JAMES W HILSINGER
121 44 5468

Introduction to Electronics

Clyde N. Herrick
Bill R. Deem
San Jose City College

Goodyear Publishing Company, Inc.
Pacific Palisades, California

Current printing (last digit):
10 9 8 7 6 5 4 3 2 1

ISBN: 0-87620-442-6

Library of Congress Catalog Card Number: 71-184130

Y-4426-6

Printed in the United States of America

PREFACE

This textbook presents a state-of-the art introduction to electricity and electronics for students with a minimum background in mathematics and physics. Although the field of electricity is comparatively stabilized, the field of electronics is progressing at a rapid rate. Electronics is becoming dominated by solid-state devices such as bipolar and field-effect transistors, gate-controlled switches, silicon controlled rectifiers, and large-scale integrated circuits. In turn, the need for relevant treatment is evident. To provide student motivation and to facilitate the learning process, the modern dual approach has been employed:

I. Functional language, topical continuity, judicious demonstrations and experiments, terminated by thought-provoking questions, serve to sustain student interest and to maximize retentivity.

II. Constructional projects that appeal to the beginning student are provided to dramatize the relation of theory to practice. These projects are comparatively inexpensive, and can be financed within the limited budgets that are often confronted by instructors and students.

This textbook is the outcome of extensive teaching experience, both on the part of the authors and of their fellow instructors at San Jose City College, who have contributed numerous constructive criticisms and suggestions. It is appropriate that we dedicate this text as a teaching tool to the instructors and students of our high schools, vocational schools, technical institutes, and junior colleges.

Clyde N. Herrick
Bill R. Deem

CONTENTS

14 Semiconductor Devices

15 Basic Printed Circuits and Integrated Circuits

16 Electronic Power Supplies

17 Electronic Oscillators

18 Radio Waves and Basic Radio Theory

19 Principles of Television

1 HISTORY OF ELECTRONICS AND ITS PLACE IN INDUSTRY

1.1 Definitions

Electronics is a branch of electrical science. Both electricity and electronics are included in the science of physics. Although electricity has had a long history, electronics was all but unknown in the nineteenth century. In 1891, Professor Stoney predicted that tiny electrical particles would be found to constitute all forms of electricity, and introduced the word *electronics* at that time. In 1896, the physicist J.J. Thomson discovered the electron while doing research on X-rays. For the next fifty years, electronics was defined as the science concerning the flow and control of electrons in vacuum tubes and similar devices. However, with the advance of science and the development of new electron devices, the definition was gradually broadened. Today, electronics is defined as a branch of physics that deals with the emission, absorption, behavior, and effects of electrons in vacuums, gases, and crystals.

1.2 Lodestone and Amber

Our recorded history of electricity and magnetism dates back to 600 B.C. in Greece, and to considerably earlier times in the Orient. The Grecian writings, however, are more detailed. Thales of Miletus wrote about certain black stones that had the ability to attract small pieces of iron. Many of these natural magnets (Fig. 1-1) were found near Magnesia in Asia Minor. Accordingly, the an-

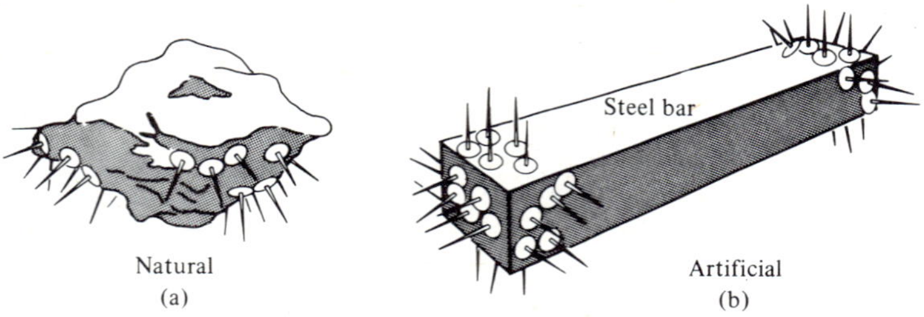

Fig. 1-1
(a) Natural magnet
(b) Artificial magnet

cients called these black stones *magnetite.* Prior to the Greeks, the Chinese had also discovered natural magnets, and knew that they would point in a north-and-south direction when floated on a light substance in a container of water. Although the date of the first navigator's compass is unknown, the common name for magnetite is *lodestone,* which has its origin in this era and means "leading stone." Magnetite is an iron ore found in the United States, Norway, and Sweden.

Thales wrote of the attraction that is exerted upon chaff by a piece of amber that has been rubbed by cloth or fur. The Greek word for amber is *elektron.* Amber is a yellow and translucent fossil resin, commonly found on seashores. We are most familiar with amber as it is used to make pipestems; it has also been used by designers of costume jewelry. When amber is rubbed by some other substance, static electricity is generated. Long before the discovery of amber, however, electricity had been experienced in the form of lightning (Fig. 1-2). Similarly, the aurora borealis (Fig. 1-3) was known in the northern hemisphere, and the corresponding aurora australis was known in the southern hemisphere.

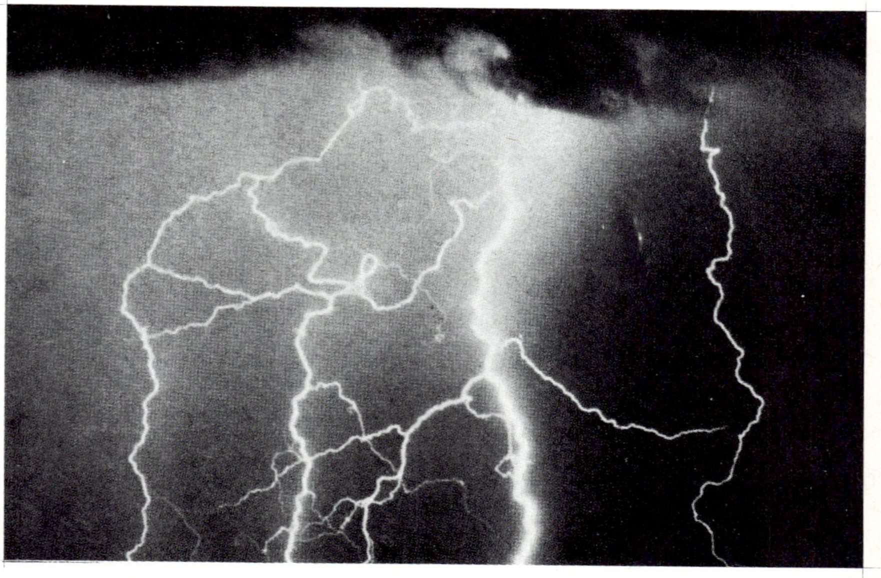

Fig. 1-2
Appearance of a lightning discharge.

Fig. 1-3
Appearance of the Aurora borealis.

(Electric eels were encountered in various parts of the world also.) None of these electrical phenomena, however, were understood until comparatively recent times.

1.3 The Magnetic Compass

A magnet has a north pole and a south pole. Although either pole will attract iron articles, a north pole of a magnet repels the north pole of another magnet, and a south pole of a magnet repels the south pole of another magnet. We can apply the rule that unlike poles attract and like poles repel.

It has been noted that the ancient Chinese devised magnetic compasses from lodestones. One of the early forms is shown in Fig. 1-4. Later, magnetized steel

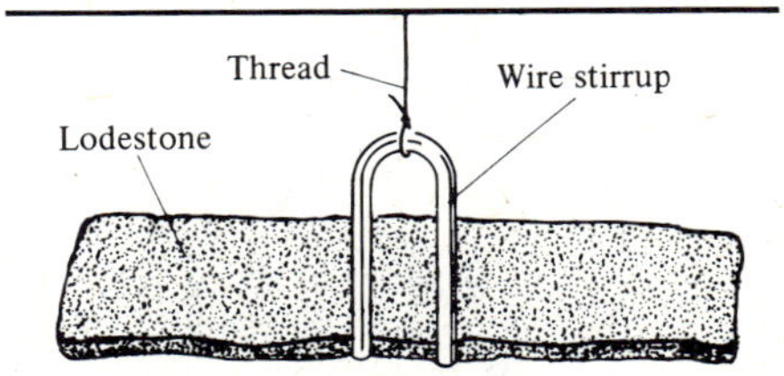

Fig. 1-4
An early type of lodestone compass.

needles were used instead of lodestones. A circular card showing the various points of the compass was mounted under the pivoted needle and served as a convenient reference in determining magnetic headings. Navigators term the course of a ship its heading. Another form of magnetic compass is constructed with the needle mounted under a circular card which rotates on a jeweled bearing. Thus, the card turns as the ship's course is changed, and a pointer fixed to the housing of the compass indicates the magnetic heading on the card. This type of compass is also designed to float in a fluid, which provides operating stability under conditions of vibration or shock.

A representation of the earth's magnetic field is seen in Fig. 1-5. Note that

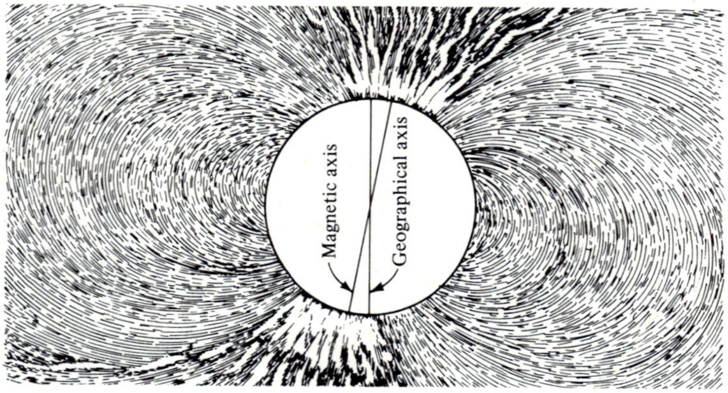

Fig. 1-5
Representation of the earth's magnetic field.

the earth's magnetic poles do not occur at quite the same locations as the geographic poles. It has been known for some time that the locations of the earth's magnetic poles are not fixed; instead, the magnetic north and south poles "wander" slowly. The strength of the earth's magnetic poles also tends to vary with the passage of time. It is evident from Fig. 1-5 that the direction of the magnetic-field forces is parallel to the earth's surface only at the magnetic equator. As we proceed toward a magnetic pole, the magnetic-field force dips downward progressively until the pole is reached, at which point the force direction is directly downward. A dipping needle, such as shown in Fig. 1-6, is used to measure the amount of dip at a given location.

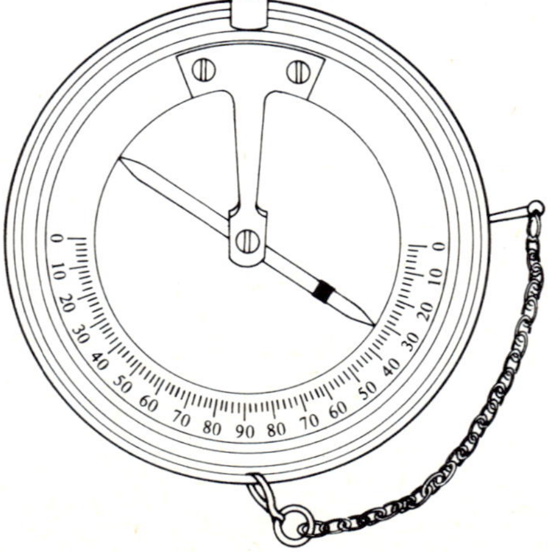

Fig. 1-6
A dipping needle.

1.4 Early Atomic Theory

Democritus, a Greek philosopher, wrote in 460 B.C. concerning his belief that all substances are built up from extremely small elementary particles, which he called *atomos.* Our modern term *atom* is derived from this Greek word. Scientific evidence for the existence of atomic elements was not available until 1808, however, when Professor John Dalton pointed out that the known facts of physics and chemistry could be most logically explained on the basis of an atomic theory. If we assume that the drop of water shown in Fig. 1-7a is halved again and again, we would eventually obtain the smallest particle of water that can be obtained: the *molecule.* According to the atomic theory, all molecules are built up from atoms.

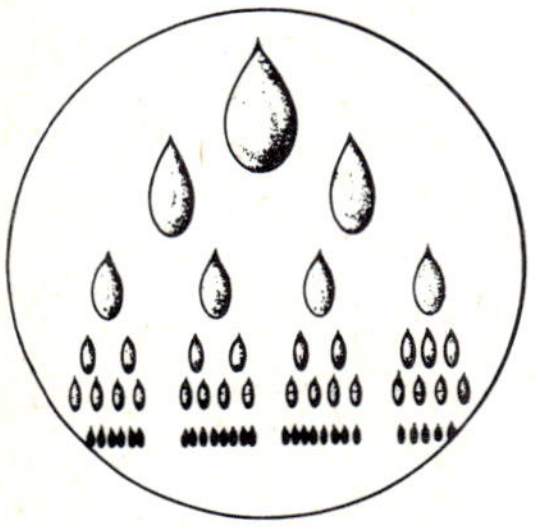

Dividing a drop of water
(a)

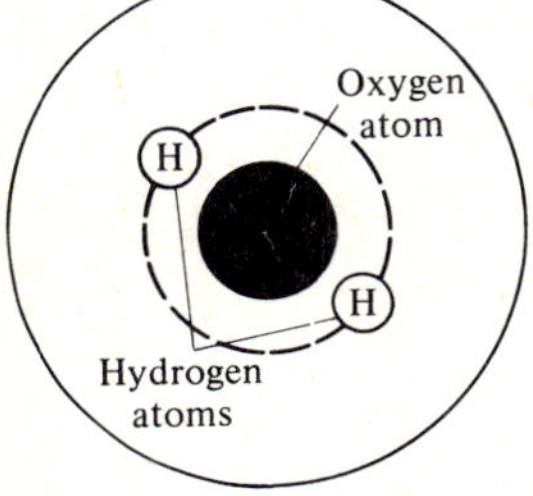

Molecule of water H_2O
(b)

Fig. 1-7
Matter is made up of molecules: (a) Dividing a drop of water. (b) Molecule of water (H_2O).

As learned in our chemistry courses, a molecule of water (H_2O) consists of one atom of oxygen (O) and two atoms of hydrogen (H), as represented in Fig. 1-7b. If the molecule of water were further subdivided, only unrelated atoms of hydrogen and oxygen would be obtained. That is, we would no longer have the substance: water. Thus, the molecule is the smallest particle into which a substance can be divided and still be called by its original name. This principle applies to all forms of matter–solids, liquids, and gases. When molecules are combined or separated without any change in their properties, they are said to undergo a *physical* change. When combining or separating results in the formation of new molecules, however, a *chemical* change is said to take place. We find that electrical action is involved in chemical changes, which brings us to the topic of subatomic structure.

Subatomic structure denotes the properties of and the relations among particles smaller than atoms. More than 100 elements (kinds of atoms) have been identified, and each element has a unique subatomic structure. Physicists have devised many experiments to study the interior of the atom. The core of an atom is called its *nucleus.* Most of the mass in an atom is concentrated in its nucleus. The nucleus of an atom contains *protons* (positively charged particles), and all atoms but one (hydrogen) have nuclei that also contain neutrons, which are electrically neutral. In turn, most of the mass in an atom is in the protons and neutrons that form its nucleus.

Whirling around the nucleus in orbits are one or more smaller particles of negative electric charge. These negative charges are called *electrons.* Normally there is one electron in orbit for each proton in the nucleus so that the net positive charge of the nucleus is balanced by an equal negative charge of the orbiting electrons. That is, an atom in its normal state is electrically neutral. Although the electrons are attracted to the nucleus, they do not fall into the nucleus because of their centrifugal force, or energy in orbital motion. This principle is analogous to the orbital motion of the planets which prevents them from falling into the sun. This general outline of subatomic structure was first described by a physicist named Niels Bohr in 1913. Although his original theory has been considerably developed, elaborated, and modified, it is of basic historical importance.

Whether an element is identified as hydrogen, oxygen, iron, or other substance depends upon the number of protons in the nucleus of the atom. Figure

1-8 shows simplified diagrams for the atoms of several elements. We observe

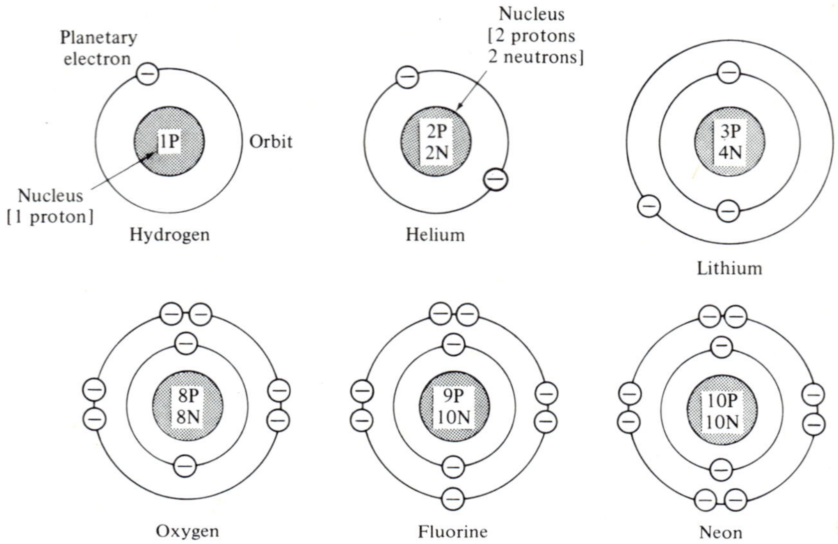

Fig. 1-8
Structure of several atoms.

that the hydrogen atom has a nucleus of one proton around which one electron orbits. Next, the helium atom has a nucleus containing two protons and two neutrons around which two electrons orbit. A heavy element such as curium, which was discovered in 1944, has a nucleus with ninety-six protons around which ninety-six electrons orbit.

1.5 The Electrostatic Machine

As noted previously, the generation of electricity is based on motion or transfer of electrons. Since an atom is normally electrically neutral, this process is also described as charge separation. For example, Fig. 1-9 illustrates how charges are separated when a hard-rubber rod is rubbed with a piece of fur. Both the fur

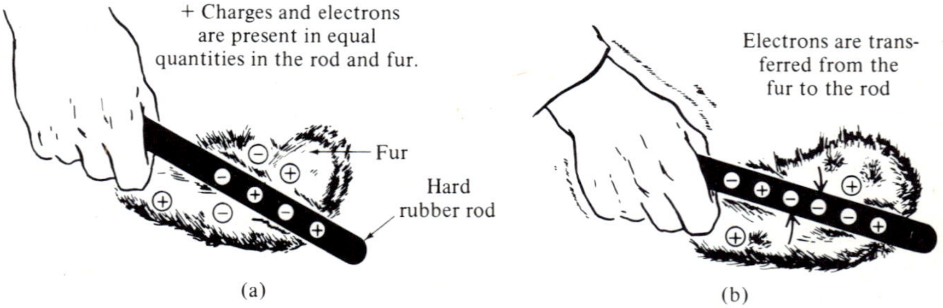

Fig. 1-9
Producing static electricity by friction.
(a) + charges and electrons are present in equal quantities in the rod and fur.
(b) Electrons are transferred from the fur to the rod.

and the rod become electrified, but with opposite polarities. That is, the fur acquires a positive charge, and the rod acquires a negative charge. This is basically the same process as that utilized by the ancients when they rubbed amber with cloth. The resulting charges are called *static electricity* because they are at rest on the fur and the rod.

In 1671, a scientist named Otto von Guericke invented the first electromechanical generator, or *electrostatic machine.* The modern version of this device, often used in classroom demonstrations, is called the Wimshurst Electrostatic Generator. A typical generator, which is illustrated in Fig. 1-10, operates by

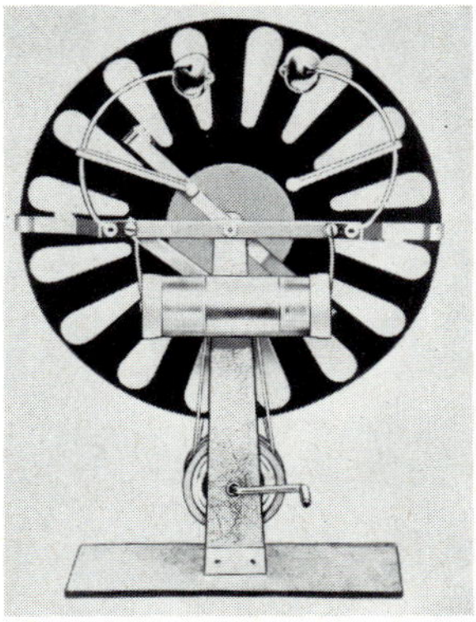

Fig. 1-10
A classroom demonstration electrostatic machine.
Courtesy: Edmund Scientific Company

changing mechanical energy into electrical energy. Two circular plastic plates are rotated in opposite directions by means of a belt-and-pulley arrangement. A number of metal-foil sectors are cemented onto each plastic disc. These sectors both generate small charges of electricity and carry these charges to the spark gap, where a small lightning discharge jumps between the pair of metallic balls. Each sector accumulates the charges derived by influence from other sectors as they revolve.

1.6 The Voltaic Cell

Although electrostatic machines are useful sources of electricity, they have several disadvantages. Aside from low efficiency, they cannot provide a steady flow of electricity, and it is impractical to generate large amounts of electricity with this type of machine. However, in 1799, a scientist named Allessandro Volta devised the *voltaic pile,* which was quickly improved and led to the development of electrical cells and batteries as they are known today. Figure 1-11 shows the plan of a voltaic pile, which consists of alternate discs of copper and zinc separated by moist cloth. Each zinc-copper pair is called a *cell,* and the pile of cells is called a *battery.*

A voltaic pile generates electricity by chemical action; it is called an *electrochemical device.* Moist cloth normally contains traces of salts among other chemical substances. In turn, the original voltaic cell made use of these traces of chemicals in its operation. Volta soon discovered that considerably more electricity was generated if the cloth separators were soaked in a strong solution of

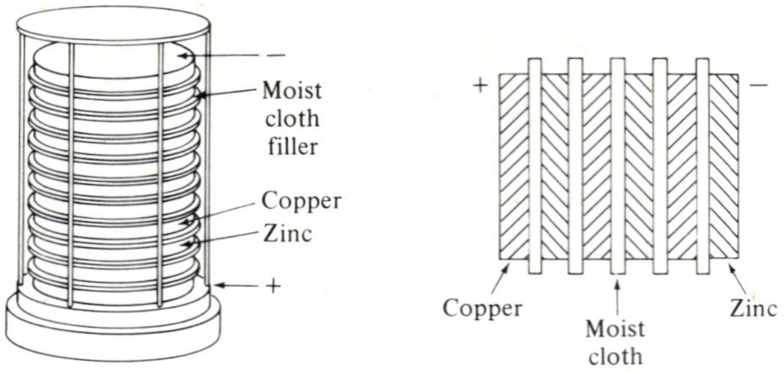

Fig. 1-11
Plan of a voltaic pile.

some salt. A voltaic cell changes chemical energy into electrical energy, and the cell eventually "goes dead." Regardless of the type of salt that is used, the zinc disc has negative polarity and the copper disc has positive polarity. Each cell generates a certain electrical pressure, and this pressure is multiplied in a battery by the number of cells that are provided.

Other early types of electric cells contained various acids instead of salts. Figure 1-12 shows the internal construction of a modern dry cell. A dry cell is a

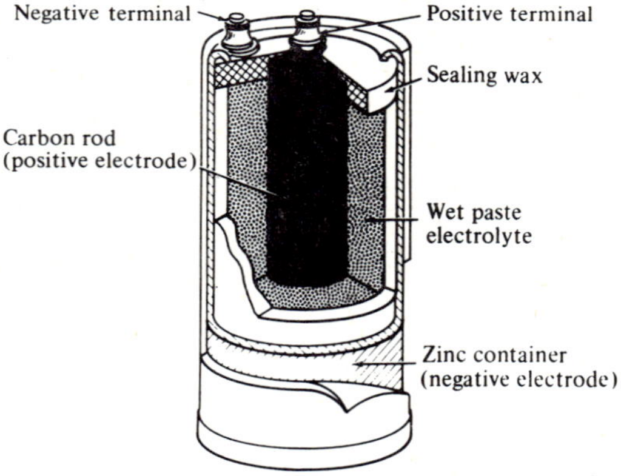

Fig. 1-12
Cut-away view of dry cell.

useful source of electricity in many applications which require portability and moderate amounts of electricity over fairly long periods of time. The efficiency of a dry cell is not very high, but this disadvantage is often outweighed by its convenience. Improved forms of dry cells, some of which are comparatively costly, have been developed within recent times and have certain advantages in special applications; they are discussed in greater detail later in the text.

1.7 Morse Telegraph

About twenty-five years after the invention of the voltaic battery, a scientist named André Marie Ampère discovered that if a current-carrying wire is wound in the form of a helix (Fig. 1-13), a concentrated magnetic field appears at each end of the coil. This arrangement is called a *solenoid*; and its magnetism disappears

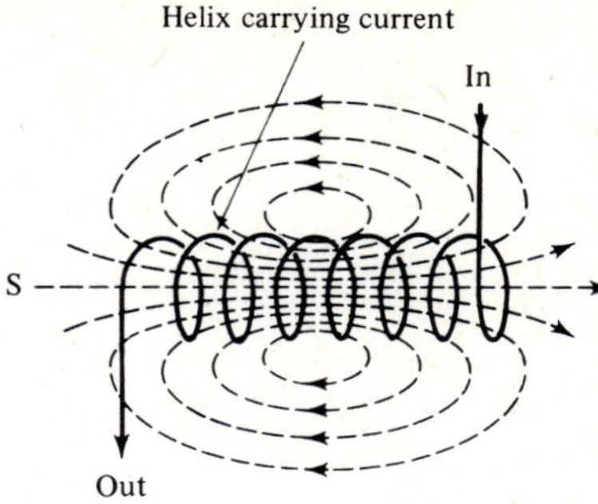

Fig. 1-13
Magnetic field around a coil.

whenever the current in the wire stops flowing. Ampère also discovered that if a core of soft iron is inserted inside a solenoid, the magnetic field becomes much stronger. This device is called an *electromagnet,* and is a basic part of many important electrical devices. The most useful feature of an electromagnet is that its magnetic force can be turned on or off by starting or stopping current flow in the coil. This is called a *control action.*

In 1832, an inventor named Samuel F. B. Morse devised the first successful electric telegraph. The principle of Morse's telegraph is illustrated in Fig. 1-14.

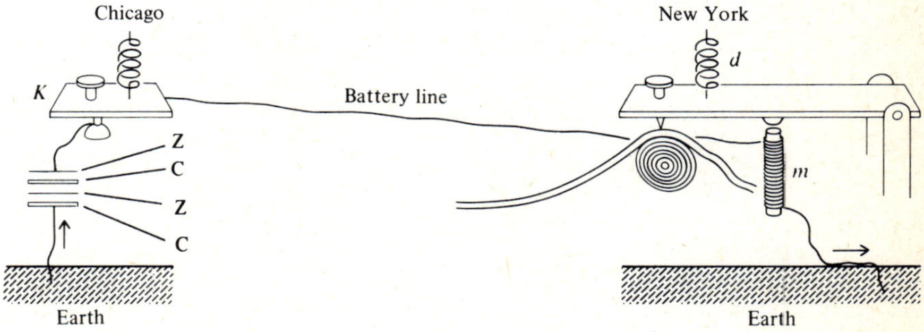

Fig. 1-14
Principle of the telegraph.

A battery line is used, with the earth serving as a return conductor for the electric current. Starting and stopping of the current is accomplished by closing and opening the switch key *K.* When the key is closed, current passes from the battery and through the key into the line, and then through the electromagnet *m.* From the coil, the electric current flows back through the earth to the battery, completing the circuit. Armature *b* is held down by the electromagnet *m* as long as key *K* is kept closed. As soon as the circuit is broken at *K,* however, the armature is pulled up by spring *d.* Morse employed a clockwork device to draw tape *c* along at a constant speed beneath the pen actuated by armature *b.* Morse also devised the standard telegraph code, whereby letters and numbers are represented by short and long strokes called dots and dashes. For many years, railroad lines used the later design of telegraph instruments shown in Fig. 1-15.

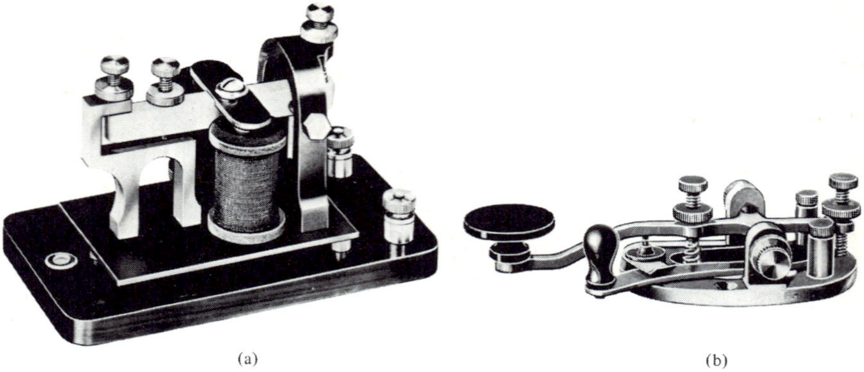

Fig. 1-15
Western Union telegraph instruments.
(a) Sounder.
(b) Key.
Courtesy: Sargent-Welch Scientific Co.

1.8 Bell Telephone

By 1875, the Western Union Telegraph Company had been in business for twenty years, and the need of speeding up communication by transmission of the human voice over wires was clearly recognized. However, no practical means of telephone communication had been found. In that year, Dr. Alexander Graham Bell, a teacher of the deaf, first succeeded in designing a device that provided voice communication over wires. The basic plan of Bell's early telephone (Fig. 1-16) was quite simple. An electromagnet was used, with its end spaced a

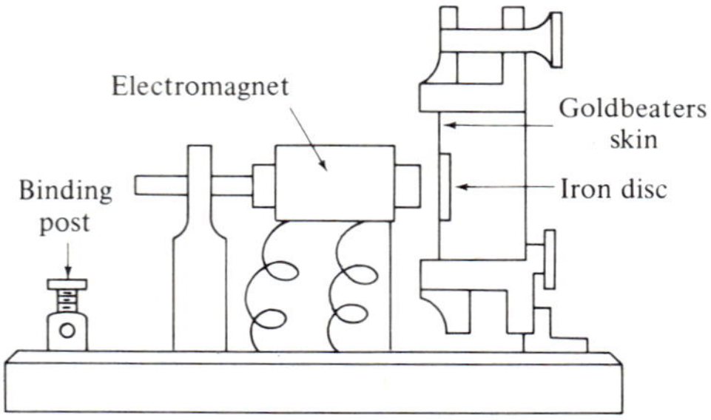

Fig. 1-16
An early Bell Telephone instrument.

short distance from a thin diaphragm of goldbeater's skin.[1] In the center of this flexible diaphragm, a thin iron disc was attached. The winding of the electromagnet was connected to the line, consisting of two wires. A battery was also connected into the line between two telephone units to produce magnetic fields in the electromagnets. The same instrument was used for both sending and receiving the signal.

Although the operation of Bell's telephone was poorly understood at the time, it gradually became recognized that the instrument action took place in the following manner. When the instrument at the transmitting end was spoken into, the diaphragm vibrated in step with the vocal tones of the speaker, carrying

1. Goldbeater's skin is the name given to the membrane of the large intestine of the ox, used to separate the leaves of gold when making gold leaf.

the iron disc with it. In turn, the spacing between the iron disc and the end of the electromagnet was changed in step with the voice vibrations. This changed the strength of the current. Since a changing electric current was flowing into one end of the line, it also flowed out of the other end of the line into the electromagnet at the receiving end. Thus the diaphragm in the receiving instrument moved in exact accordance with the diaphragm motion in the transmitting instrument. This method of electrical-current generation is called *electromagnetic induction,* and is explained in greater detail in Chapter 8. A somewhat later form of Bell's telephone receiver is shown in Fig. 1-17.

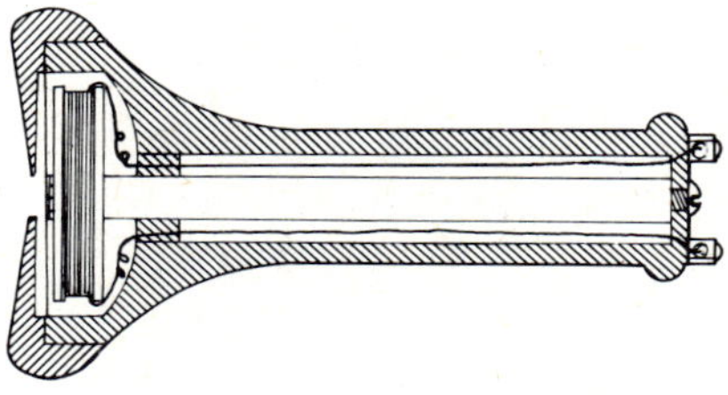

Fig. 1-17
A later form of Bell's telephone receiver.

1.9 Edison Electric Light

Thomas Alva Edison constructed the first successful incandescent lamp in 1879. His lamp consisted of a carbon filament which was sealed into an evacuated glass bulb, as shown in Fig. 1-18. The filament was connected to an external source

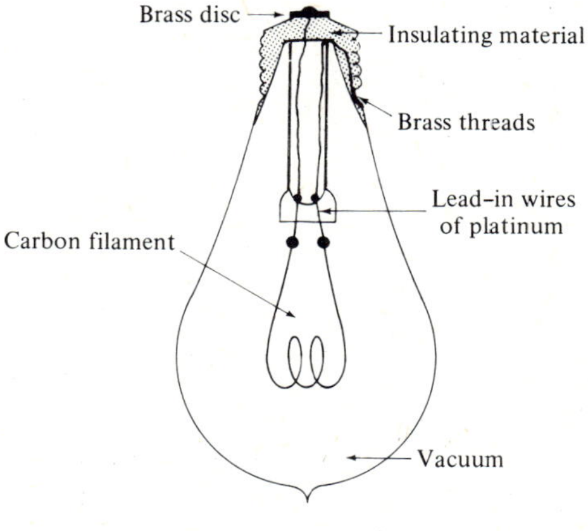

Fig. 1-18
A diagram of the old carbon lamp invented by Edison.

of electricity through lead-in wires. Although carbon filaments were a useful source of light, they had several disadvantages. For example, the light that was produced was not white, but had a yellow cast. Also, carbon filaments were somewhat fragile, and tended to break if the lamp was jarred. Finally, as the lamp aged, the carbon filament slowly vaporized and settled as a dark film on the inside of the bulb; this reduced the effective light output considerably. Today, most incandescent lamps employ tungsten filaments.

Although the incandescent lamp was a very important invention, its descendants were almost of equal importance. For example, the radio vacuum tube evolved from the incandescent lamp, as did the television picture tube. By 1882, the larger cities were turning to electric lighting, leading to the development of large generating stations and substations. At about this same time, the Edison Electric Company installed the first power plant in New York City to meet the demands of the new electrical industry. Thereafter, electric power developments proceeded rapidly, and within a few years electric transmission lines were a familiar sight throughout the country.

1.10 Marconi Radio System

Radio waves were unknown until 1887 although their existence had been predicted by James Clerk Maxwell in 1865. Radio or electromagnetic waves are electric and magnetic fields traveling through space at the velocity of light (186,500 miles per second). Heinrich Rudolf Hertz, a physicist, made the experiment illustrated in Fig. 1-19 to demonstrate electromagnetic waves. The spark coil was similar to the ignition coil used in automobiles. When the circuit was closed

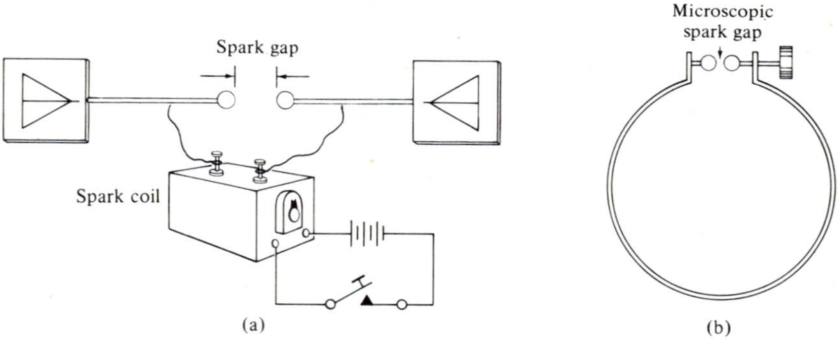

Fig. 1-19
Hertzian electromagnetic-wave apparatus:
(a) Transmitter
(b) Receiver

with the key switch, a spark jumped across the gap. This made up the transmitter arrangement. At some distance away, a loop of wire with a very narrow spark gap was used as a receiver of the electromagnetic waves. Wave reception was indicated by minute sparks jumping across the narrow gap. Although this demonstration arrangement was very crude, it paved the way for development of the Marconi radio system.

Guglielmo Marconi was a businessman who was interested in the commercial possibilities of wireless telegraphy, as it was called at that time. He was also a practical engineer who was capable of developing a laboratory experiment into a useful communication system, such as his invention of the elevated aerial now termed antenna. By greatly increasing the transmitter power, by using large antenna structures, and by developing comparatively sensitive receiving devices, Marconi succeeded in spanning the Atlantic Ocean in 1901. Soon thereafter, the Marconi Wireless Telegraph Company of America was founded, thus introducing the Marconi radio system in the United States.

1.11 DeForest Vacuum Tube

Edison's invention of the incandescent lamp led to important developments in the field of electronics. While he was trying to develop improved filaments, Edison discovered that a metal plate mounted in the light bulb near the filament became negatively charged. He could not explain this fact, and it was not recognized for several years that the hot filament was emitting electrons, which were collected by the plate and charging it negatively. However, in 1904, Professor John Fleming found that an incandescent lamp with a filament and plate operated as an electric "valve" in a circuit. That is, the device permitted electric current to flow in one direction, but blocked its flow in the other direction. This *Fleming valve* was the first radio tube, and was used by Marconi in his later types of receivers. Figure 1-20 shows the plan of a Fleming valve.

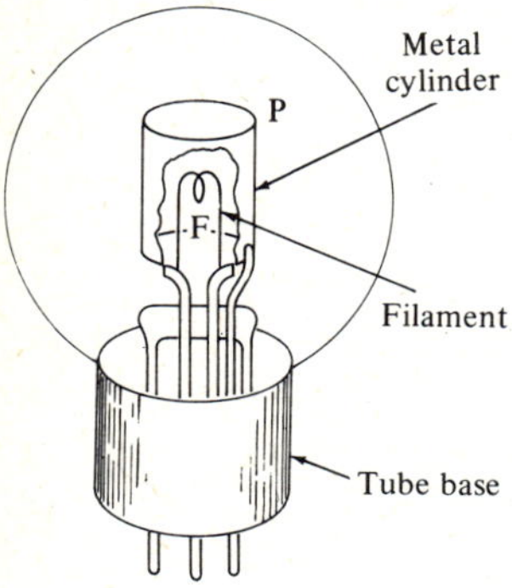

Fig. 1-20
Plan of a Fleming valve.

Although the Fleming valve was an important invention, there was a need for a device or method that could be used to step up (amplify) weak radio signals that were otherwise inaudible. In 1905, Dr. Lee De Forest succeeded in devising a new type of radio tube that had amplifier action. This was the famous *audion* vacuum tube, illustrated in Fig. 1-21. Although it has a basic resemblance to the Fleming valve, the audion has a striking distinction in that a grid electrode is placed between the filament and the plate. This grid has an open structure so that electrons can flow through it in their travel to the plate. Because the grid is placed in the path of electron flow, it has a control action. That is, a very weak electrical signal applied to the grid will control the flow of a similar but much stronger electrical signal in the plate-filament circuit.

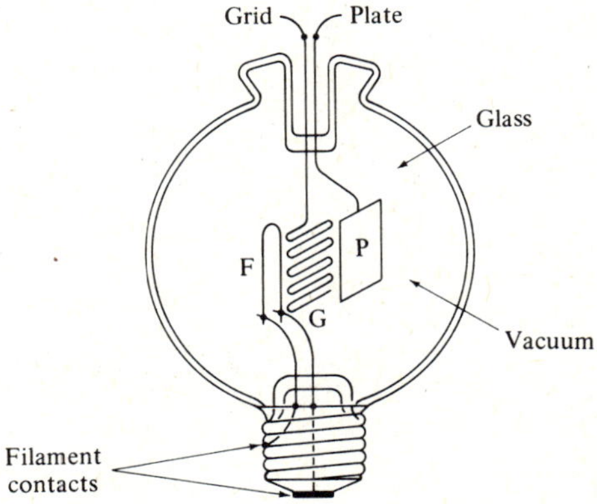

Fig. 1-21
Plan of the DeForest audion.

1.12 Development of Television

Television technology entails the conversion of light into electricity, and reconversion of electricity into light. In 1888, a scientist named Wilhelm Hallwachs discovered the photoelectric effect wherein certain kinds of metals enclosed in an evacuated glass bulb emit electrons when struck by light rays. When this *phototube* is connected into a circuit, electric current flows in the circuit, caused by light rays entering the tube. Figure 1-22 shows an early type of phototube.

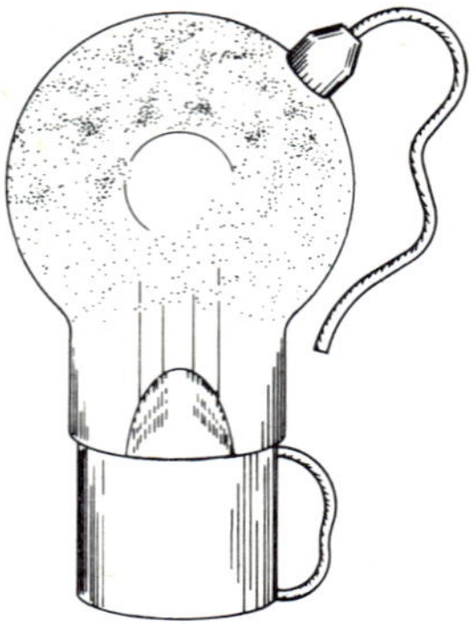

Fig. 1-22
An early type of phototube.

This type of tube was one of the basic parts in the first television transmitters. At the receiver end of the television system, a neon glow tube was used to change electrical signals into light rays. Professor Julius Plücker invented the glow tube in 1858, and named it a Geissler tube. Modern neon signs are an example of glow tubes that are very similar to the original Geissler tubes.

One line of development of glow tubes led to the neon glow lamp (Fig. 1-23)

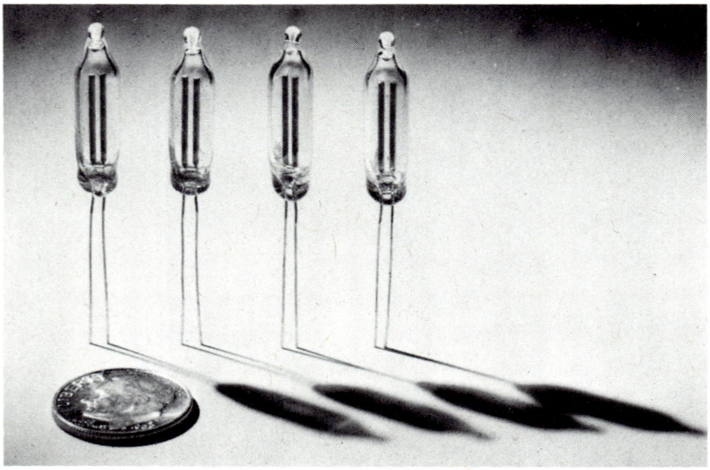

Fig. 1-23
A neon glow lamp used in early television receivers.
Courtesy: Signalite, Inc.

in which the anode (the positive electrode) had a comparatively large area. Early television receivers used neon glow lamps with anodes approximately one inch square. This meant that the reproduced picture was limited to this size unless a large magnifying glass was mounted in front of the tube. Another essential device in early television was the *scanning disc* (Fig. 1-24) used at both the transmitter and receiver. The scanning disc was invented in 1884 by a scientist named Paul Gottlieb Nipkow. When the disc was rotated, each opening in the disc passed a strip on the image so that the light from each small area of the image could be processed or viewed one after another. That is, any television system must "ana-

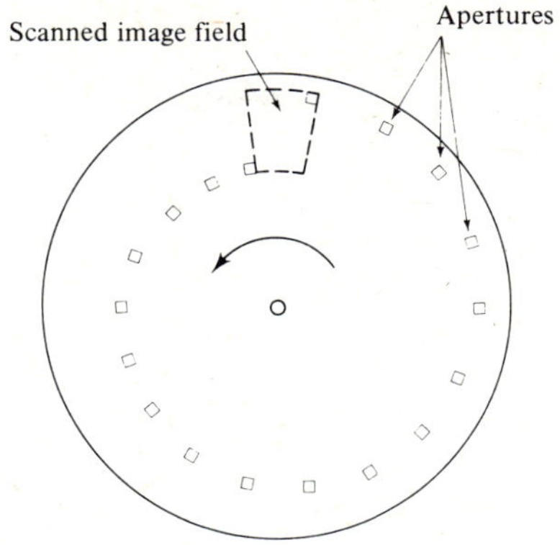

Fig. 1-24
The Nipkow scanning disc.

lyze" the light from an image, point by point and line by line. For this reason, the scanning process is called a *sequential* operation.

Figure 1-25 shows the essentials of the first television system. The televised

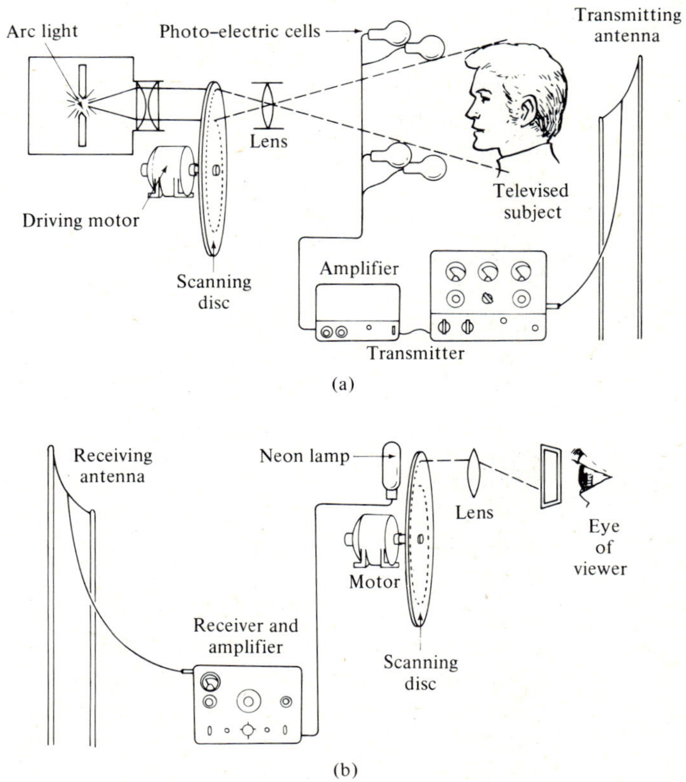

Fig. 1-25
The first television system.
(a) Transmitter
(b) Receiver

subject was seated in a room that was lighted only by a flying spot of light from the transmitting mechanism as shown in Fig. 1-25a. This flying spot of light was produced by a rotating scanning disc with an arc-light chamber mounted behind the disc. In turn, the face of the televised subject was rapidly scanned from left to right and from top to bottom by the flying spot of light. The reflected light was picked up by a bank of photoelectric cells located in front of the televised subject. This produced an electric current which was proportional at any instant to the intensity of the reflected light from the flying spot. This television signal was then amplified and radiated by a transmitter as electromagnetic waves from an antenna.

At the receiver (Fig. 1-25b), the television waves were picked up by the receiving antenna, amplified, and applied to a neon lamp. The scanning disc in front of the neon lamp produced a flying spot of light similar to the flying spot at the transmitter. By this method, the viewer saw a light image that reproduced that of the televised subject. It is evident that the scanning disc at the receiver had to be kept closely in step, or *synchronized,* with the scanning disc at the transmitter. Otherwise the received image would become "jumbled" and meaningless. Therefore, a motor speed control was used at the receiver. Because this control required readjustment often, a special synchronizing signal was later transmitted for this purpose.

Although the scanning-disc television system was crude, it provided picture transmission on an experimental basis. Commercial television, however, awaited the development of new electronic devices that could provide large and high-quality images. In 1897, Professor Karl F. Braun invented the basic *cathode-ray tube* (CRT), shown in Fig. 1-26. This tube consisted of a cathode that emitted

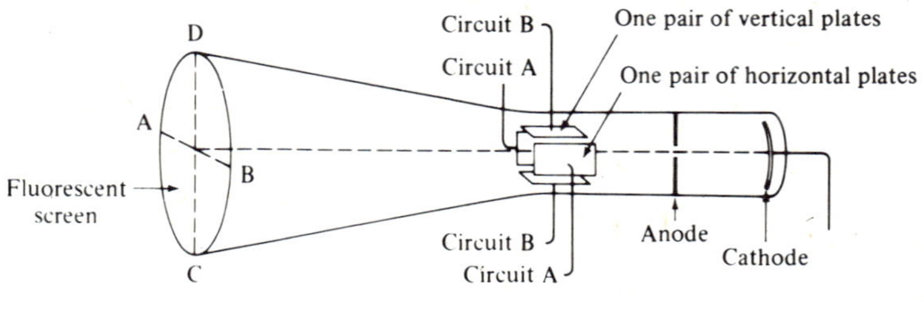

Fig. 1-26
An early form of cathode-ray tube.

electrons, and an anode with a small aperture to form a thin beam of electrons that traveled through the horizontal and vertical plates to the fluorescent screen. This screen was coated with a mineral that produced a glowing spot where it was struck by the electron beam. Note that the vertical plates when energized by circuit A produce spot deflection from C to D on the screen, and that the horizontal plates when energized by circuit B produce spot deflection from A to B on the screen. Thus, by the use of suitable deflecting signals, the screen could be scanned as required for reproduction of a television image. Then, by varying the amount of electric potential on the cathode, or anode, the brightness (intensity) of the spot could be controlled at any point in the image.

Modern television picture tubes are considerably more complicated than the Braun tube shown in Fig. 1-26. Their basic function, however, is the same. The modern television system also uses an electronic camera tube, called the *image orthicon.* This camera tube replaces the historical scanning disc at the television transmitter and generates high-quality picture signals. Color television requires the use of considerably elaborated cathode-ray tubes and electronic camera tubes. Basically, camera tubes and picture tubes used in a color-television system must break down visible light into its three primary colors, and process each of the primary colors as a separate electrical signal. Details of the technology that is used are covered in Chapter 20.

1.13 Electronic Computers

In many ways an electronic computer can be compared with an adding machine. However, there are various important differences between electronic and mechanical calculators. An electronic computer operates at tremendous speed in comparison to a mechanical calculator, and also operates on the basis of binary numbers instead of the decimal numbers used by ordinary adding machines. Students who have taken courses in the new mathematics are usually familiar with binary arithmetic. As an example of binary to decimal conversion, we may write

Decimal		Binary
540	=	1000011100

Since an electronic computer operates on the basis of binary numbers, some means of conversion between binary and decimal numbers must be used at the start and at the end of a problem. This is often the task of a *programmer* who translates from an ordinary arithmetic problem into "machine language." After the electronic computer has obeyed the orders that have been placed on the tape, the programmer may have to note the answer that has been given in binary numbers, and change the answer to decimal notation. However, various input-output devices are used which automatically convert between binary and decimal numbers. In turn, an electronic computer can be operated by comparatively unskilled operators.

A modern electronic computer is shown in Fig. 1-27. The first electronic computer was constructed in 1946 by Professors Carl Eckart and John Mauchly at the Moore School of Electrical Engineering. This computer was called the ENIAC, which comes from the name *E*lectronic *N*umerical *I*ntegrator *A*nd *C*omputer. Instead of using a punched tape, this first electronic computer was programmed by means of switches and a plugboard. The second electronic computer was developed by International Business Machines (IBM), and the punched-card program, now in wide use, was introduced in 1948. Punched tapes and cards are a form of program (or information) storage which are external to the computer. In 1945, Dr. John von Neumann had suggested the use of internal program storage. Today, all electronic computers have this feature in addition to the internal storage of numerical data used in the computation process.

Fig. 1-27
A modern electronic computer.
Courtesy: IBM Corp.

1.14 Industrial Electronic Devices

Since World War II, electronics has played an increasing role in all branches of industry. The first applications were simple. For example, photoelectric cells were used with indicators to automatically count the number of objects that passed by on a conveyor, as shown in Fig. 1-28. As each object passes in front

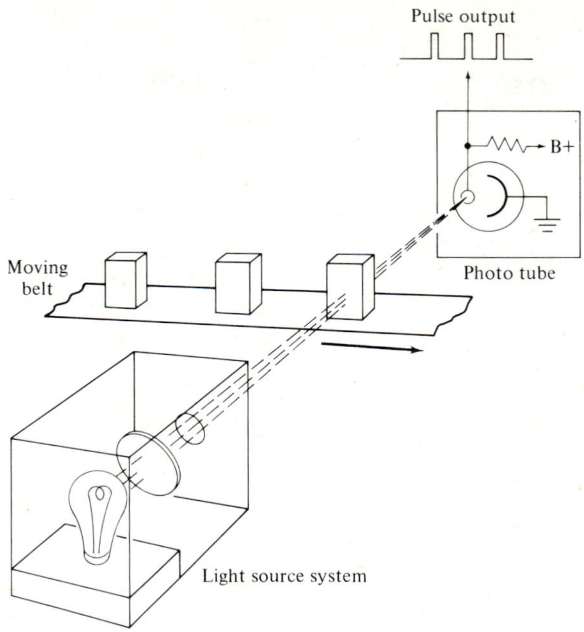

Fig. 1-28
Interruption of light beam produces an electrical pulse output.

of the light beam, an electric pulse is produced. The pulses energize an indicator that records the number of pulses that have been applied. In turn, an accurate count is kept of the day's production or of a particular production run. Phototubes have also been used for many years to control processes, such as illustrated in Fig. 1-29. In this example, the material from the feed roll is kept in a precise

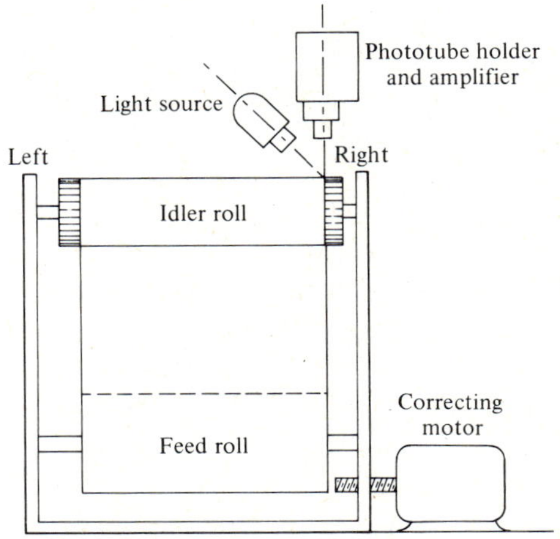

Fig. 1-29
An electronic side-register control.

path in passing over the idler roll. A correcting motor is used to shift the feed roll to the left or right, as needed. The motor is controlled by an electric current from the amplifier (an amplifier steps up the weak electrical output from the phototube). Note that a light beam is directed at the edge of the moving material. The idler roll has a contrasting color. Thus, if the edge of the material tends to move from the center of the light beam, an electrical correcting signal is produced and applied to the motor.

1.15 X-Ray

X-ray tubes have well-known applications in dentistry and medicine. X-rays are also used in industrial plants, mainly for automatic inspection and gauging. A basic form of X-ray tube invented in 1895 by Professor Wilhelm Konrad Roentgen is shown in Fig. 1-30. In this tube, a high electric potential is applied between

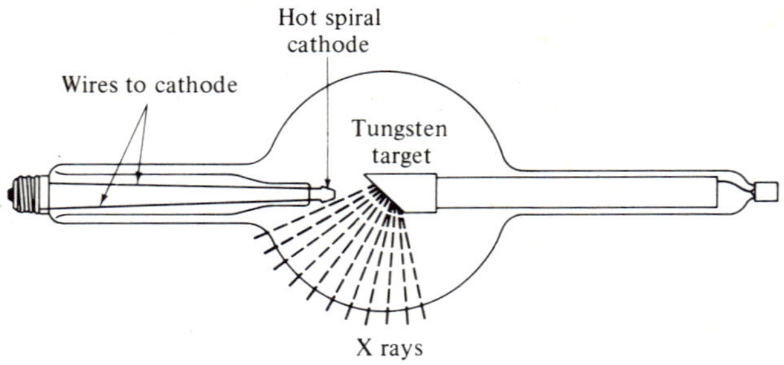

Fig. 1-30
Plan of an X-ray tube.

the cathode and the target. Electrons are emitted by the heated cathode and are attracted to the target. Since the electrons strike the target at very high speed, considerable shock is given to the tungsten atoms, which respond by radiating X-rays. These rays are basically similar to radio waves except that they have a very much shorter wavelength (a higher rate of vibration). In use, a manufactured product (such as a casting) is passed through the X-ray beam, and the intensity of the beam that emerges from the other side is measured with an X-ray detector. Various flaws, cracks, or contained bubbles cause an increase or decrease in the intensity of the emerging X-ray beam. Any change in the X-ray beam is used either to record the test result or to turn on an automatic rejector.

An X-ray detector is usually designed on the plan of a Geiger-Mueller tube, as shown in Fig. 1-31. This tube was introduced in 1910 and is basically a metal-and-glass enclosure containing a gas at low pressure. A wire is mounted along the axis of the tube, and serves as one of the electric terminals. The metal portion of the enclosure serves as the other electric terminal. When X-rays enter the Geiger-Mueller tube, some of the gas molecules are *ionized.* Ionization means that one or more of the electrons orbiting around the nucleus of a gas atom is ejected from the atom as a result of being struck by the X-rays. If an electric potential is applied between the terminals of such a Geiger-Mueller tube, a small electric current will flow each time that the positive terminal collects an electron. It is obvious that a strong X-ray beam will produce a greater current flow than that of a weak beam.

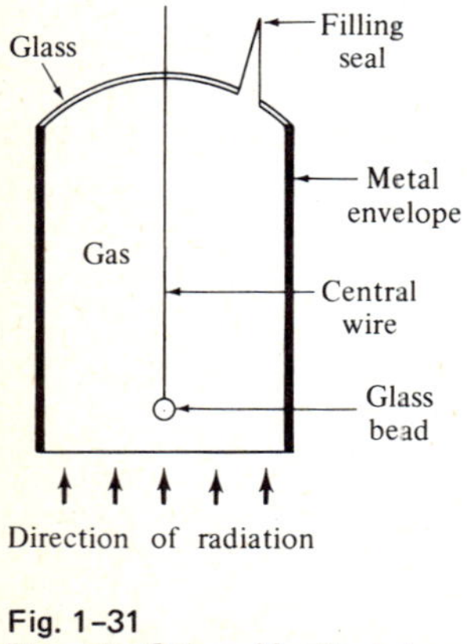

Fig. 1-31
Plan of a Geiger-Mueller tube.

1.16 Oscilloscope

The cathode-ray tube was discussed earlier. This device is widely used in industry for quality-control applications, and is the basic component of an *oscilloscope,* shown in Fig. 1-32. At this point we will regard the oscilloscope as an electronic

Fig. 1-32
An industrial oscilloscope.
Courtesy: Hewlett-Packard Co.

device that traces graphs on the screen of a cathode-ray tube. A *display* on the screen of a cathode-ray tube is always produced by the application of a varying electric potential. However, this potential variation may be caused by almost any physical relationship, such as air vibrations, cylinder explosions, fluid phenomena, and so on. That is, an oscilloscope generally provides an indirect indication of physical relationships by means of suitable *transducers.* A transducer is a device, such as a telephone transmitter, that changes one form of energy into some other form of energy. In the case of an oscilloscope, a transducer might change an explosive shock wave into an electrical impulse which in turn produces a *waveform* on the screen of the cathode-ray tube.

1.17 Principles of Automation

Industrial electronic devices find their widest usage in automated systems. The term *automation* was introduced in 1936 by D. S. Harder of General Motors and refers to an automatic system of working. We will find that the difference between an automated process and a mechanized process is basically a matter of complexity. For example, a steam engine usually includes a governor to control the speed of the engine automatically. A simple form of governor is shown in Fig. 1-33. This is a typical example of an automatic controlled mechanized sys-

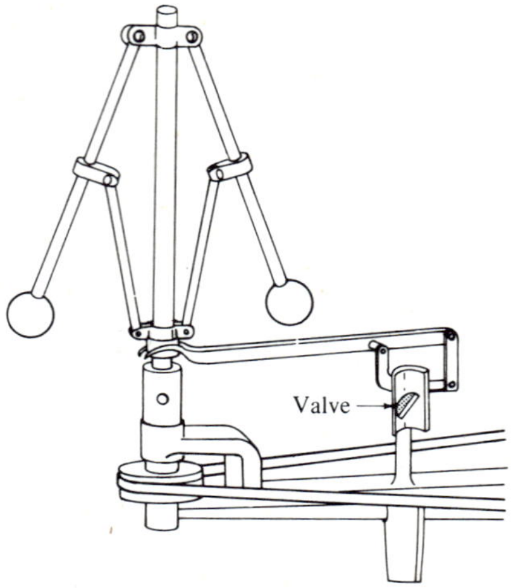

Fig. 1-33
A typical governor.

tem. Now let us consider a piston factory in an automobile plant that is fully automated. In the piston factory, raw material (metal bars) is fed into an input device, and quality-controlled finished pistons are delivered by an output device. This is a typical example of industrial automation.

Automated processes can be basically described in terms of processing, mechanical handling, sensing, and control operations. *Processing* is the fabrication procedures that are necessary to change the raw material into a finished product. *Mechanical handling,* as the term suggests, is concerned with physical transport, locating, securing, and controlled movements. *Sensing* refers to automatic checking and correction (when required) of fabrication processes. For example, a sensor might register a slight change in the diameter of a piston while it is being turned automatically in a lathe operation. In turn, the sensor will actuate a corrective mechanism that brings the chisel to its precise reference position. A corrective operation involves *control.* With reference to Fig. 1-34a, a human

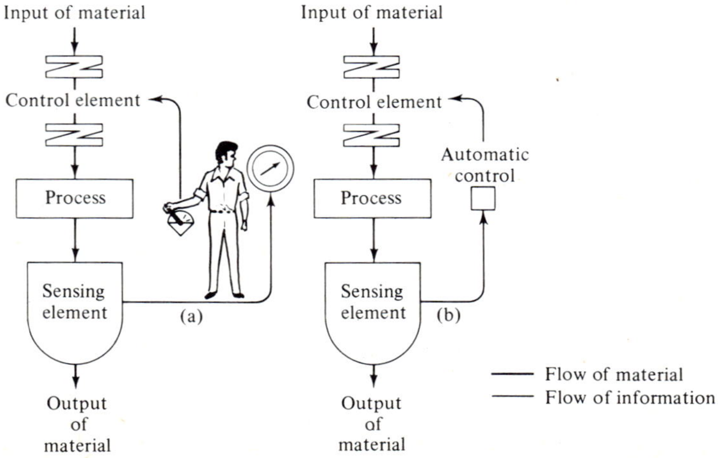

Fig. 1-34
Plan of control systems.
(a) With human operator.
(b) With automatic feedback.

operator may observe an indicator and adjust machine actions accordingly; or as shown in Fig. 1-34b, an automatic control arrangement may be used.

Note in Fig. 1-34 that the two basic requirements concern the flow of material and the flow of information. Automatic control replaces human control by means of feedback from output to input, wherein the necessary information provided by the sensing element is converted into the correct form for actuation of the control element. In other words, the main difference between a mechanized system and an automated system is found in a feedback loop, as shown in Fig. 1-35. We find that feedback loops often use electronic components. The

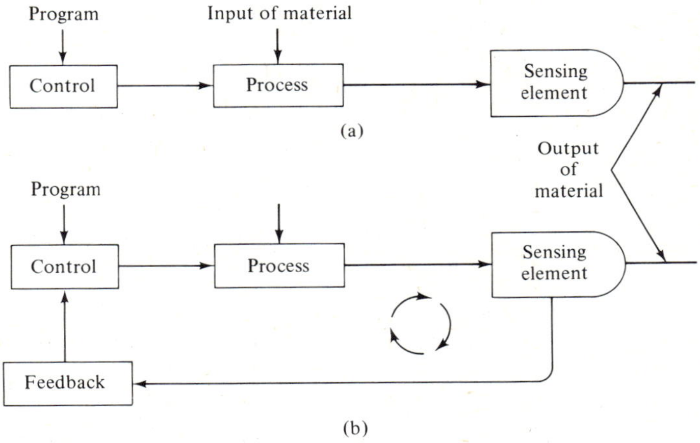

Fig. 1-35
Basic manufacturing systems.
(a) Mechanized.
(b) Automated.

program indicated in Fig. 1-35 may be as simple as a punched card or a punched tape; however, it may represent a large electronic computer. Figure 1-36 illustrates an installation of this type. As we proceed with our study of electronics, we will develop a better understanding of automated systems.

Fig. 1-36
Portion of an automated system.
Courtesy: The Foxboro Company

Questions
Chapter 1

1. Give a definition of electronics as it was discussed in this text.
2. A natural magnet is called a lodestone. What does this term mean?
3. What was the origin for the word "electron"?
4. Explain the operation of a compass.
5. What is the rule for the interaction of one magnet to another?
6. Describe a dip needle.
7. State the name of the smallest part of a material, such as water, that can be obtained.
8. What is the name given to the center of the atom?
9. Explain the principle that determines whether or not a material is identified as iron, oxygen, and so forth.
10. What is the purpose of a voltaic cell?
11. Describe the construction of a dry cell.
12. What is a solenoid and what is another name for this device?
13. Draw a simple diagram of a telegraph system and explain its operation.
14. What is the basic difference between a telegraph and a telephone system?
15. Who invented the incandescent lamp?
16. What was the name given to the first vacuum tube? VALVE
17. State the name of Dr. De Forest's triode vacuum tube. AUDION
18. What is the photoelectric effect? LIGHT TO ELEC & BACK
19. Describe some of the problems of the first television system. SYNCRONIZATION
20. What was the importance of the invention of the cathode-ray tube to the development of television? ELIMINATED MOTOR
21. State the number system upon which digital computers operate.
22. Give a simple example of the use of a photocell as a control device.
23. How are X-rays used to test welds in a metal?
24. What is an oscilloscope?
25. Give the definition of the term "automation."
    REPLACE OPERATOR WITH A MACHINE

2

OCCUPATIONAL FIELDS FOR THE ELECTRONICS TECHNICIAN

2.1 Introduction

An electronics technician has a wide variety of occupational fields from which to choose. The widest choice is available to the best qualified group of technicians. At the start of his career, the technician should consider whether he prefers to be self-employed or to work for some electronics organization. For example, a man who desires to "be his own boss" may embark upon a television and radio repair business. A technician, however, who wishes to have maximum economic security may take a job with one of the government services or work in the laboratory or in the field activity of a large industrial plant.

From a very broad viewpoint, the job opportunities in electronics technology can be divided into the federal government market, the industrial electronics market, and the consumer market. It may be noted that the Department of Defense is the largest purchaser of electronics equipment. The industrial electronics market is second in dollar volume, followed by the consumer market. It is helpful to note that the consumer market is the most stable field of employment and that the federal government market is the least stable from the viewpoint of job security. That is, activity in the Department of Defense is marked by rapid rise-and-fall periods in accordance with changing military requirements.

2.2 Defense Plant Activities

Laboratories and factories that are concerned with government electronics contracts have activities that can be classified into research and development, production design, manufacturing, calibration and test, and such associated operations as equipment maintenance, defective-products troubleshooting, and stockroom management. Each research-and-development engineer usually requires several technicians to assist in model building and testing. Similarly, each of the other activity classifications requires a staff of technicians. It is evident that some technicians will work on higher levels than others and that a given job classification may require specialized training and experience.

After military and naval electronics equipment has been placed in service, there is a continuing problem of maintenance and repair. Since there has seldom been a sufficient number of trained technicians available to maintain and repair existing military and naval electronics equipment, a greater burden falls upon the design laboratory. For example, a warship may be provided with $20 million worth of electronics equipment and devices. To ease troubleshooting and repair by comparatively unskilled technicians, design engineers usually include automatic-failure indicators in electronic systems. These special circuits and devices indicate the section or subsection in which a failure has occurred within the system.

2.3 Industrial Electronics Plant Activities

Although less research and development is found in industrial electronics organizations than in defense plants, this is nevertheless an important occupational area for technicians. Industrial electronics work includes computers, process-control and operating equipment for factories or commercial installations, test and measuring equipment, medical electronic devices and systems, products for nuclear analysis and instrumentation, office dictating equipment, closed-circuit television units, and many related devices and systems. Approximately one half of the electronics technicians employed by industry are assigned to process-control activities. Automation is the most important part of this branch of industrial electronics.

Electronic computers are widely used in control systems for milling machines, various types of lathes, and general metalworking machinery. Electronic process control has made considerable progress in the automation of the garment industry. However, the full potential of the computer market has not been realized owing to the lack of trained programmers to manage computer operation. Another factor is a shortage of qualified computer-maintenance technicians. The technical knowledge required to troubleshoot a large computer exceeds the capabilities of some junior engineers. That is, the training required of a general technician in this area may actually be that of a qualified engineer. A computer troubleshooter, however, has a very specialized training background.

Communications is a general term that means information transfer, and includes telephone, telegraph, teletype, radio, telemetry, television, and microwave relay systems, for example. These fields require large numbers of technicians to install and to maintain the equipment in proper working condition. Industrial or commercial radio installations use two-way systems that need extensive technical knowledge and training for proper maintenance. Some industrial and commercial closed-circuit television systems are quite simple, while others are very complicated. Some of the most elaborate color-television installations are found in medical colleges. The performance of such television systems greatly exceeds that of ordinary home-entertainment television receivers.

Another large employment area for electronics technicians is provided by electrified railway systems. Telecommunication facilities, signaling equipment, control devices, and train sensors require maintenance by qualified technicians. This is a field in which new electronics equipment is introduced at intervals. That is, older electrified railway systems periodically go through modernization programs. These programs involve both new signaling equipment and new telecommunication equipment. However, improvement of signaling equipment requires the largest cost and provides the most jobs for technicians.

Industrial electronics technicians may be most often employed for the installation and maintenance of test and measuring devices. Although this is a rather minor field of employment, it represents a career possibility for technicians who prefer to work with electronic instruments, such as oscilloscopes, recorders, generators, and meters. A worker in this area is often called an electronics metrologist. Metrology is the science of weight and measurements. Employment opportunities in electronics metrology are about equally divided between qualified engineers and technicians. The more demanding responsibilities are assigned to graduate engineers with experience in this field. Routine installation and maintenance jobs are usually given to technician metrologists.

An expanding employment area for laboratory technicians is offered by the medical electronics field. This is a branch of industrial electronics that has been developed by the medical profession. For example, X-ray machines are designed and their installation is planned under the guidance of physicians. The technical details of construction are carried out by engineers. A sizable number of X-ray technicians are employed by factories to work with the engineering staff, both inside the plant and at various field locations. Other medical electronics products include patient-monitoring equipment for respiratory and pulse rates, blood pressure, and body temperature. As in the X-ray field, technicians are required to work with design, production, and field engineers.

Another important employment opportunity for electronics technicians is provided by the nuclear instrumentation industry. Although this is a rather new field, it has a good growth rate. Radiation detectors are used in all nuclear laboratories and plants that process radioactive materials or that have installed atom-smashing equipment. Radiation detectors are also used by prospectors for radioactive ore and mining operations. Color-television technicians occasionally employ radiation detectors to check the operation of high-voltage supplies. Offices and residences provided with bomb shelters against nuclear attack usually have one or more centrally located warning devices which are basically radiation detectors. Other main products in this field include radiometers, spectrometers, and spectroradiometers. A radiometer measures radiation intensity, and a spectrometer measures its wavelength. A spectroradiometer measures the energy distribution in a sample of radiation.

Although the nuclear instrumentation field is a specialized branch of the industrial electronics industry, there is quite a bit of overlap. Thus, meters, oscilloscopes, and electronic computers are generally types of devices which are used by both process-control and nuclear-plant engineers. Remote-control microscopes and manipulators are in general use wherever a radiation hazard exists. These devices and systems serve to position and move radiation samples from the human operator. Closed-circuit television is also commonly used to observe the remote operation. Thus, a technician who specializes in this area of metrology may find employment in more than one type of industrial electronics activity.

2.4 Consumer Market Activities

The consumer electronics market includes such products as television receivers, radios, high-fidelity units, record players, tape players and recorders, garage-door openers, intercommunication systems, walkie-talkies, and so on. There are more self-employed technicians in the consumer electronics field than in any other area. In turn, the electronics servicing market is highly competitive, and the average income in this activity has not kept pace with the general cost of living. There has been a trend toward reduction in the number of small service shops and establishment of large service companies that employ large numbers of technicians. Technicians in this field are mainly concerned with servicing of black-and-white and color-television receivers, although most operators in this area will service radio receivers.

Other branches of the consumer electronics market are more highly specialized. For example, a technician who specializes in servicing electronic organs will seldom take a request for servicing a television or a radio. The reason for this specialization is that modern electronic organs are complicated units, and extensive training and experience on the part of the technician is needed to understand their operation. For example, electronic computer techniques are used in organ design. In turn, it is much easier for an electronic-organ technician to branch out into general computer work than for a radio-and-television technician to do so.

Another rather specialized branch of the consumer electronics market is that of servicing high-fidelity and stereo sets. A technician in this area is qualified to service record players, tape recorders and players, and high-fidelity FM radio receivers. Considerably more technical knowledge is required here than in the servicing of radio and television receivers. For example, complex mechanical designs are used in automatic record changers. Also, specialized test equipment is needed to check the operation of any high-fidelity unit. For example, a high-fidelity unit is rated for power output, harmonic-distortion percentage, and occasionally for intermodulation-distortion percentage. A high-fidelity technician must also know how to use specialized audio test equipment.

A very important consideration of the self-employed technician is that he is responsible for the entire operation of his business. He is the purchasing agent, the bookkeeper, the salesman, the cashier, the employer, and the technician. His success depends upon his ability in these areas as much as on his technical ability.

2.5 Allied Occupational Fields

Some technicians prefer commercial employment instead of technical occupations. For example, a technician may prefer to become a salesman for one or more electronic products. The technician has a decided advantage over the nontechnical person in merchandising elaborate equipment such as computers. Even the component salesman in an electronic parts supply store must have some basic knowledge of electricity and electronics. Individuals who have a highly social nature, and who become bored when working alone, are well advised to consider some form of technical sales work. A talented commission salesman can often earn a much larger income than a routine technician.

Another field related to electronics technology is drafting. Many draftsmen are employed by industry to prepare schematic diagrams, production drawings, and similar material. Draftsmen are also employed by technical publishers to prepare diagrams and drawings for books and magazines. The production of operating manuals for military and naval electronics equipment also requires the services of draftsmen. It is very difficult for a mechanical draftsman to produce an acceptable schematic diagram, for example, unless he has at least a basic understanding of electricity and electronics. Although drafting is not one of the highly paid fields, it offers more security in many cases than other occupations.

Still another field allied to electronics technology is technical editing and publication production. Books and magazines on electronics are planned and prepared for production by technical editors and artists. This is basically a processing operation, in which "raw" manuscript and illustrations are prepared for the printing plant. An editor selects material that he believes will be of interest to prospective purchasers of the publication, reviews manuscripts for technical accuracy and correct grammatical expression, and decides which illustrations, if any, shall be included with the text. Thus, the technical editor is a middleman between an author and the printer of a publication. In turn, a technical editor should be knowledgeable in the fields of electronics technology and English grammar and composition. Like the drafting field, technical editing is not one of the highly paid fields, but it usually offers substantial security.

One of the most attractive, though one of the most difficult, of these related occupational fields is that of creative writing. A creative electronics writer is typically a highly individualistic worker, who conceives and executes literary products, such as new and better school textbooks, improved operating manuals for electronics equipment, and technical books that shed new light on familiar but incompletely understood subjects. A creative writer is usually self-employed, and works on a contract basis with one or more book publishers. If he contracts with several publishers, he is classified as a free-lance writer. Although creative writing is a hopelessly difficult field for most technicians, the talented person will find it among the most lucrative of occupations.

Questions Chapter 2

1. List five types of jobs that are available to electronics technicians in your area.
2. What are the educational requirements for each of the jobs listed in Question 1?
3. State the approximate starting salaries of the jobs listed in Question 1.
4. What are the ideal working conditions of a job that you would like for yourself?
5. Pick one of the five jobs in Question 1 that comes closest to qualifying as your ideal job.
6. What preparation is necessary to qualify for the job that you selected in Question 5?
7. List the special school courses and training opportunities available in your area to help you qualify for the job that you selected in Question 5.
8. Study your job selection and the educational requirements for that job very carefully. Now, estimate the length of time that would be required to prepare yourself for that job and the total cost of that preparation.
9. Are you prepared to put forth the effort and expense for the training that you have determined to be necessary for your future job selection? If so, good luck. If not, you should select another job and repeat the last three questions.

3 STATIC ELECTRICITY IN NATURE AND IN THE LABORATORY

3.1 Lightning and Franklin's Experiment

Lightning was not understood before 1751, although many different theories had been suggested. One of these theories was that lightning was an electrical discharge. Benjamin Franklin believed that lightning was a form of electricity, and devised an experiment to prove this theory. This experiment, which was simple although quite dangerous, consisted of flying a kite with a tinsel string in low-lying thunderclouds. A key was suspended at the end of the string, from which Franklin drew sparks when he brought his knuckles near the key. (Fig. 3-1).

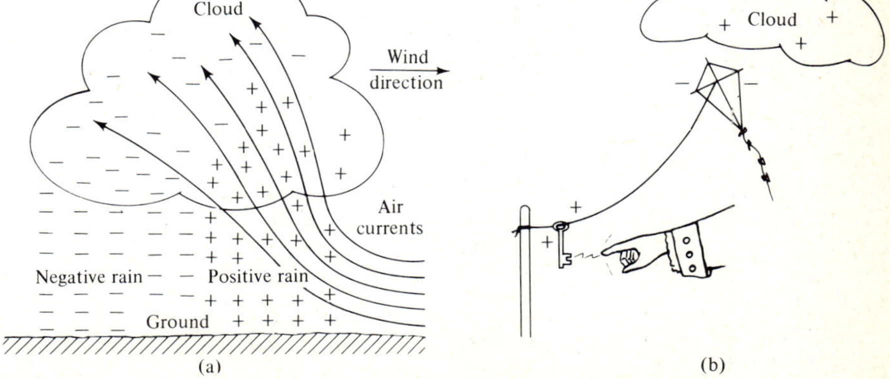

Fig. 3-1
Examples of static charge.
(a) Distribution of electric charges in a thundercloud.
(b) Franklin's experiment.

When the kite was flown in the negative portion of a thundercloud, and a person was standing on the positive area of ground beneath, a potential difference existed between the tinsel string and the person's body. Whenever the potential difference happens to be sufficient, a spark can be drawn from the string.

This is a dangerous experiment because the electric potential occasionally jumps up to a very high value and can kill the kite flyer. Several scientists were killed while attempting to repeat Franklin's experiment. It is important to understand why the experiment proved that lightning is an electrical discharge. The proof consisted in demonstrating that, when the tinsel string is in a thundercloud, electric charges flow down the string and can be taken from the string at the ground location. Proof that charges were being taken from the string was shown by producing an electric spark, in the form of a miniature lightning discharge, from the key to Franklin's knuckles. Franklin also showed that if a stick of sealing wax was touched to the kite string, the wax rod became charged as if it had been rubbed with cat's fur.

3.2 Laws of Positive and Negative Charges

We have learned that positive and negative charges are produced by ionization of atoms. That is, if a neutral atom loses an electron, the atom becomes an ion; the ion has a net positive charge, and the electron that has been lost has an equal negative charge. We know that ions can be produced by mechanical friction, such as that produced when a glass rod is rubbed with a silk cloth. Atoms can also be ionized by chemical action, as in a dry cell. It has been noted that atoms can be ionized (electrons emitted) by heating a metal filament to incandescence. When an electron is emitted from a glowing filament, it will be captured by any gas molecule that happens to be in its path. If a neutral gas molecule captures an electron in this manner, the molecule then has a net negative charge. In turn, the gas molecule becomes an ion. Thus, ions are formed when atoms or molecules capture electrons.

Static electricity means "charges on electricity at rest." In Fig. 3-1, the cloud is charged with static electricity which is due to friction of the air currents on the water vapor. If a bolt of lightning jumps between the cloud and the ground beneath, the charges are no longer at rest. Instead, the charge on the cloud suddenly discharges or moves along a column of ionized air between the cloud and the ground

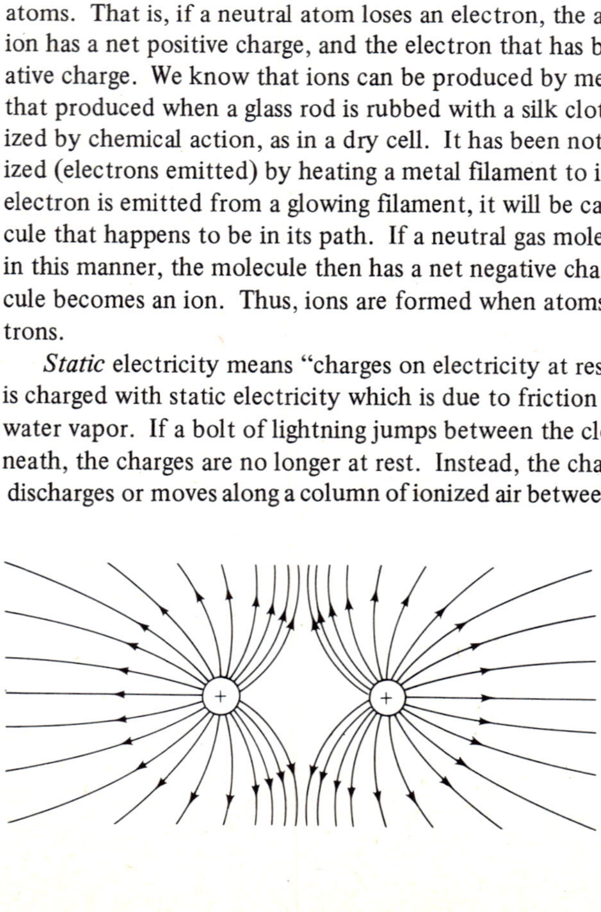

Fig. 3-2
Patterns of electric fields for charges of like polarity.

beneath. The air becomes ionized as a result of a very high electric potential, or pressure, that has built up on the cloud because of the friction from air currents. The basic electrical law that is at work in this situation is called the *Law of Charges,* and it is stated as follows:

1. ***Like charges repel each other.*** A pair of negative charges repel each other, and a pair of positive charges repel each other, as shown in Fig. 3-2.
2. ***Unlike charges attract each other.*** A positive charge and a negative charge attract each other, as shown in Fig. 3-3.

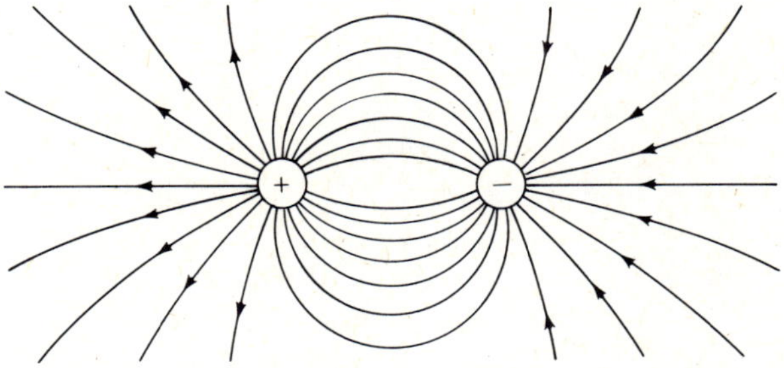

Fig. 3-3
Patterns of electric fields for charges of unlike polarities.

3. ***A charge of either polarity attracts a neutral body.*** Thus, a positive charge attracts a neutral body, and a negative charge attracts a neutral body. This is because a charge of opposite polarity is *induced* in the neutral body, as explained in greater detail at a later point.

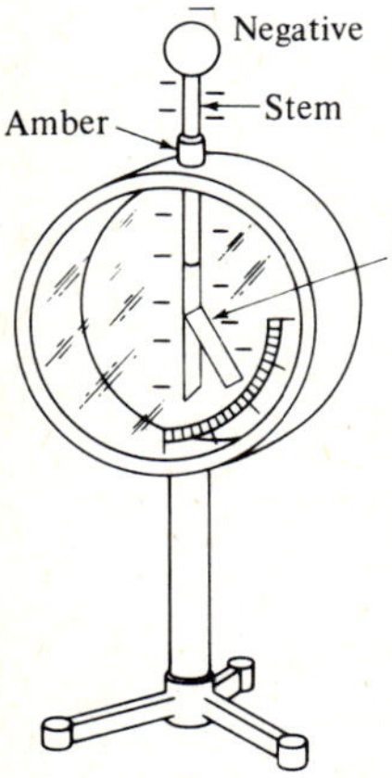

Fig. 3-4
A classroom electroscope, with negative charge indicated.

An electrical instrument called the *electroscope* (Fig. 3-4) is often used to show the presence of static electricity, and to measure its potential difference. This instrument operates upon the principle that like charges repel each other. It consists of a pair of very thin metal leaves (often gold-leaf foil) suspended from a metal rod. If a negative charge is applied to the rod, each leaf repels the other, and the leaves open; if a positive charge is applied to the rod, each leaf repels the other, and the leaves open. The amount of opening is a measure of the potential difference (electric pressure or tension) that is forcing the leaves apart.

An Electrostatic Experiment

The objectives of this experiment are:

The process of electrostatic charging by induction.
The process of electrostatic charging by contact.
Like charges repel each other.
Unlike charges attract each other.

Materials:

Vulcanite (hard-rubber) rod and cat's fur, or the equivalent.
Glass rod and silk handkerchief, or the equivalent.
Laboratory electroscope.
Single and double pith balls suspended by silk threads from stands.

Procedure:

To demonstrate the process of electrostatic charging by induction, rub the vulcanite rod with the cat's fur. The rod is thereby charged negatively. When the rod is brought near the metal ball atop the electroscope, the gold-foil leaves open. This action occurs because the negatively charged vulcanite rod repels electrons from the metal ball, and the electrons flow down the metal stem to the gold-foil leaves (Fig. 3-5). The leaves are thereby *charged by induction.* Since both leaves are charged negatively, they repel each other and the leaves open.

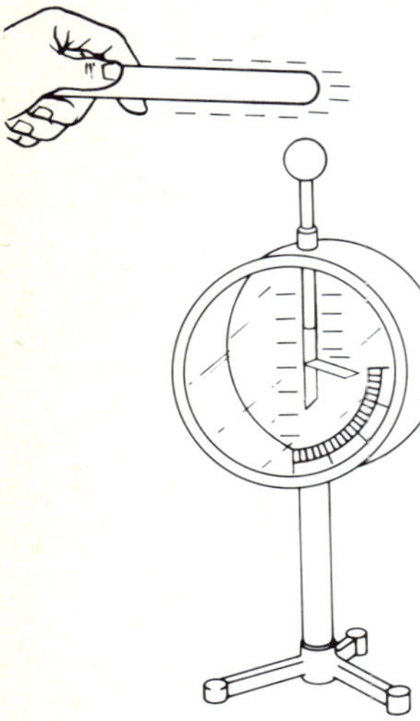

Fig. 3-5
Charging an electroscope by induction.

Next, when the vulcanite rod is withdrawn from the area of the electroscope, the gold leaves fall back together. This action occurs because the electrons that were concentrated on the gold leaves now flow back up the metal stem, and the electroscope becomes neutral once again. In turn, the leaves no longer repel each other, and the force of gravity causes them to fall together. Notice that the amount of opening of the leaves depends upon how far the vulcanite rod is spaced from the metal ball. Thus, the amount of opening indicates how much charge has been induced in the electroscope.

To demonstrate electrostatic charging by contact, rub the vulcanite rod with cat's fur, and touch the negatively charged rod to the metal ball atop the electroscope. The gold leaves open as before; however, when the vulcanite rod is withdrawn, the leaves remain open, as shown in Fig. 3-6. This action occurs because electrons are transferred from the charged vulcanite rod to the metal ball upon contact. The result is that the metal stem and the gold leaves are charged negatively. When the vulcanite rod is withdrawn, the electrons that have been transferred remain trapped in the electroscope. For this reason, the gold leaves remain open.

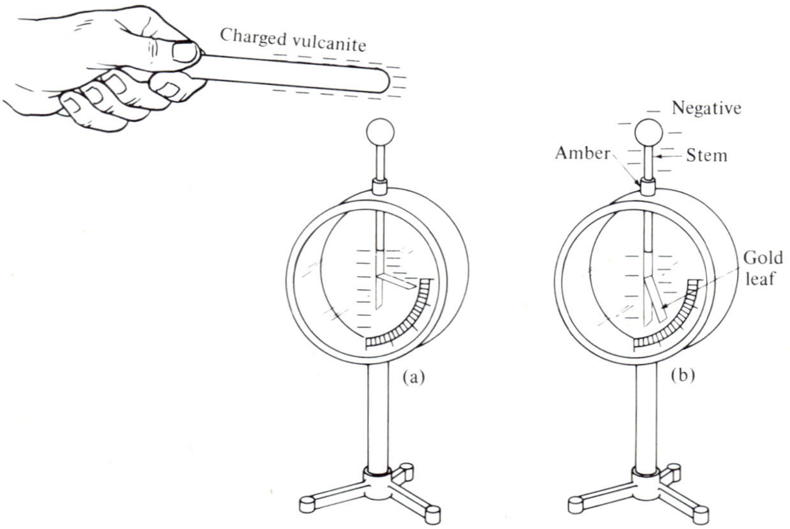

Fig. 3-6
Charging an electroscope by contact. (a) Electrons are transferred to the metal ball. (b) Charge remains after vulcanite rod is withdrawn.

To demonstrate that an induced positive charge cancels a negative charge, rub the glass rod with the silk cloth. The glass rod thereby becomes charged positively. When the charged glass rod is brought near the metal ball atop the electroscope (which is charged negatively from the foregoing experiment), the leaves fall back toward each other. As shown in Fig. 3-7, the positively charged glass rod

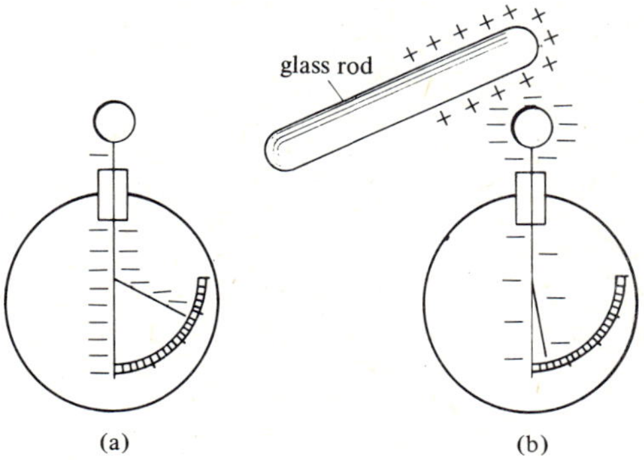

Fig. 3-7
Cancellation of negative charge by induced positive charge. (a) Negatively charged electroscope. (b) Leaves fall back as positively charged glass rod approaches.

attracts electrons from the gold leaves to the metal ball. This causes the leaves to lose their negative charge and to fall back toward each other. If the glass rod is withdrawn, the leaves again open. On the other hand, if the glass rod is touched to the metal ball, the leaves do not open when the rod is withdrawn.

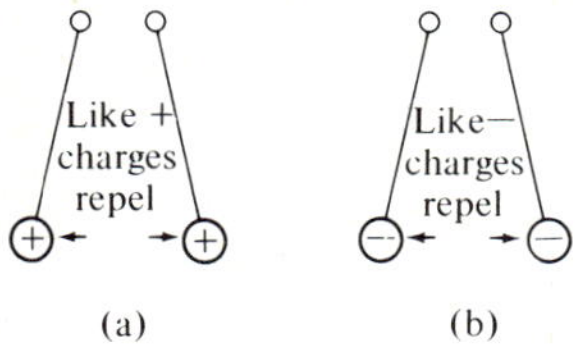

Fig. 3-8
Repulsion of like polarities. (a) Pith balls charged positively. (b) Pith balls charged negatively.

To demonstrate that like charges repel each other, use the glass rod to charge each pith ball of a pair positively by contact. As shown in Fig. 3-8a, the balls repel each other and swing apart. Next, use the vulcanite rod to charge each pith ball of another pair negatively by contact. As shown in Fig. 3-8b, the balls repel each other and swing apart.

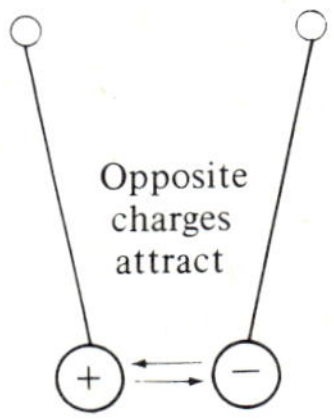

Fig. 3-9
Attraction of like polarities.

To demonstrate that unlike charges attract each other, charge a single pith ball positively by contact with the glass rod. Then charge another single pith ball negatively by contact with the vulcanite rod. As the stands are brought near each other, the unlike charges attract each other and the pith balls tend to swing together (Fig. 3-9).

If the foregoing experiment is repeated and each step is carefully observed, we can note that when the vulcanite rod is brought near the neutral pith ball, the ball is at first attracted to the rod, as shown in Fig. 3-10a. Then, after the pith ball has been charged negatively by contact, the ball is repelled by the rod, as shown in Fig. 3-10b. Figure 3-11 may help us to understand why a charged

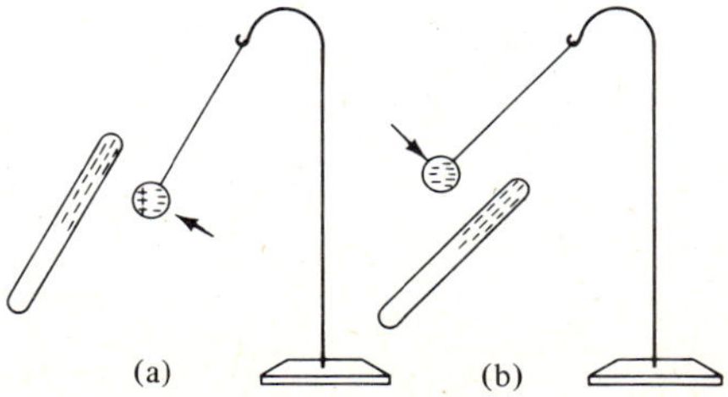

Fig. 3-10
Reaction of polarities. (a) A metallic coated pith ball is attracted by a charged rod. (b) After contact, the pith ball is repelled.

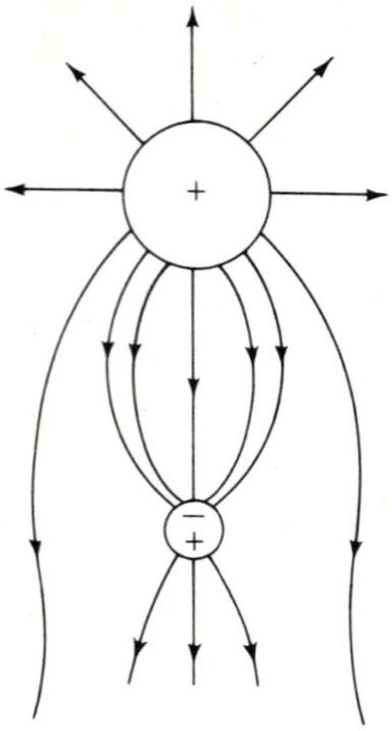

Fig. 3-11
A positively charged body induces an adjacent negative charge on a neutral body.

body attracts a neutral body in its vicinity. We notice that the neutral body is charged by induction; and since the negative induced charge is nearest to the positively charged body, a force of attraction results.

3.3 How an Electrostatic Machine Operates

We know that a glass or hard-rubber rod becomes charged when rubbed by some other substance. Scientists use this principle in electrostatic machines to generate large quantities of static electricity. One of the most common arrangements is the Van de Graaff generator. A classroom model of this generator is shown in Fig. 3-12. We can explain its operation with reference to Fig. 3-13. A

Fig. 3-12
A classroom Van de Graff demonstrator.
Courtesy: Edmund Scientific Co.

large-diameter metal electrode is mounted on an insulating column (usually made from porcelain), so that charges on the electrode cannot leak to ground. The electrode is charged by a motor-driven insulation belt, made of fabric. In operation, the moving belt transports electrons from a source of electricity *E* up to the electrode at the top of the generator. As the belt continues this action, the electric potential on the electrode will build up any desired potential, or until an artificial-lightning discharge occurs.

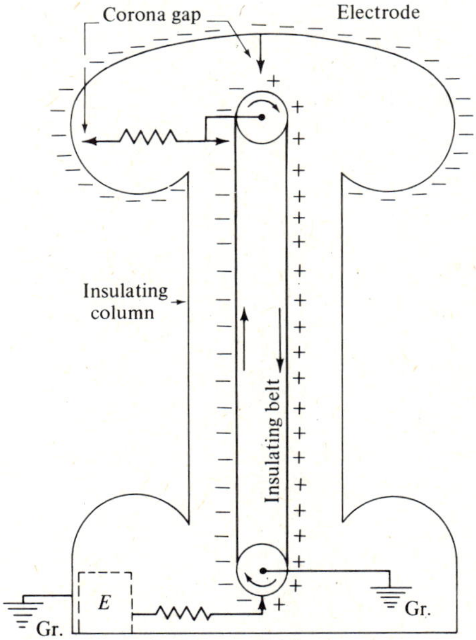

Fig. 3-13
Plan of Van de Graaff generator.

In Fig. 3-13, the source *E* has a sufficiently high potential so that a glow discharge occurs between the belt and the pointed electrode at the bottom. This causes the portion of the belt moving upward to be charged negatively. At the top, the negative charge on the belt produces a glow discharge to the top pointed electrode, transferring the negative charge to the inside surface of the electrode. From the inside surface, this negative charge immediately flows to the outside surface of the electrode. We find that the inside surface of a metal object cannot hold an electric charge. Thus, the potential on the outside of the electrode will continue to build up until the surrounding air ionizes and an artificial-lightning discharge occurs.

3.4 The Coulomb Unit

Various units are used for purposes of measurement. For example, the inch is a unit of length, and the ounce is a unit of weight. Similarly, the *coulomb* is a unit of electric charge. A coulomb may have either positive polarity or negative polarity. One coulomb consists of 6.28×10^{18} electrons, or 6,280,000,000,000,000,000 electrons. The unit of electric charge is named after Charles de Coulomb, a French scientist who made many important discoveries.

3.5 The Volt Unit

Another basic electrical unit is used to measure potential difference, or electric pressure. This unit is called the *volt* (V), and is named after Alessandro Volta, who invented the electric battery. It has since been shown that high voltages can be measured with reasonable accuracy by determining how far a spark will jump between needle points. Figure 3-14 shows that about 28,000 volts are required to make a spark jump across a one-inch gap. The ignition system of an automobile operates at high voltage to make a spark jump across the gap in a spark plug, whereas the headlamps of an automobile operate at 12 volts. This low voltage cannot make a spark jump across a gap unless its spacing is microscopic.

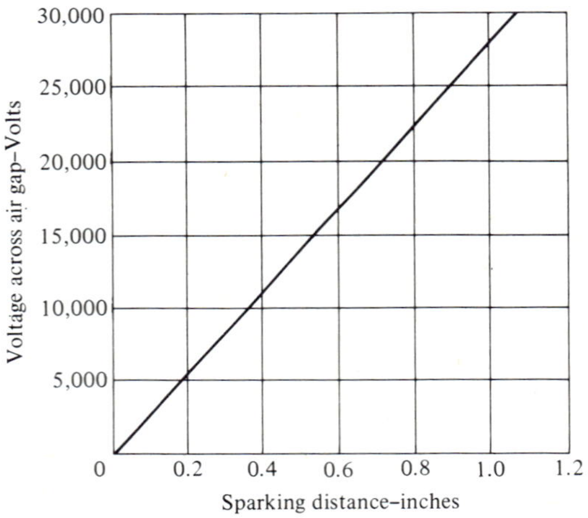

Fig. 3-14 Voltage required to produce a spark between sharp metal points.

3.6 The Leyden Jar and the Farad Unit

In 1746, Professor Pieter van Musschenbroek made various experiments in an attempt to devise a means of storing the electricity produced by a static machine. These investigations led to the invention of the *Leyden jar.* A model of the Leyden jar is shown in Fig. 3-15. The Leyden jar was the first *condenser,* or *capacitor.* Although both terms are used, "capacitor" is preferred by engineers. A Leyden jar consists of two metal surfaces separated by glass. Modern capacitors generally use flat metal plates. Figure 3-16 shows the operating principle of a capacitor. When one plate is charged positively, and the other plate is charged negatively, the resulting electrostatic field is concentrated in the space between the plates. This is stored field energy, which is retained by the capacitor after it is disconnected from the source of electricity. However, if the plates of the capac-

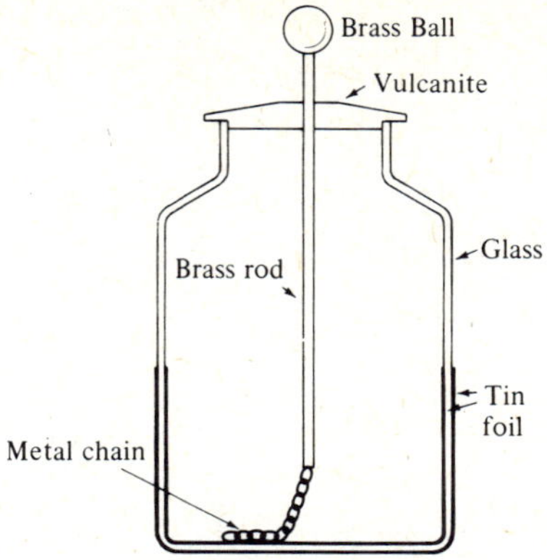

Fig. 3-15
Leyden Jar.

itor are then connected together, the stored electric charge rushes out and often produces a bright and loud spark.

Capacitance is measured in *farad* units. With reference to Fig. 3-16b, a capacitor has a capacitance of one farad (F) if it stores one coulomb (C) of charge when it is connected across a one-volt source of electricity. In other words, a one-farad capacitor will store 6,280,000,000,000,000,000 electrons when one volt is applied across the plates. Since a farad is a very large unit, we usually work with microfarad units. A microfarad is equal to one-millionth of a farad. We also work with picofarad units. A picofarad is equal to one-millionth of a microfarad. We will consider capacitor action in greater detail as we continue our study of electronics.

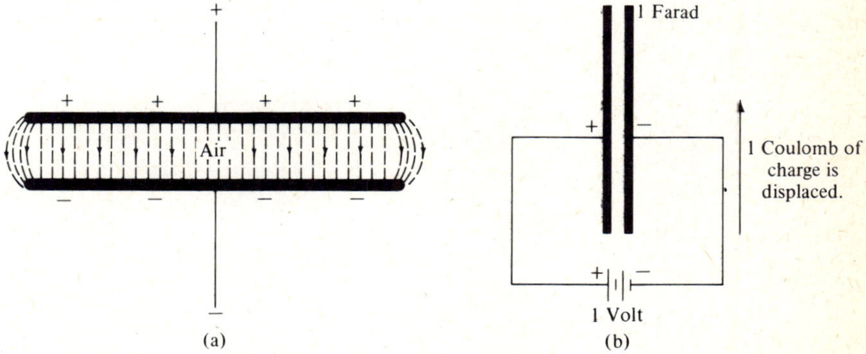

Fig. 3-16
The capacitor principle. (a) Concentration of electrostatic field. (b) A 1-farad capacitor stores 1 coulomb of charge at 1 volt.

3.7 Industrial Applications of Static Electricity

Static electricity has wide application in industry. For example, Cottrell *precipitators* (Fig. 3-17) have been used for many years. One precipitator acts to separate smoke and dust particles from gases, which are then collected at the bottom of the unit. As the gas enters the unit the suspended particles are electrically neutral. These neutral particles are attracted to the 30,000-volt wires. When the particles touch the electrified wires, they are charged in the same polarity as the wires

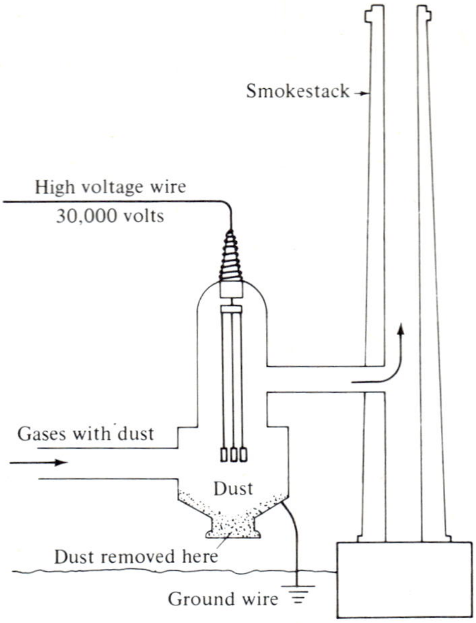

Fig. 3–17
A Cottrell precipitator.

and are rapidly repelled. Since the particles are now electrified, they are attracted to the grounded (neutral) collector at the bottom of the unit. The electrically "cleaned" gases then pass into the smokestack, and the atmospheric pollution is reduced.

Electrostatic painting is another important industrial application of static electricity. In Fig. 3–18, an object to be painted is moved through a strong electrostatic field. The object is grounded, with the result that the field is most intense at the surface of the object. A cloud of paint spray is directed at the object from

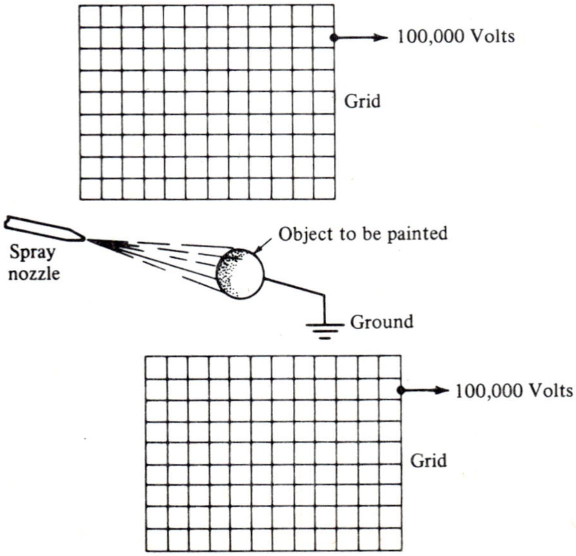

Fig. 3–18
Plan of an electrostatic painting unit.

one or more spray nozzles. Electrostatic attraction of the charged paint droplets by the grounded object causes the paint to adhere very tightly, and this produces a better coating than ordinary spray painting. In mass-production operations, another advantage is the reduction in wasted paint that is realized by the electrostatic process.

Electrostatic deposition of sand or emery particles has been used in the manufacture of sandpaper for many years. Figure 3-19 shows the plan of the equipment used for this process. The operation is as follows: Paper or cloth is moved

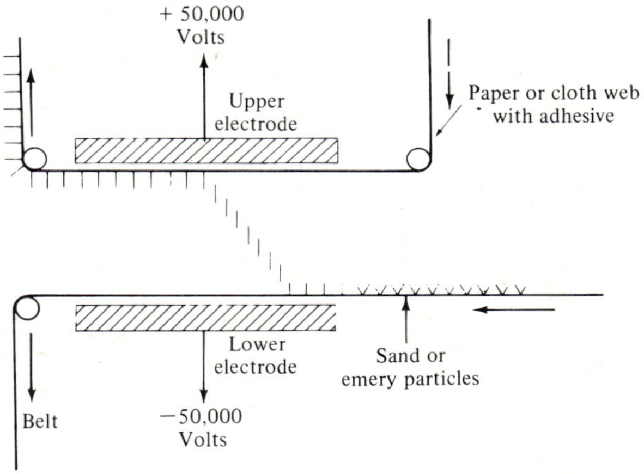

Fig. 3-19
Sandpaper manufacture by electrostatic deposition.

from a continuous roll with an adhesive applicator over the face of an electrode connected to a high-voltage source. At the same time, a belt that conveys the particles is moved over the face of a lower electrode. The particles are charged negatively by the lower electrode, causing their attraction to the positive upper electrode. The charged particles strike the adhesive at high speed and are firmly embedded, producing a high-quality type of sandpaper. One of the advantages of the electrostatic method of producing sandpaper is that the sharpest edges of the particles point out from the surface, providing a better cutting action.

Electrostatic sorting of various materials is used in many industries. For example, if a mixture of two different materials falls by gravity through an electrostatic field, one of the materials may be attracted more strongly than the other material. Whenever this difference in attractive force is strong enough to make one substance fall in a separate stream, the method can be used for sorting, as shown in Fig. 3-20. We find this method used in ore and mineral processing, in the food industry, and so on. The unsorted material passes over a grounded conveyor roller and starts its fall in an electrically neutral condition. As the materials fall through the strong electrostatic field, they become charged by induction and are attracted to the positive electrode. However, the force of attraction will be stronger for one material than for the other, causing that material to drop into the right-hand hopper.

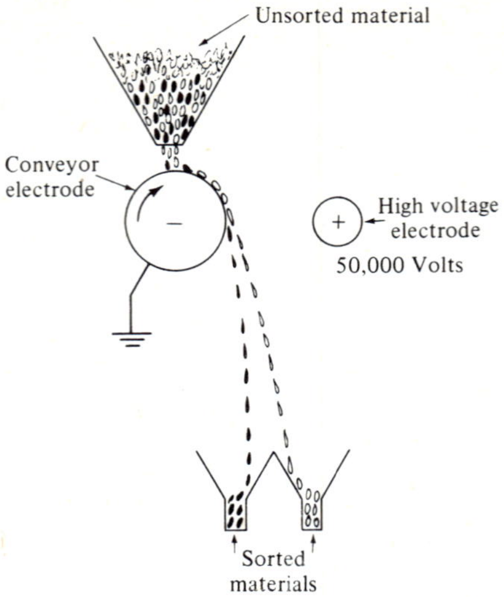

Fig. 3-20
Principle of electrostatic sorting.

Electrostatic fields are also used to remove excess paint from manufactured products that are dip-painted. When an object is dip-painted, excess paint tends to collect in the form of "teardrops" that tend to spoil the appearance of the painted surface. However, these "teardrops" can be automatically removed by passing the freshly dipped object over an electrode charged to about 100,000 volts. The strong electrostatic field set up between the electrode and the painted object pulls the excess paint off the surface of the object. This process is called *electrostatic detearing.*

3.8 Protection Against Static Electrical Discharges

Buildings, airplanes, trucks, and electric power lines must often be protected against fire and damage by static electrical discharges. These discharges may be in the form of lightning strokes or caused by air friction against the surface of a moving object. Benjamin Franklin invented the lightning rod, which has since been used for protection against lightning strokes. As shown in Fig. 3-21, the

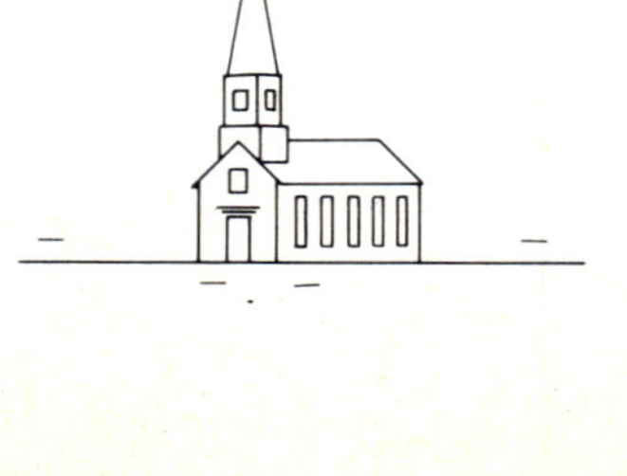

Fig. 3-21
A lightning rod protects a building against lightning strokes.

drawing demonstrates the manner in which a charged cloud will discharge to the tip of the lightning rod and be conducted safely to ground. To conduct this charge, a thick copper wire is usually connected to the lightning rod and is run down the side of the building to a pipe that is driven into moist earth.

As an airplane flies through particles of dust, sleet, snow, or rain, it often collects electrical charges from the particles as they strike the metal fuselage. The sum of all the small charges that are collected is added, and the airplane may be charged to a potential of several hundred thousand volts. This intense charge will gradually escape in the form of a glow discharge called St. Elmo's fire. This type of discharge produces radio interference and may also cause a fire if gasoline vapor happens to be leaking out. To prevent this problem, a special discharging device is generally used on an airplane. One type of discharger called a "cat's tail," consists of a five-foot length of carbonized plastic, terminated by a five-foot bundle of extremely fine wires with sharp points. This arrangement produces a smooth discharge at comparatively low voltage.

High-speed trucks often become charged to high voltages by air friction. These high voltages occasionally cause discharges to take place as sparks in the same manner as from a static machine. These sparks impose a fire hazard when vapor is escaping from the gasoline tank or from the fuel line. To provide a continuous discharge of static electricity as fast as it is generated, truckers may attach a short chain to the rear frame. This chain drags on the pavement and serves to conduct the charges to ground. Although the truck wheels are in contact with the pavement, rubber is a good insulator and prevents the flow of static electricity from the truck frame to ground.

Plastic seat covers will often generate high-potential static electricity when a person moves or slides across the plastic surface. The high potential causes a disagreeable spark to jump from the charged person's hand to the door handle as he starts to get out of the car. To eliminate this nuisance, conductive sprays are available which are invisible, but provide a conductive coating on the plastic surface. These coatings provide protection against build-up of static electricity by conducting the charges away as fast as they are generated. Similar conductive sprays are also used to keep the glass or plastic faceplates of electrical meters neutral. This is necessary because a static charge collecting on the faceplate of an electrical instrument causes the pointer to be attracted to the area of strongest charge and the scale indication to be incorrect.

Questions: Chapter 3

Multiple choice: choose the best answer.

1. Benjamin Franklin proved that
   a. a key should be made of metal
   b. lightning is a form of electricity
   c. a string will conduct electricity
   d. lightning is negatively charged
2. Franklin proved that electricity was being taken from the kite string by
   a. using a voltmeter
   b. using electroscopes
   c. the sound of the electric arc
   d. producing an electric spark
3. An atom or molecule that has gained or lost an electron is called a/an
   a. proton
   b. neutron
   c. ion
   d. nucleus
4. The principle of the Law of Charges can be demonstrated with an instrument known as a/an
   a. charge meter
   b. electroscope
   c. static meter
   d. ergometer
5. The unit of electric charge is the
   a. coulomb
   b. volt
   c. watt
   d. franklin
6. The unit of electric pressure is called the
   a. volt
   b. coulomb
   c. farad
   d. watt
7. The unit of capacitance is called the
   a. volt
   b. coulomb
   c. franklin
   d. farad
8. A microfarad is ________ farads.
   a. 0.001
   b. 0.000,001
   c. 0.000,000,001
   d. 0.01

9. Static electricity may be used in industry to aid in
   a. cleaning the air
   b. painting
   c. making sandpaper
   d. all of the above
10. An example of an undesirable effect of static electricity
   a. St. Elmo's fire
   b. the static charge on a gasoline truck
   c. static charge on a meter face
   d. all of the above

True–False:

1. Benjamin Franklin invented lightning.
2. Franklin's kite experiment was very dangerous.
3. Static electricity means charges at rest.
4. A molecule that has gained or lost an electron is called an ion.
5. The Law of Charges states that like charges attract.
6. An electrostatic machine is used in the classroom to demonstrate the production of static electricity.
7. The volt is the unit of electric charge.
8. The volt is the unit of electric pressure.
9. The Leyden jar was the first capacitor.
10. A capacitor stores charges in the form of field energy.
11. Static electricity has little industrial application.
12. Static electricity can produce discharges from planes and trucks that can be dangerous.

4

CURRENT ELECTRICITY—PHYSICAL AND CHEMICAL SOURCES

4.1 Electric Current and Charge Flow

We have learned that when a spark jumps between two conductors, an electric current is flowing through the ionized air between the electrodes. This is just another way of saying that electric charges are flowing; in this example, the charges are electrons. When an electric lamp is connected to a battery, electrons flow in the connecting wires as indicated in Fig. 4-1. As the electrons flow through the filament in the lamp, electrical energy is changed into thermal (heat) energy and into light energy. The heat causes the lamp to glow and become warm. If we use a higher voltage battery, the current flow is greater and the lamp glows more brightly. If too much voltage is applied to the lamp, however, its filament will burn out and the lamp will no longer be useful.

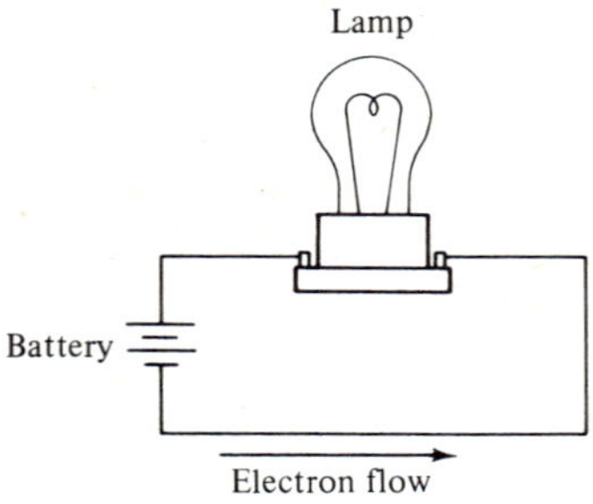

Fig. 4-1
Electrons flow as indicated by the arrow.

4.2 Unit of Electric Current: the Ampere

Current means *charge flow per second.* The unit of electric current is called the *ampere.* One ampere is equal to one coulomb passing a given point in one second. That is, an ampere is equal to the passage of 6.28×10^{18} electrons past a point in one second. For example, if the current in the lamp filament (Fig. 4-1) happens to be one-fourth ampere, there are 1.57×10^{18} electrons passing through the filament each second.

We must have some sort of energy source or converter to cause a current or charge flow; therefore, we will study some of the available energy sources or converters in this chapter.

4.3 Basic Batteries

The operation of all *primary* batteries is based on the voltaic cell.[1] A primary battery changes chemical energy into electrical energy; its chemical reactions are irreversible. That is, a primary battery cannot be recharged, but must be replaced after it "goes dead." The dry cell is in very wide use and is manufactured in several different sizes, ranging from the No. 6 ignition dry cell to the small penlight cell. If the same lamp bulb is operated by a large dry cell, and then by a small dry cell, the large cell will supply electricity for a comparatively long time before it must be replaced. In other words, the *current capacity* or current capability of a large dry cell is greater than that of a small dry cell. Figure 4-2 shows various sizes of dry cells.

Fig. 4-2
Dry cells are manufactured in various sizes.
Courtesy: Burgess Battery Co.

We have learned that a dry cell (Fig. 4-3) uses zinc and carbon electrodes. This combination of electrode materials provides an electromotive force (electric potential) of approximately 1.5 volts. The electromotive force (emf) of a dry cell is the same whether it is a large-size No. 6 cell or a small penlight cell. That is, the emf of a cell is determined by its chemicals and not by its size. However, a large cell has a greater current capability, as mentioned before. Not only will a

1. See Chapter 1.

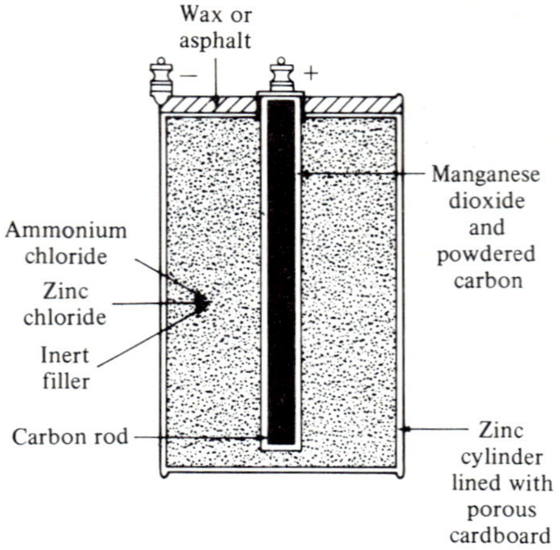

Fig. 4-3
Sectional view of a cry cell.

large dry cell have a longer life than a small dry cell, but the large cell can operate a powerful lamp, whereas the small cell can operate a small lamp only. Let us see why this is so.

A large cell has more electrode area. Since a small area can produce only a certain number of amperes, we must double this area if we wish a cell to produce twice as many amperes. Thus, a very large cell can supply a very large number of amperes. The ampere capability of cells that are the same in size may also be different because of a difference in speed of chemical action.

The chemical reactions in a cell are one-way reactions. This means that, when one or more of the chemicals in the cell is completely used up, the cell will "go dead." We find that a cell will also go dead after a long period of time, even if it is standing idle. This is called the *shelf life* of a cell. The length of this shelf life depends upon the amount of *local action* that takes place in the cell. A No. 6 dry cell has a shelf life of about one year. As shown in Fig. 4-4, local action consists of a chemical reaction between the zinc electrode and any impurities, such as carbon, iron, copper, or other conductive elements. Although the cell is standing idle, this local action eats away at the zinc electrode. Local action can be stopped by using pure zinc; however, this is an expensive metal. Therefore, in most cells amalgamated zinc is used in which there is a small amount of mercury applied to the zinc to form a very thin film over its surface. The mercury prevents any foreign particles from attacking the zinc.

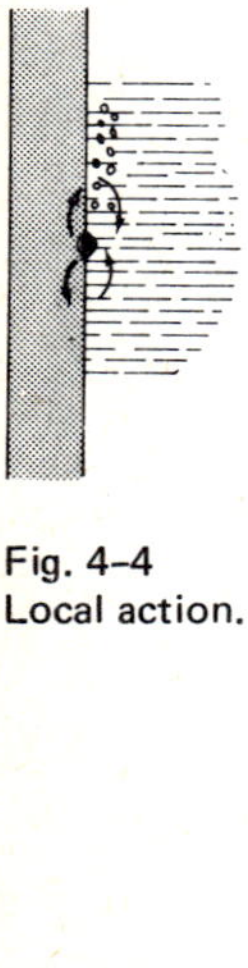

Fig. 4-4
Local action.

4.4 Mercury Batteries

When more current and longer life are required than can be provided by a dry cell, the more costly mercury cell is often used. Figure 4-5 shows a mercury battery made up of seventeen mercury cells, each cell having an emf of approximately 1.34 volts. This is another example of a primary cell, which cannot be recharged. A mercury cell can supply about five times as much current as a dry cell of the same size.

Fig. 4-5
Mercury battery.
Courtesy: P. R. Mallory & Co. Inc.

4.5 Alkaline-Manganese Batteries

Another type of primary cell that supplies more current than a dry cell is the alkaline-manganese cell—often called an alkaline cell or battery. The alkaline cell, which is used in portable radios, tape recorders, shavers, and electric toys, has an appearance that is the same as that of a flashlight cell. The alkaline cell (Fig. 4-6)

Fig. 4-6
Alkaline-manganese cells.
Courtesy: Burgess Battery Co.

has an emf of approximately 1.2 volts. In Fig. 4-6, we can see an example of a cell that is labeled as a battery. Although a single cell is popularly termed a battery, we find that by strict definition a battery consists of two or more cells connected together.

4.6 Rechargeable Batteries

Various types of rechargeable batteries are in wide use. Within recent years, the nickel-cadmium battery (Fig. 4-7) has become very popular. It has a very long

Fig. 4-7
Typical nickel-cadmium batteries.
Courtesy: Edmund Scientific Co.

life, and is not damaged by overloads although an extremely large current may be drawn. A nickel-cadmium cell has an emf of approximately 1.25 volts, and is an example of a *secondary* cell. That is, when discharged, the nickel-cadmium cell can be connected to a source of electricity and recharged in the same basic manner as an automobile storage battery. (A rechargeable or secondary cell is often called a *storage* cell). The action of a secondary cell can be compared in some ways with the action of a Leyden jar or capacitor. However, a capacitor stores electric charge directly, whereas a secondary battery stores electric charge indirectly by means of electrochemical action. Nickel-cadmium batteries are used in cordless electric shavers, electric toothbrushes, camera flash units, flashlights, slide viewers, portable radio and television receivers, and various battery-powered novelties and toys.

Almost everyone is familiar with the lead-acid storage battery used in automobiles. To understand how it works, two lead plates may be placed in a water solution of sulfuric acid, as shown in Fig. 4-8. When these two plates are connected

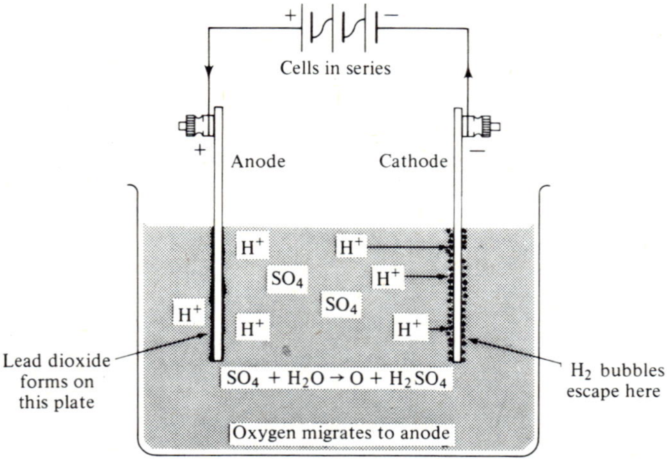

Fig. 4-8
The chemical action that occurs during the charging of a storage cell.

to a source of electricity (such as a battery, for example), the acid solution is decomposed and hydrogen bubbles escape from the negative electrode. The negative electrode is often called the cathode. At the positive electrode (anode), lead and oxygen combine to form lead dioxide. If the lead-acid cell is now disconnected from the source of electricity, we find that the cell operates as a primary cell, except that its terminal emf is approximately two volts. The plate with the lead dioxide is positive.

As a charged lead-acid cell discharges, the lead dioxide changes into lead sulfate, and water is formed. The other plate also becomes coated with lead sulfate. After the cell is discharged, it can be recharged. The result is that the lead sulfate combines with water to reform lead dioxide on the positive plate and to reform pure lead at the negative plate.

The chemical action in a lead-acid battery is *reversible.* Since lead sulfate and water are changed into sulfuric acid during the charging process, water is replaced by sulfuric acid allowing the acid solution to become more concentrated. (Sulfuric acid is heavier than water.) In turn, its specific gravity increases. This change in specific gravity is used to check the condition of a storage battery. Figure 4-9 shows the use of a hydrometer. A standard lead-acid battery has a

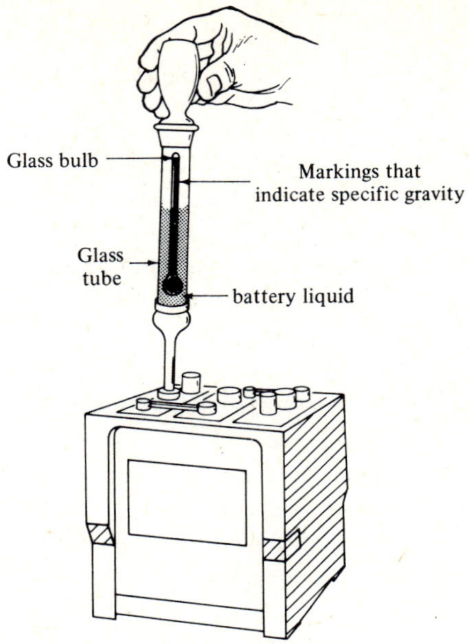

Fig. 4-9
Taking the specific gravity of the electrolyte.

specific gravity of 1.3 when fully charged, and a specific gravity of 1.15 when nearly discharged. The hydrometer is a glass ball inside a glass tube. The bulb floats in the liquid that is drawn into the glass tube. The marking on the bulb indicates the specific gravity of the liquid in which the bulb is floating. This type of battery should never be completely discharged, and should not be stored in a nearly discharged condition; otherwise, harmful reactions will occur that damage the plates. It is good practice always to keep a lead-acid battery reasonably well charged, and to fully charge a battery that is placed in storage. Since sulfuric acid can *burn your clothes or skin severely,* use care when checking the electrolyte.

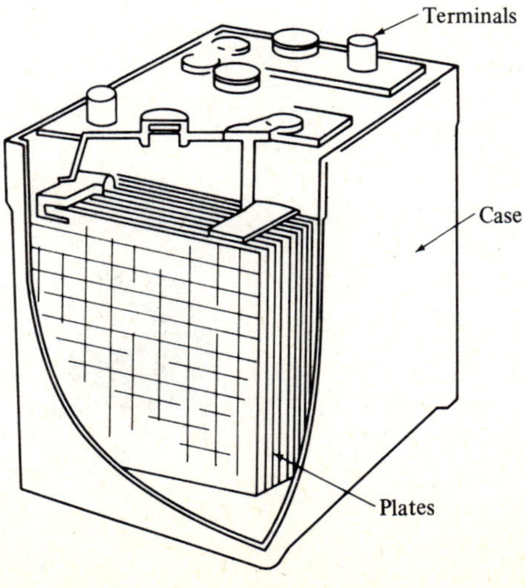

Fig. 4-10
Cutaway view of a lead-acid storage cell.

Remember also, that hydrogen gas is *highly explosive*—never light a match near a lead-acid battery that is being charged.

A commercial lead-acid cell has basically the same construction as that seen in Fig. 4-8. The positive plate, however, has a large number of recesses that form a grid. Each of these recesses is packed with lead dioxide. The negative plate also has a large number of recesses that are filled with spongy lead. A cutaway view of a commercial cell is shown in Fig. 4-10. Separators of rubber, wood, or plastic are placed between adjoining plates in order that one plate cannot touch the next plate. The use of a large number of plates in a storage cell permits a very large current capability to be obtained. This is the chief reason for using this type of battery power in an automobile—when the car is started, several hundred amperes will be drawn by the starter motor for a short length of time. Most automobiles use a battery consisting of six cells.

The current capability of a storage battery is expressed in terms of a unit called the *ampere-hour.* This unit is defined as the supply of 1 ampere for 1 hour, or ½ ampere for 2 hours, or 4 amperes for ¼ hour, and so on. Although the ampere-hour rating of a storage battery gives a general idea of its current capability, the rate of discharge must also be taken into consideration. That is, if a 60 ampere-hour battery is discharged at the rate of 1 ampere, we will usually obtain considerably more than 60 hours of active use. On the other hand, if the same battery is discharged at the rate of 120 amperes, we will usually obtain considerably less than ½ hour of active service.

4.7 Conversion of Light Energy to Electrical Energy

We have learned that phototubes, as well as other devices, are used to change light energy into electrical energy. For example, a silicon solar cell is shown in Fig. 4-11. This type of device is called a *photovoltaic cell,* and is used in solar

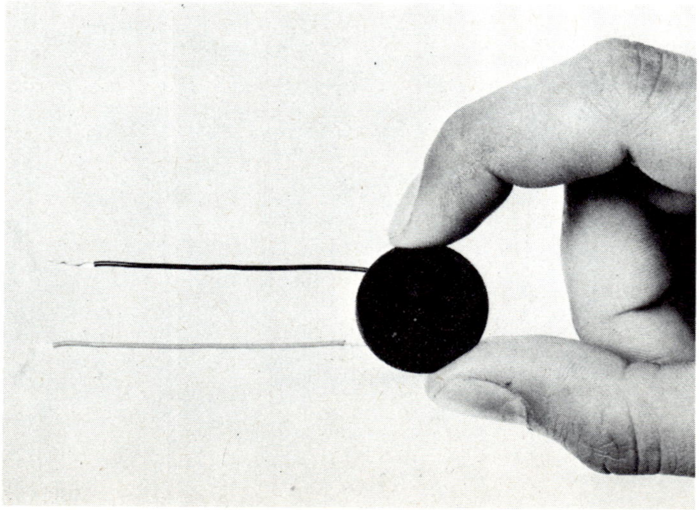

Fig. 4-11
A silicon solar cell.
Courtesy: Edmund Scientific Co.

power applications to power satellite radio transmitters, model boats, transistor radios, telephone stations, burglar alarms, automatic lighting controls, and photo counting devices. A large bank of silicon solar cells has been constructed to operate a 1912 Baker electric automobile. A unit such as shown in Fig 4-11 will supply about 0.1 ampere of current at 0.5 volt when illuminated by direct sunlight.

Architects have been developing and experimenting with solar houses. A model solar house, which is illustrated in Fig. 4-12, operates by collection of solar heat in water flowing down the heat collector on the roof. This heated water is

Fig. 4-12
A model solar house.
Courtesy: Edmund Scientific Co.

then stored in a tank placed in a bed of stones; in turn, the stones are warmed by the tank. Air is circulated through the warm stones and admitted into the interior of the house for heating in the same manner that heated air is used from a furnace. Silicon solar cells are connected together to form a solar battery when a substantial amount of electric power is needed. A silicon solar cell is defined as a photovoltaic cell.

4.8 Conversion of Heat Energy to Electrical Energy

Thermocouples are used to change heat into electricity. The operation of a basic arrangement (Fig. 4-13) is explained in terms of the rates at which electrons diffuse from one kind of metal into another kind of metal. When the metals bismuth and antimony are brought into contact, electrons diffuse faster from the bismuth into the antimony than from the antimony into the bismuth. This produces a small potential difference across the junction between the metals. If a complete path for electron flow is provided, and one junction is heated, electric current flows because the hot junction generates a higher voltage than the cold junction.

Although any two dissimilar metals will generate thermoelectricity, antimony and bismuth are the two most efficient of the more common metals. It would seem that, if the junction is heated to higher and higher temperatures, more current would flow as a result. However, there is a certain optimum temperature for any thermocoupled junction; the current rises to a maximum at this temperature

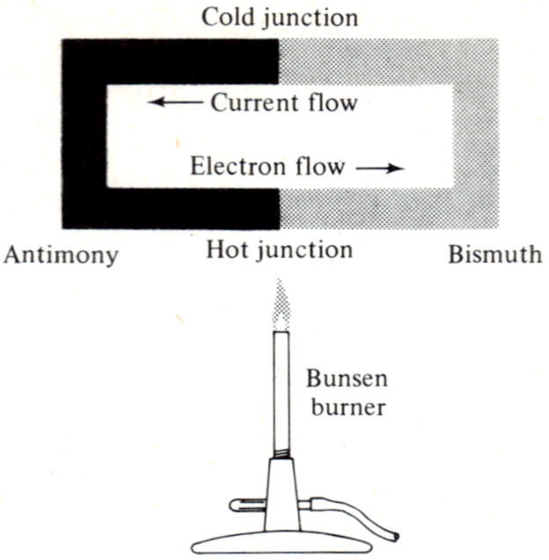

Fig. 4–13
A basic thermocouple.

and then decreases at still higher temperatures. Although a thermocouple generates only a small voltage, a very large current will flow if the conductors have a large diameter, and if the circumference of the complete path for current is fairly short. When higher voltages are needed, a number of thermocouples are connected together. This arrangement is called a *thermopile*. Thermocouples and thermopiles are used in some types of electrical meters, and in electrical thermometers. When designed to measure extremely high temperatures, an electrical thermometer is called a *pyrometer.*

4.9 Piezoelectricity

Electricity is generated when certain crystals are compressed, twisted, or vibrated. This is called the *piezoelectric effect*. The most widely used piezoelectric crystals are quartz, Rochelle salt, and barium titanate. Within recent years, special ceramics have been developed that produce piezoelectricity. Figure 4–14a shows

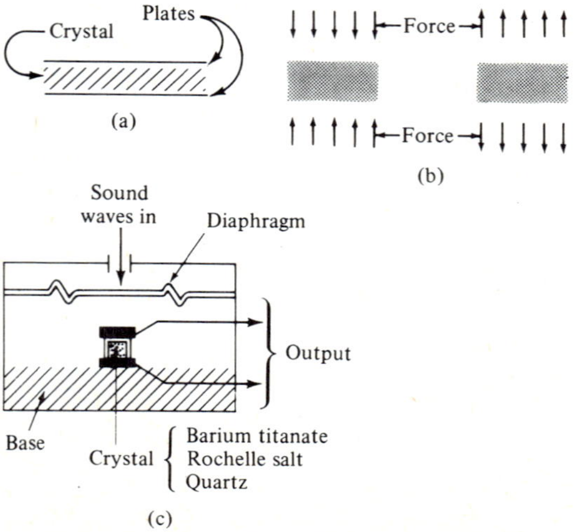

Fig. 4–14
Generation of piezoelectricity.
(a) Basic crystal arrangement.
(b) Development of electric charges.
(c) Plan of a piezoelectric microphone.

the basic plan of a piezoelectric device; a crystal wafer is ground with mirror-flat surfaces, and mounted between a pair of polished metal plates. If the crystal is forced together (compressed), its upper and lower surface develop positive and negative charges; if the crystal is pulled apart (decompressed), the charges are reversed, as shown in Fig. 4–14b. Although this potential difference, or voltage, is very small, devices producing piezoelectricity are widely used in the electronics technology.

We find piezoelectric devices used in phonograph pickups, tape-recorder microphones, industrial flaw testers, pressure gauges, combustion analyzers, and many other applications in which mechanical energy is changed into electrical energy, or electrical energy is changed into mechanical energy. Figure 4–14c shows the plan of a piezoelectric microphone. In this unit, sound waves strike the diaphragm, making it vibrate. This mechanical energy is applied to the upper crystal electrode. This causes electric charges to be generated at the upper and lower surfaces of the crystal. The output voltage is then taken from the crystal electrodes by a pair of output leads.

4.10 Electrets

Although electrets were invented many years ago, it is only within recent time that they have been put to practical application in electronics devices. An electret (Fig. 4–15) is a specialized source of static electricity. There is a piezoelectric effect associated with the operation of an electret. When resin, carnuba wax, or beeswax is rubbed with a cloth, it develops static electricity, as we have previously learned. However, if resin is melted and placed between a pair of charged metal plates, it acquires electric charges throughout as it cools. After the resin has solidified, it shows an induced charge. That is, the surface of the resin next to the positive plate has been charged negatively, and the surface of the resin next to the negative plate has been charged positively.

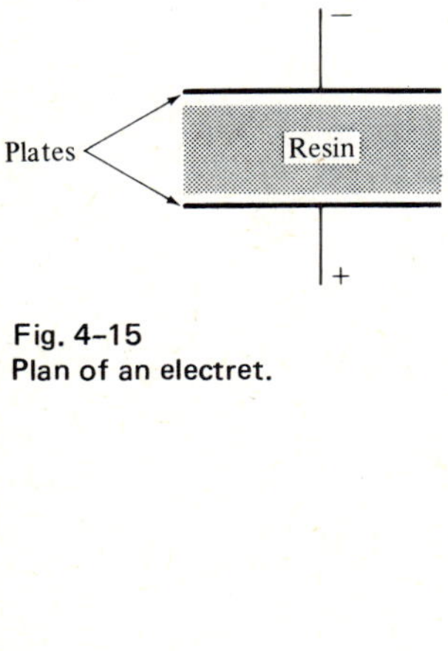

Fig. 4–15
Plan of an electret.

To show that the charge exists all the way through the resin, the outer surfaces can be removed, and the new surfaces also show a charge. One of the surprising facts in the manufacture of an electret is that this original charge reverses its polarity within a few days. The reversed charge condition is stable, and usually lasts for many years. Scientists explain this electret charge as a piezoelectric effect resulting from the internal mechanical strains that are set up in the solidified resin. Although an electret can supply substantial voltage, it has an extremely small current capability. Therefore, electrets are used only in special applications that need a potential difference with practically no current drain.

4.11 Generation of Electricity from Magnetism

We find that electricity can also be generated from magnetism by means of mechanical energy. This is one of the main methods of generating electricity, and is used in large electric power-generating stations, such as the installation at Niagara Falls. However, the details that are involved will be explained after we have learned more about the basic principles of electricity.

Questions and Problems Chapter 4

True-False:

1. When an electric current flows through the filament of a lamp, energy is being converted.
2. The light given out by a lamp bulb does not change if the battery voltage is increased.
3. An ampere of current in a circuit is equal to a certain number of electrons passing a given point in one second.
4. A dry cell is supposed to be recharged.
5. The emf of a dry cell is approximately 2.2 volts.
6. To produce twice as many amperes from a cell, we must double the area of the electrodes.
7. When one of the chemicals in a cell is used up, the cell will "go dead."
8. A cell will stay charged for many years while in storage.
9. A cell has a rating called its "shelf life."
10. The shelf life of a cell is determined by the local action within the cell.
11. Mercury is sometimes used with zinc in a dry cell to extend its shelf life.
12. A mercury cell supplies less current than a zinc cell.
13. A mercury cell supplies an emf of approximately 1.5 volts.
14. By definition, a battery consists of two or more cells connected together.
15. The nickel-cadmium battery is a very popular rechargeable battery.
16. A nickel-cadmium cell is a primary cell.
17. A rechargeable cell is often called a storage cell.
18. The lead-acid battery can be recharged.
19. Any lead-acid battery should be handled with care because of the danger of a burn from the sulfuric acid in the battery.
20. The emf of a lead-acid cell is 1.5 volts.
21. The current capability of a storage battery is expressed in terms of the ampere-hour rating.
22. A solar cell converts chemical energy to electrical energy.
23. Solar cells could be used to power an electric car.
24. A thermocouple changes heat energy into electrical energy.
25. A thermocouple is formed by combining an insulator and a metal into a junction.
26. A pyrometer is a high-temperature thermocouple.
27. Electricity is produced by a piezoelectric effect when certain crystals are compressed, twisted, or vibrated.
28. A piezoelectric device can be used in a microphone.
29. An electret is a type of battery that can be recharged.

Multiple choice: choose the best answer.

1. When an electric current flows through the filament of a lamp,
   a. the filament is used up
   b. electrical energy is being converted to heat energy
   c. heat energy is being converted to electrical energy
   d. it remains constant when the supply voltage changes.
2. An ampere is a measurement of
   a. current
   b. voltage
   c. wattage
   d. electrical power
3. When a dry cell goes dead,
   a. one of the chemicals in the cell is used up
   b. it should be recharged
   c. new chemicals should be added
   d. liquid should be added
4. Local action within a cell
   a. increases the output current
   b. is desirable
   c. increases the shelf life
   d. reduces the shelf life
5. A very popular rechargeable battery used in portable equipment is the
   a. dry-cell battery
   b. mercury-cell battery
   c. nickel-cadmium battery
   d. solar battery
6. The current rating of a storage battery is expressed in terms of the
   a. volume of the battery
   b. ampere-hour rating
   c. voltage-ampere rating
   d. ampere rating
7. A rechargeable cell is often called a
   a. wet cell
   b. portable cell
   c. storage cell
   d. dry cell
8. A device that converts solar energy into electrical energy is called a
   a. photo cell
   b. pyrometer
   c. thermocouple
   d. solar cell
9. A device that converts heat energy into electrical energy is a
   a. crystal
   b. battery
   c. thermocouple d. solar cell

10. A thermocouple is made by forming together
    a. an insulator and a metal
    b. two like metals
    c. two unlike metals
    d. two insulators
11. A device that can be used in a microphone is a
    a. piezoelectric crystal
    b. thermocouple
    c. electret
    d. pyrometer
12. An electret can be used as a
    a. constant voltage supply
    b. constant current supply
    c. solar cell
    d. thermocouple

5 DIRECT – CURRENT CIRCUITS

5.1 Definition of a Circuit

In the previous chapter we learned that the basic requirements for a circuit are conductors, an energy source, and an energy converter. For example, when an incandescent lamp is connected to the terminals of a dry cell, an electric current flows. The electrical energy from the battery converts to light and heat energy. A circuit is defined simply as a complete path for electric current. As we progress in our study of electronics, we will find that there is more than one kind of electric circuit. We will also discover that the basic definition of a circuit is not quite as simple as it appears at this point in our studies. With these words of caution, let us examine a simple circuit, and consider its operation.

Considering the circuit in Fig. 5-1 there are three new facts to be observed. First, Fig. 5-1a is called a *pictorial diagram*. Although a pictorial diagram is the

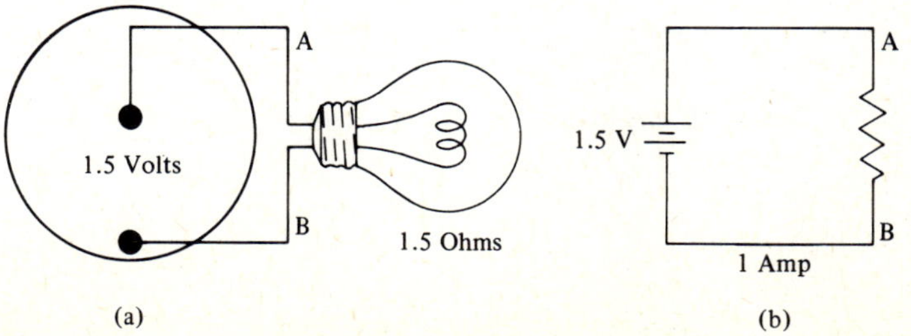

Fig. 5-1
A battery and lamp circuit.
(a) pictorial diagram.
(b) schematic diagram.

most descriptive representation of an electric circuit, it is time-consuming to draw. Therefore, engineers and technicians prefer to use a form of technical shorthand, which is called a *schematic diagram*. It is evident from the diagram in Fig. 5-1b that a schematic diagram can be drawn quicker and easier than a pictorial diagram. Of course, a schematic diagram is meaningless if we do not understand the *symbols* that are used. We note that a dry cell is represented by a symbol consisting of a short line and a long line; the short line denotes the negative terminal, and the long line denotes the positive terminal. The lamp filament is represented by a *resistance* symbol (resistance is explained in the following paragraph). Conducting wires are represented by straight interconnecting lines between the cell symbol and the resistance symbol.

5.2 The Meaning of Resistance

Resistance means *opposition to current flow*. To understand this, consider that when an electric current flows in a wire, electrons are passed along from one atom to the next, as shown in Fig. 5-2. Each atom in this conductor has an outer electron that is easily detached, and which moves along the wire when a voltage is applied. These electrons make up an electric current and are called *free electrons*. However, we must not think that an electron is entirely free to move, since it takes a certain amount of work to move an electron along a wire. This is just another way of saying that metals have an *opposition* or resistance to current flow. Resistance is measured in *ohms*, or *ohm units*. We define an ohm as the resistance that is present when a voltage of one volt causes a current of one ampere to flow.

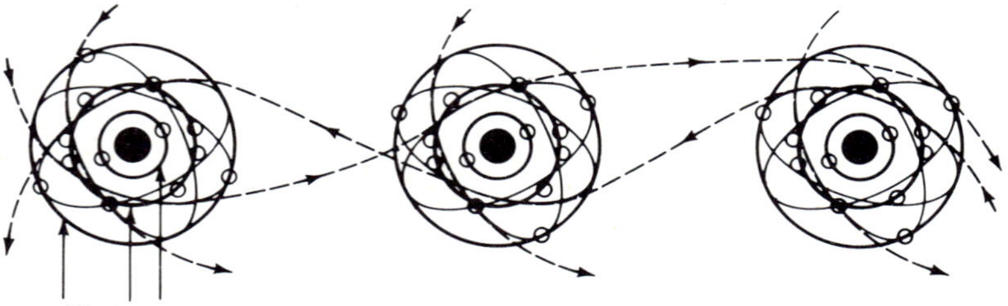

Fig. 5-2
Atoms in a metal wire.

To describe the relationship between resistance, voltage, and current in an electrical circuit, we use *Ohm's law*, named after George Simon Ohm. Ohm's law is the most important law in the science of electricity. This law states the relation between voltage, current, and resistance, as follows:

$$\text{Current } (I) = \frac{\text{voltage } (E)}{\text{resistance } (R)}$$

or

$$I = \frac{E}{R}$$

Note that we denote current by I, voltage by E, and resistance by R. In terms of electrical units, we write

$$\text{Amperes} = \frac{\text{volts}}{\text{ohms}}$$

Resistance is often denoted by the Greek letter omega (Ω) instead of R; similarly, voltage is often denoted by V instead of E. For example, 100 Ω means 100 ohms and 12 V means 12 volts. Note also that a *kilohm* means 1,000 ohms and that a *megohm* means 1,000,000 ohms.

In Fig. 5-1, we note that the lamp filament has a resistance of 1.5 ohms; the dry cell applies 1.5 volts to the filament, and a current of 1 ampere flows through the filament. These relations follow from Ohm's law:

$$I = \frac{E}{R} = \frac{1.5\ \text{V}}{1.5\ \Omega} = 1 \text{ ampere}$$

5.3 Wire Sizes and Electrical Properties

Most electrical conductors are made of copper, although aluminum is also used in applications where light weight is an advantage. We find that copper is a somewhat better conductor than aluminum. Silver is a better conductor than copper, but is too expensive except for certain critical applications. Most electrical conductors or wires are manufactured from commercial-grade copper. Electric wire is usually round; however, square or rectangular wire is sometimes used. We also find copper strip used for winding specialized coils. In recent time, printed or plated conductors in the form of very thin strips have come into wide use (Figure 5-3).

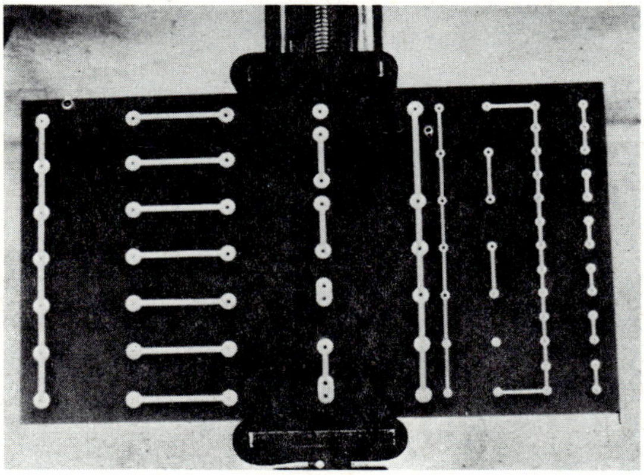

Fig. 5-3
A printed wiring, or printed circuit; the wiring is bonded to the insulator board.

Since the most common conductors are round, we will now examine their characteristics. Round copper conductors may be either bare or insulated. Many types of insulation are in general use. The most basic description of round copper wire is its diameter, which is usually expressed as a gauge number. A gauge number refers to the diameter of the wire only, and disregards any insulation that the wire may be covered with. The standard wire gauge is called the American Wire Gauge (AWG), and is sometimes known as the Browne & Sharpe gauge (B & S). Table

5-1a lists gauge numbers of copper wires with corresponding diameters, areas, and

(a)

Gage no.	Diam, mils	Circular-mil area	Ohms per 1,000 ft of copper wire at 25°C	Gage no.	Diam, mils	Circular-mil area	Ohms per 1,000 ft of copper wire at 25°C
1	289.3	83,690	0.1264	21	28.46	810.1	13.05
2	257.6	66,370	0.1593	22	25.35	642.4	16.46
3	229.4	52,640	0.2009	23	22.57	509.5	20.76
4	204.3	41,740	0.2533	24	20.10	404.0	26.17
5	181.9	33,100	0.3195	25	17.90	320.4	33.00
6	162.0	26,250	0.4028	26	15.94	254.1	41.62
7	144.3	20,820	0.5080	27	14.20	201.5	52.48
8	128.5	16,510	0.6405	28	12.64	159.8	66.17
9	114.4	13,090	0.8077	29	11.26	126.7	83.44
10	101.9	10,380	1.018	30	10.03	100.5	105.2
11	90.74	8,234	1.284	31	8.928	79.70	132.7
12	80.81	6,530	1.619	32	7.950	63.21	167.3
13	71.96	5,178	2.042	33	7.080	50.13	211.0
14	64.08	4,107	2.575	34	6.305	39.75	266.0
15	57.07	3,257	3.247	35	5.615	31.52	335.0
16	50.82	2,583	4.094	36	5.000	25.00	423.0
17	45.26	2,048	5.163	37	4.453	19.83	533.4
18	40.30	1,624	6.510	38	3.965	15.72	672.6
19	35.89	1,288	8.210	39	3.531	12.47	848.1
20	31.96	1,022	10.35	40	3.145	9.88	1,069

(b)

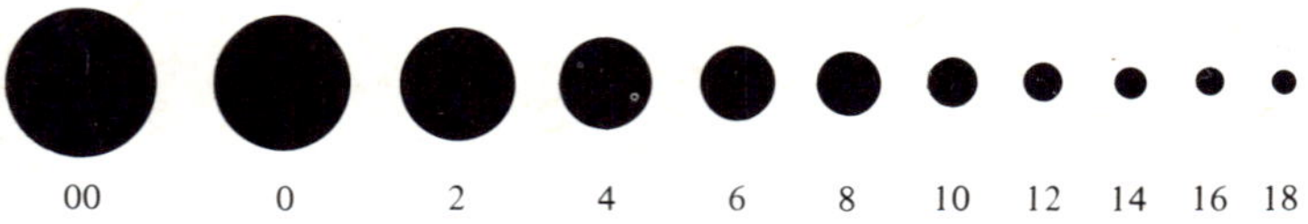

Table 5-1
Wire sizes.
(a) AWG copper wire gauges.
(b) Actual diameters of different sizes of copper wire, without the insulation.

resistance values per 1,000 feet. Diameters are given in *mils*, and areas are given in *circular mils.* A mil is equal to 0.001 inch. A circular mil is defined as the area enclosed by a circle with a diameter of one mil.

Figure 5-4 shows comparisons of the *square mil* and the *circular mil.* Note that a square mil is equal to the area of a square with sides one mil in length. When we are working with round wires, it is easier to consider cross-sectional areas in circular-mil units. On the other hand, when we are working with rectangular or square conductors, it is easier to consider cross-sectional areas in square-

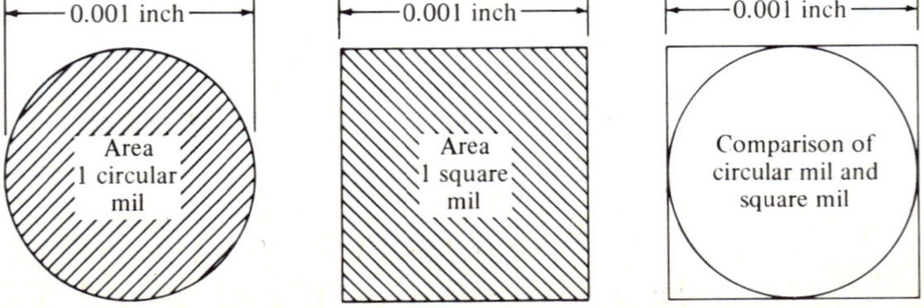

Fig. 5-4
Enlarged comparison of circular mil and square mil.

mil units. Let us examine an example of the circular-mil unit in Table 5-1. The diameter of the No. 36 wire is five mils, or its area is equal to twenty-five circular mils. In other words, *to find the cross-sectional area of a wire in circular-mil units, we square the diameter expressed in mil units.*

Next, let us see how to change circular-mil units into square-mil units by continuing with the foregoing example. Since No. 36 wire has an area of twenty-five circular mils, the number of square mils (SM) that corresponds to twenty-five circular mils (CM) is found by multiplying circular mils by 0.7854

$$0.7854 \text{ CM} = \text{SM}$$

or

$$0.7854 \times 25 \text{ CM} = 19.64 \text{ SM}$$

If we were given the cross-sectional area of a conductor in square mils, we would change this value into circular mils simply by dividing 0.7854 into the square mil area

$$\frac{\text{SM}}{0.7854} = \text{CM}$$

Another important unit of wire measure is the *circular-mil-foot*, as shown in Fig. 5-5. This unit is used to describe the resistance of the metal used in manufacturing the wire. The resistance of ordinary commercial wire is expressed as about 10.4 ohms per circular-mil-foot. This term is often abbreviated to *mil-foot*. A mil-foot consists of one foot of the wire, with a diameter of one mil. Next, let us observe the relations shown in Fig. 5-6. If a mil-foot of a given wire has a resistance of 10.4 ohms, then two of these wires side by side will have only one-half as much resistance, or the double conductor will have a resistance of 5.2 ohms. This is just another way of saying that if a certain wire has twice as much cross-sectional area as another wire, its resistance will be one-half as much.

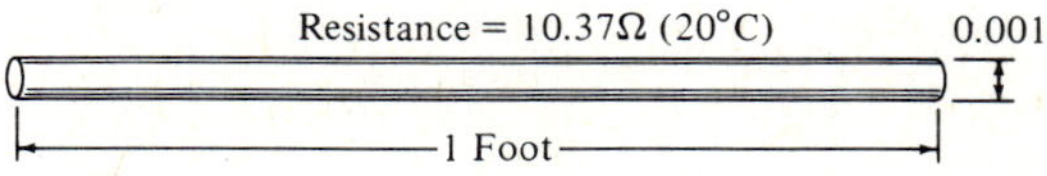

Fig. 5-5
Illustration of dimensions and resistance of a circular mil foot of copper.

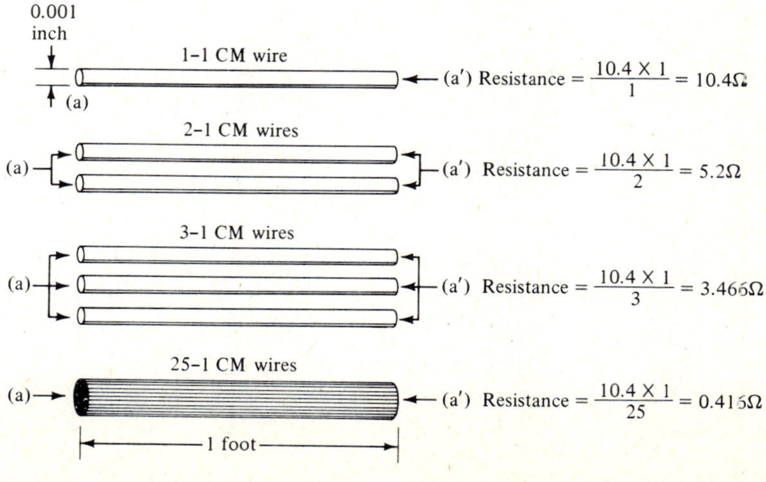

Fig. 5-6
The resistance of a conductor decreases with an increase of the area through which the current flows.

Various means can be used to measure the diameter of a wire. A typical wire gauge is shown in Fig. 5-7. This type of gauge will check AWG gauge sizes from No.0

Fig. 5-7
Standard wire gauge.
Courtesy: Sargent-Welch Scientific Co.

to No. 36. As we stated earlier, it is important to remember that the wire must be bare when a gauge is used. That is, gauge numbers are based on the diameter of a wire without any insulation. When checking stranded wires, such as shown in Fig. 5-8, we must count the number of conductors in addition to measuring the gauge of the individual wire. In the example shown, there are thirty-seven strands,

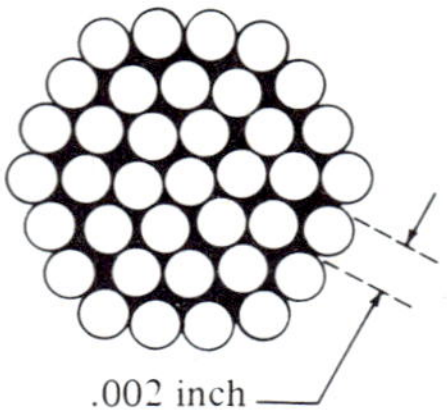

37 Strand conductor

Diameter of each strand = .002 inch
Diameter of each strand = 2 mils
Circular mil area of each strand = D^2 = 4 CM
Total CM area of conductor = 4X37 = 148 CM

Fig. 5-8
Example of a stranded wire.

each with a diameter of 2 mils. Therefore, the circular-mil area of each conductor is 4 circular mils, and the total circular-mil area of the stranded wire is 148 circular mils.

In any electrical installation, the conductor size must be sufficient to carry the required current without having too large resistance losses. Since the resistance losses generate heat, a conductor that is much too small for the given current will burn out, and possibly start a fire. The current-carrying capacity of a wire increases with an increase in its diameter. Ordinary house-lighting circuits generally use No. 12 or No. 14 wire. Table 5-2 lists the normal current-carrying capacities of various sizes and types of copper wire. Notice that the current-carrying capacity depends upon the type of insulation that covers the wire in addition to the size of the wire itself. An insulation, such as asbestos, that will stand a high temperature will allow the wire to carry a greater current safely.

Size	Rubber or thermoplastic	Thermoplastic asbestos, var-cam, or asbestos var-cam	Impregnated asbestos	Asbestos	Slow-burning or weatherproof
0000	300	385	475	510	370
000	260	330	410	430	320
00	225	285	355	370	275
1	195	245	305	325	235
2	165	210	265	280	205
3	140	180	225	240	175
4	120	155	195	210	150
6	105	135	170	180	130
8	80	100	125	135	100
10	55	70	90	100	70
12	40	55	70	75	55
14	25	40	50	55	40
	20	30	40	45	30

Table 5-2
Current-carrying capacities of copper wires of various sizes and types.

5.4 Operation of Simple Circuits

We have learned that a circuit is a complete path for electric current. Therefore, the lamp in Fig. 5-9 glows while the circuit is closed. On the other hand, if we

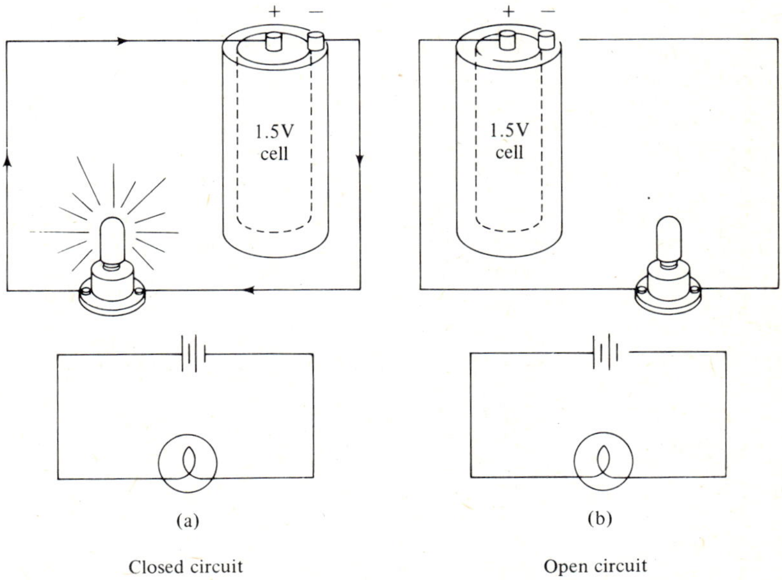

Fig. 5-9
Simple electric circuit.
(a) Closed.
(b) Open.

disconnect the dry cell, the lamp does not glow because the circuit has been opened. Logic tells us that breaking the circuit at any point will cause the lamp to stop glowing. An example of the circuit in Fig. 5-9 is a flashlight. To turn the flashlight off or on, a *switch* is provided to open and close the circuit. There are many types of switches in general use, but they all operate on the same principle. That is, any switch operates to open or close a circuit. The most basic type of switch is the *knife switch* shown in Fig. 5-10a. The symbol for a switch is shown in the schematic diagram in Fig. 5-10b.

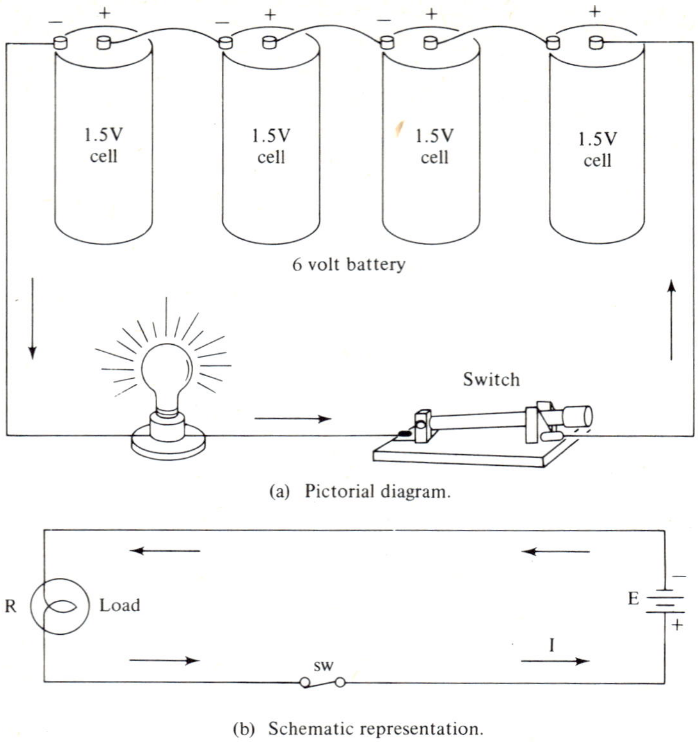

Fig. 5-10
Illustrating a simple electric circuit. (a) Pictorial diagram. (b) Schematic representation.

Another important part of the circuit in Fig. 5-10 is the series connection of four dry cells to form a battery. The positive terminal of one cell is connected to the negative terminal of the next cell. This causes the voltages of the cells to be added. That is, each cell has a voltage of 1.5 volts, and the battery has a voltage of 6 volts. Therefore, 6 volts are applied to the lamp when the switch is closed. In the circuit of Fig. 5-1, however, only 1.5 volts was applied to the lamp. Lamps are rated for operation at a certain voltage. For example, a flashlight lamp may be rated for operation at 1.5 volts, 3 volts, 4.5 volts, or 6 volts. A lamp must be operated at its rated voltage. If we place a 1.5-volt lamp in a 6-volt circuit, the lamp will burn out. On the other hand, a 6-volt lamp placed in a 1.5-volt circuit will glow very dimly.

5.5 Electric Power

Electricity is useful because it does *work*. For example, electricity can operate the starter of an automobile, run a circular saw, or operate a washing machine. We often measure work in terms of foot-pounds, as shown in Fig. 5-11. If we lift a one-pound weight 3 feet above its resting position, we have done three foot-pounds of mechanical work. Work is also called *energy*. The weight in the above example has three foot-pounds of *potential energy*[1] after it has been

1. Potential energy is the energy that an object has due to its position, whereas kinetic energy is the energy that an object has due to its motion.

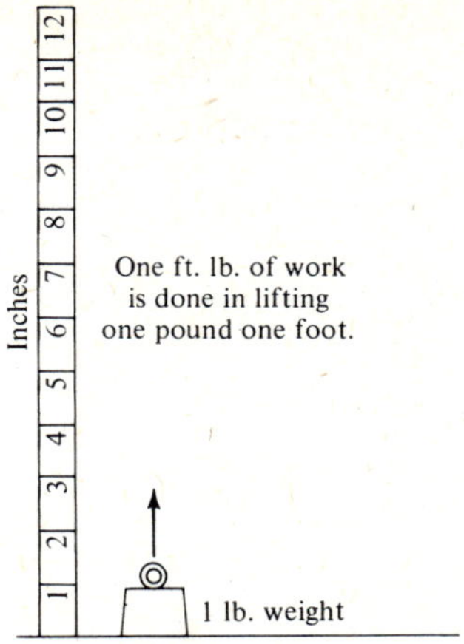

Fig. 5-11
Work done in lifting a weight equals weight times vertical distance.

raised. Notice that we have made no reference to time. This is because the same amount of mechanical work has been done in the previous example, whether we lift the weight in a second, a minute, or an hour.

Another important subject in electrical circuits is power. Power is the time rate of doing work. We often measure power in terms of foot-pounds per second. For example, if we take one second to lift a one-pound weight 3 feet above its resting position, the power that we have used is three foot-pounds per second. Or, suppose that we take three seconds to lift a one-pound weight 3 feet above its resting position; the power that we have expended is one foot-pound per second. The horsepower is a unit of work that is often used to describe the rate at which an electric motor can do work. For example, a washing-machine motor is usually a one-fourth-horsepower motor. As shown in Fig. 5-12, one horsepower is equal to moving a one-pound weight 550 feet in one second.

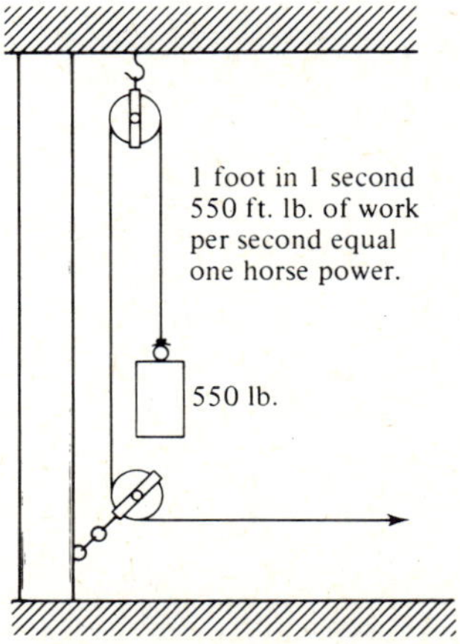

Fig. 5-12
How to determine horsepower.

In an electrical circuit, we use the *watt* as the unit of electrical power. When one volt moves one coulomb of electrons past a point in one second, the rate of doing work is one watt. Therefore, the power in an electrical circuit is equal to voltage times current

$$E \text{ (volts)} \times I \text{ (amperes)} = P \text{ (watts)}$$

Using these relationships between horsepower and power in an electric circuit, we can find how much mechanical power is produced by a certain amount of electrical power

$$1 \text{ horsepower} = 746 \text{ watts}$$

For example, if a motor is rated at one-fourth horsepower, this amount of mechanical power is equal to 186.5 watts of electrical power.

Let us now take a practical example of an electrical circuit, and see how Ohm's law and the power law are used. Figure 5-13 shows a simple series circuit, with a voltmeter and an ammeter connected. The ammeter indicates the current

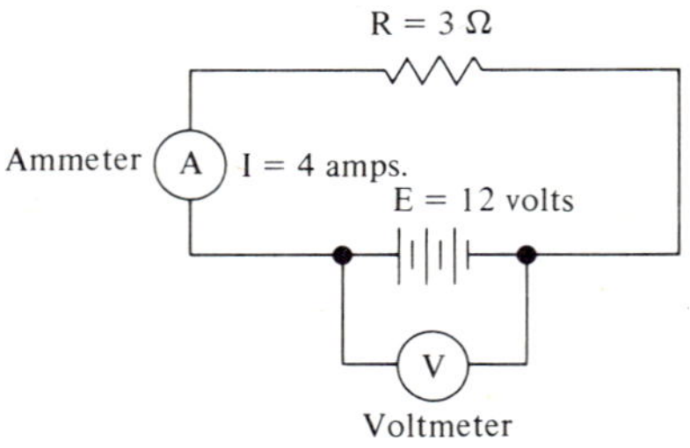

Fig. 5-13
Simple electrical circuit.

in the circuit, and the voltmeter indicates the voltage across the battery. Notice that the ammeter is connected in series with the circuit, and that the voltmeter is connected across the battery. In this example, the meters indicate that the circuit current is 4 amperes, and that the battery voltage is 12 volts. Let us set this circuit up as a problem and we will apply Ohm's law.

Problem: The load resistance R is 3 ohms. Find the current that will flow on the basis of the meter readings and Ohm's law.

$$E = 12 \text{ volts (indicated by the voltmeter)}$$
$$R = 3 \text{ ohms (given in the problem)}$$
$$I = \frac{E}{R} \text{ (Ohm's law)}$$
$$I = \frac{12}{3} \text{ or 4 amperes}$$

Another form of the problem would be to find the load resistance R on the basis of the meter indications and Ohm's law.

$$E = 12 \text{ volts (indicated by the voltmeter)}$$
$$I = 4 \text{ amperes (indicated by the ammeter)}$$
$$I = \frac{E}{R} \text{ (Ohm's law)}$$
$$R = \frac{E}{I} \text{ (rearranging Ohm's law)}$$
$$R = \frac{12}{4} \text{ or 3 ohms.}$$

The third form of the problem would be to find the battery voltage, knowing that the current is 4 amperes and the load resistance R is 3 ohms.

$$I = \frac{E}{R} \text{ (Ohm's law)}$$

$$E = IR \text{ (rearranging Ohm's law)}$$

$$E = 4 \times 3 = 12 \text{ volts}$$

A simple way to remember the three forms of Ohm's law is to keep the diagram of Fig. 5-14 in mind. To use the diagram, place a finger over the unknown, and then read the relationship of the other two terms.

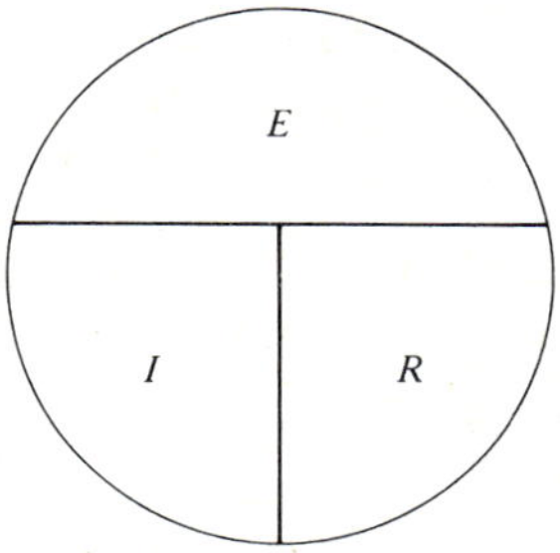

OHM'S LAW
Current = $\frac{\text{Electromotive Force}}{\text{Resistance}}$ $I = \frac{E}{R}$ Amperes = $\frac{\text{Volts}}{\text{Ohms}}$
Resistance = $\frac{\text{Electromotive Force}}{\text{Current}}$ $R = \frac{E}{I}$ Ohms = $\frac{\text{Volts}}{\text{Amperes}}$
Electromotive Force = Current X Resistance $E = IR$ Volts = Amperes X Ohms

Fig. 5-14
Ohm's law in diagram form.

Next, let us see how we would apply the power law to find the power in the circuit of Fig. 5-13.

Problem: Find the power from the meter indications.

$$P = EI \text{ (power law)}$$

$$P = 4 \times 12 \text{ or } 48 \text{ watts}$$

Another form of the problem could be to find the power in the circuit, knowing that the load resistance R is 3 ohms, and that the current is 4 amperes.

$$P = EI \text{ (power law)}$$

$$I = \frac{E}{R} \text{ (Ohm's law)}$$

$$E = IR \text{ (rearranging Ohm's law)}$$

$$P = I^2R \text{ (substituting in Watt's law)}$$

$$P = 4^2 \times 3 = 16 \times 3 = 48 \text{ watts}$$

The third form of the problem would be to find the power in the circuit, given that the load resistance R is 3 ohms, and that the battery voltage is 12 volts.

$$P = EI \text{ (power law)}$$

$$I = \frac{E}{R} \text{ (Ohm's law)}$$

$$P = \frac{E^2}{R} \text{ (substituting in Watt's law)}$$

$$P = \frac{12 \times 12}{3} = \frac{144}{3} = 48 \text{ watts}$$

A handy way to solve electrical-circuit problems is to use the diagram shown in Fig. 5-15. This diagram shows that there are twelve forms of expressions for

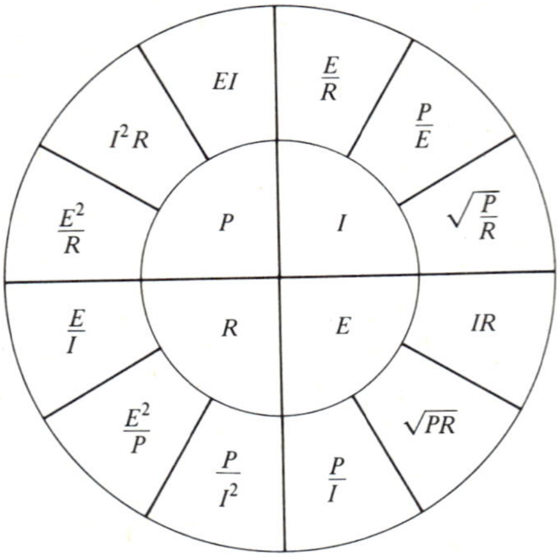

Fig. 5-15
There are 12 expressions for *I*, *E*, *R*, and *P*.

I, E, R, and P. We see that there are three forms each of E, I, R, and P.

5.6 Types of Resistors

There are only four basic electronic components: resistors, capacitors, inductors, and active devices. We will begin our study of the basic components with an explanation of the important types of resistors. Resistors are the most widely used of the components in electronic equipment, and the easiest to understand. Resistors may be used to divide voltages, to obtain a voltage drop by current flow through the resistor, or to limit the amount of current that can flow in a circuit. Many other uses for resistors will be explained as we continue our study of electronics.

Figure 5-16 shows the construction of the most common type of molded composition-type resistor. Composition resistors are manufactured in many shapes and sizes. Their resistance values are from one ohm to many million ohms (megohms). The physical size of a resistor is related to its ability to radiate heat or to its power-dissipation ability. The power-dissipation ability of a resistor is given in watts. A resistor should not be required to carry more current than is permitted by its power rating. Otherwise, the resistor will overheat and may be damaged. For example, commonly used resistors have power ratings of one-fourth watt, one-half watt, one watt, and two watts, as shown in Fig. 5-17. As an example of the power rating of resistors, suppose that a resistor to be used in a circuit has a resistance value of 2,500 ohms, and that fifty volts will be applied across its terminals. What power rating will be assigned to the resistor?

$$P = \frac{E^2}{R}$$

$$P = \frac{50 \times 50}{2{,}500} = \frac{2{,}500}{2{,}500} = 1 \text{ watt}$$

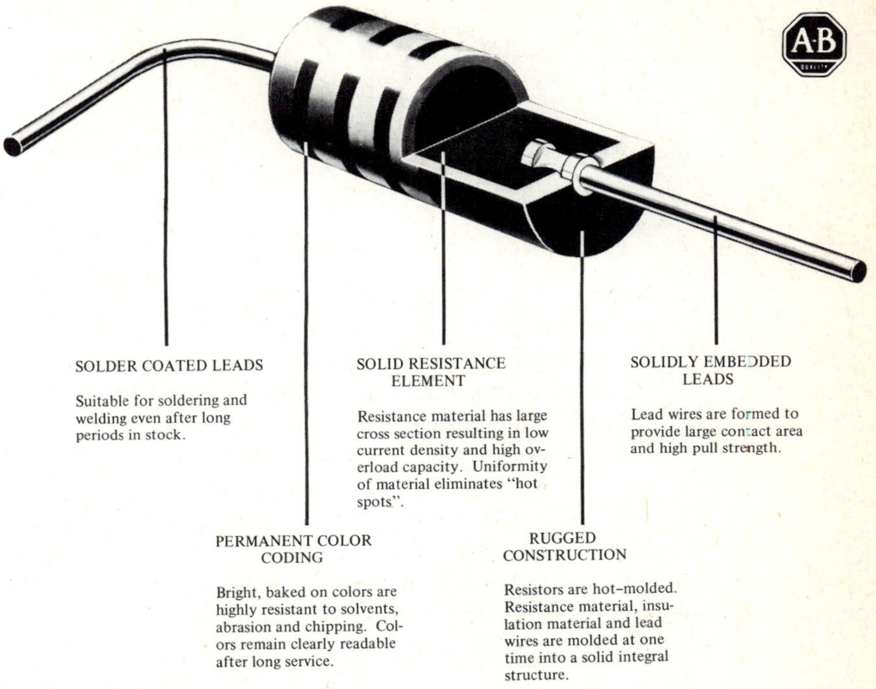

Fig. 5-16
Construction of a composition resistor.
Courtesy: Allen-Bradley Co.

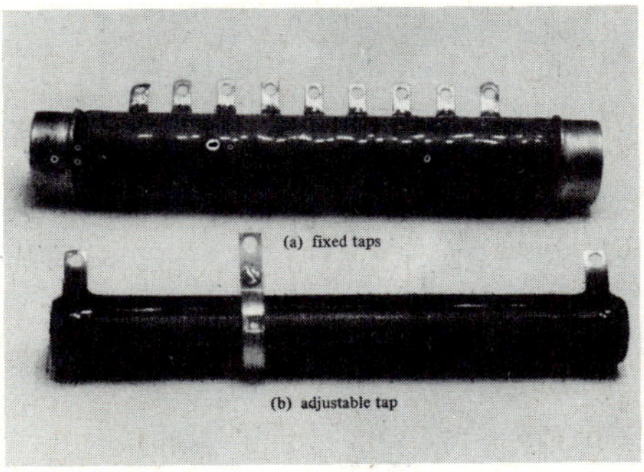

Fig. 5-17
Adjustable resistors.
(a) fixed taps
(b) adjustable tap.

Therefore, we must use a resistor that has a power rating of at least one watt. For safety, we always employ resistors with power ratings two to four times the power dissipation.

Composition resistors are color coded to indicate their ohmic value. Instructions are given in the appendix for reading the color code. Some resistors that have a resistance of several ohms are wirewound, although they look like composition resistors. Most resistors with a resistance less than one ohm are wirewound. High-current (power) resistors are also wirewound, regardless of their resistance value. Figure 5-18 shows the appearance of typical wirewound resistors. Wire-

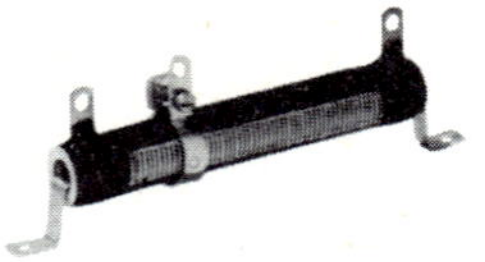

(a)

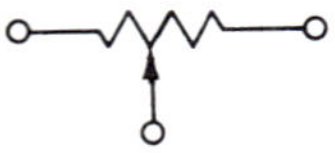

(b)

Fig. 5-18
An adjustable wire-wound resistor. (a) Appearance. (b) Symbol.

wound resistors are formed by winding the wire on a ceramic core, and covering the windings with a coating of enamel for insulation. Wirewound resistors with power ratings from 5 watts to 200 watts are often used in electronic equipment. Connections or *taps* may be provided in between the end terminals of the fixed wirewound resistor to tap off different resistance values.

The resistors that have been described are called *fixed* resistors, because their resistance values cannot be changed. However, many *variable* or *adjustable* resistors are also used in electronic equipment. An adjustable resistor (Fig. 5-19) has

(a)

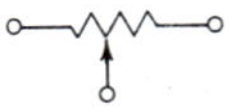

(b)

Fig. 5-19
Typical variable resistors. (a) Appearance. (b) Symbol.
Courtesy: Ohmite Co.

one or more movable contacts which are secured in place, but which can be relocated by loosening a nut or bolt. This type of adjustable resistor is often used in power supplies. The variable resistor shown in Fig. 5-20 has a sliding contact

Fig. 5-20
Variable resistor (potentiometer)

which is moved around a molded composition ring by a knob-and-shaft assembly. This type of variable resistor is often called a *potentiometer.* The resistive element may have a resistance value of less than 100 ohms up to many megohms. Larger potentiometers are rated at two watts. When high currents must pass, wirewound resistive rings are used instead of composition rings.

Many other types of resistors are used in electronic equipment, and we will become familiar with them as we continue our study of electronics.

Questions Chapter 5

True–False:

1. A schematic diagram is a form of technical shorthand used by the electronics technician.
2. The filament of a lamp is represented by a voltage in a schematic diagram.
3. In a schematic diagram, conductors are represented by straight lines that connect all parts of the circuit.
4. The unit of voltage is the ohm.
5. The unit of current is the ampere.
6. The symbol for resistance is the Greek letter Ω.
7. A correct statement of Ohm's law is given as $I = \frac{E}{R}$.
8. In a given circuit, increasing the value of the resistance will cause the current to increase.
9. The short cut for writing 1,000 ohms is 1 kilohm.
10. Connecting a 100-ohm resistor across a 200-volt battery causes a current of 0.1 ampere.
11. Most electrical conductors are square in shape.
12. The diameter of wire is usually given in mils.
13. The current carrying capacity of a wire is determined by its cross-sectional area.
14. A battery is made up of one or more cells.
15. A dry cell is not designed to be recharged.
16. A secondary cell is sometimes called a dry cell.
17. Electricity is useful because it can be used to do work.
18. Potential energy is the energy that a body has due to its motion.
19. Kinetic energy is the energy that a body has because of its motion.
20. Current in an electrical circuit is the result of carriers in motion.
21. Mechanical power in horsepower can be compared to electrical power in watts.
22. An electrical motor that used 746 watts of electrical power would have an output of one horsepower, if the motor had no losses because of friction.
23. A correct statement of Ohm's law is $E = IR$.
24. A correct statement of the power law is $P = IR$.
25. A current of two amperes through a four-ohm resistor would produce a power dissipation of sixteen watts.
26. There are only five basic electronic components.
27. The power rating of a resistor determines the amount of heat that the resistor will safely dissipate.
28. High-wattage resistors are usually made of carbon.

29. A potentiometer is a variable resistor.
30. A tap is sometimes used to vary the resistance of high-wattage variable resistors.

Multiple choice: choose the best answer.

1. Electronics technicians use a form of technical shorthand when drawing an electrical circuit, called a
   a. pictorial diagram
   b. schematic diagram
   c. block diagram
   d. circuit board
2. The unit of resistance is the
   a. ohm
   b. volt
   c. ampere
   d. watt
3. The unit of voltage is the
   a. ampere
   b. volt
   c. watt
   d. ohm
4. The unit of current is the
   a. ampere
   b. volt
   c. ohm
   d. watt
5. The symbol used for resistance is
   a. A
   b. P
   c. I
   d. Ω
6. The unit of current is the
   a. ampere
   b. volt
   c. ohm
   d. watt
7. Connecting a 50-ohm resistor across a 200-volt battery results in a current of ____________ amperes.
   a. 0.25
   b. 2
   c. 4
   d. 0.5
8. A lamp connected across a 6 volt battery causes a current of 0.25 ampere; the resistance of the lamp is __________ ohms.
   a. 1.5

b. 24
c. 12
d. 5

9. Increasing the voltage to 12 volts in Problem No. 8 would cause the current to (assuming the resistance remains the same)
a. increase to 48 amperes
b. decrease to 12 amperes
c. increase to 3 amperes
d. decrease to 6 amperes

10. A correct statement of Ohm's law is
a. $I = \frac{E}{R}$
b. $E = \frac{I}{R}$
c. $R = EI$
d. $E = ER$

11. The current through a 12-ohm resistor is 3 amperes; the power dissipated across the resistor is ________ watts.
a. 4
b. 36
c. 108
d. 432

12. The power dissipated by an electric coffee pot is 1,100 watts when it is connected into the 110-volt electrical outlet; the current through the line is ________ amperes.
a. 20
b. 0.1
c. 10
d. 5

13. A 1,500-watt electric iron is plugged into the 110-volt electrical outlet; the current through the line is ________ amperes.
a. 1.36
b. 13.6
c. 5
d. 10

14. The current-carrying capacity of a conductor is determined by its
a. insulation
b. shape
c. insulation and cross-sectional area
d. length

15. A primary cell
a. is not designed to be recharged
b. is designed to be recharged
c. uses no electrolyte
d. has no metal in its construction

16. A storage battery used in an automobile
    a. is made of secondary cells.
    b. is made of primary cells
    c. cannot be recharged
    d. is made of a single cell
17. The energy that a body has owing to its position is called __________ energy.
    a. heat
    b. potential
    c. light
    d. kinetic
18. The energy that a body has owing to its motion is __________ energy.
    a. potential
    b. kinetic
    c. motion
    d. light
19. One horsepower of mechanical energy is equivalent to __________ watts of electrical energy.
    a. 1,000
    b. 33,000
    c. 525
    d. 746
20. High-voltage resistors are usually
    a. made of carbon
    b. wirewound
    c. tapped
    d. made of ceramic
21. A variable resistor placed in a unit with three leads and a control shaft is called a
    a. variometer
    b. tapped resistor
    c. potentiometer
    d. carbon resistor

6

SERIES, PARALLEL, AND SERIES – PARALLEL CIRCUITS

6.1 Operation of Series Circuits

We find series circuits in many types of electrical and electronics equipment. For example, Christmas-tree light bulbs are sometimes connected in series, as shown in Fig. 6-1a. This string of lights has eight bulbs; all the bulbs are the same.

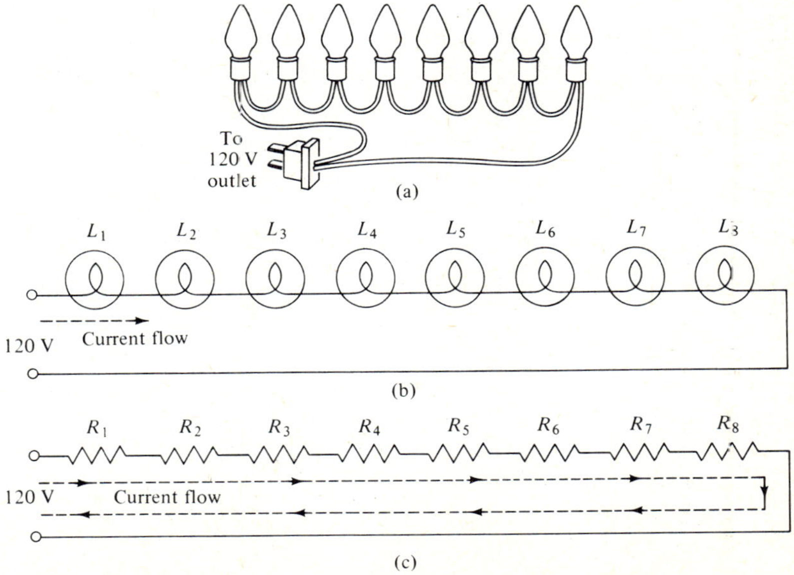

Fig. 6-1
Series string of Christmas-tree lights. (a) Appearance of string. (b) Schematic diagram. (c) Equivalent circuit.

When the plug is inserted into a 120-volt outlet, all eight bulbs light up. A schematic diagram for the series string of lights is shown in Fig. 6–1b. This diagram shows how electric current enters the filament of L_1, then passes through the filament, enters the filament of L_2, and so on. The filaments are connected in series, and a complete path for current flow is provided.

Because each of the filaments has resistance, we can draw an *equivalent circuit* for the series string, as shown in Fig. 6–1c. In this equivalent circuit, each lamp filament is represented by a resistor. Since all the bulbs are the same, all the resistances in the equivalent circuit are the same. Notice that the current enters at one terminal of the circuit and leaves at the other terminal. Now, let us consider an experiment that shows an important fact about a series circuit.

Instructor Demonstration or Student Experiment:

Plug a series string of Christmas-tree bulbs into a power outlet. Remove one of the bulbs from its socket, and note how all the bulbs become dark immediately. Can you explain why this happens?

Questions:

1. If a filament happens to burn out in one of the bulbs in a series string, will all the bulbs become dark?
2. Explain how you can find a burned-out bulb in a series string.
3. Do you think that there is likely to be more than one bulb burned out when a series string becomes dark?
4. Why is it very unlikely that all the bulbs in a series string will burn out at the same time?

Next, let us consider how many volts there will be across each of the light bulbs in Fig. 6–1. Since there are eight bulbs in the string, and all of the bulbs are the same, we know that there will be one-eighth of the supply voltage across each bulb. That is, there will be 15 volts across each bulb, because 8 times 15 is equal to 120 volts. Let us observe a demonstration which proves this fact.

Instructor Demonstration or Student Experiment:

Connect a voltmeter, as shown in Fig. 6–2, to measure the voltage across each filament in the series string. Observe that the voltmeter reads the same voltage for each filament. Also, observe that the voltage across each filament is equal to one-eighth of the supply voltage.

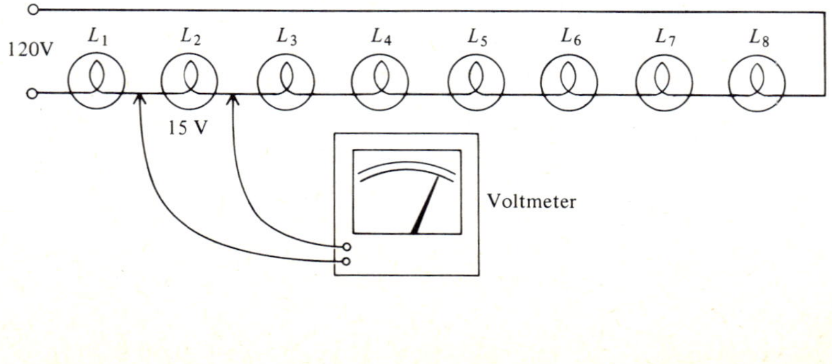

Fig. 6–2
The voltmeter measures the voltage across L_2, in this example.

Questions:

1. If the voltmeter were connected across L_2 and L_3 in Fig. 6-2, how many volts would we read on the voltmeter scale?
2. If the voltmeter were connected across L_3 and L_4 in Fig. 6-2, how many volts would we read on the voltmeter scale?
3. If the voltmeter were connected across the eight lamps L_1-L_2-L_3-L_4-L_5-L_6-L_7-L_8 in Fig. 6-2, how many volts would we read on the voltmeter scale?

When a lamp filament burns out in a series string, an *open circuit* occurs. The effect is the same as if there were a switch in the circuit, and the switch had been opened. Figure 6-3 shows how a switch is connected into a series string of light bulbs. Notice that the switch is in series with the lamp bulbs. When the switch is

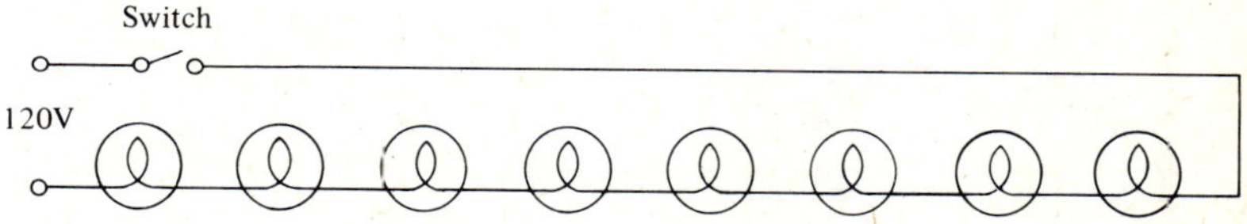

Fig. 6-3
How a switch is connected into the series-string circuit.

open, there are 120 volts across the switch, and there is 0 volt across each lamp filament. On the other hand, when the switch is closed, there is 0 volt across the switch, and there are 15 volts across each lamp filament. Let us observe a demonstration which proves this fact:

Instructor Demonstration or Student Experiment:

Connect a series string of lamp bulbs in series with a switch, as shown in Fig. 6-4. Connect a voltmeter across the switch, and another voltmeter across one

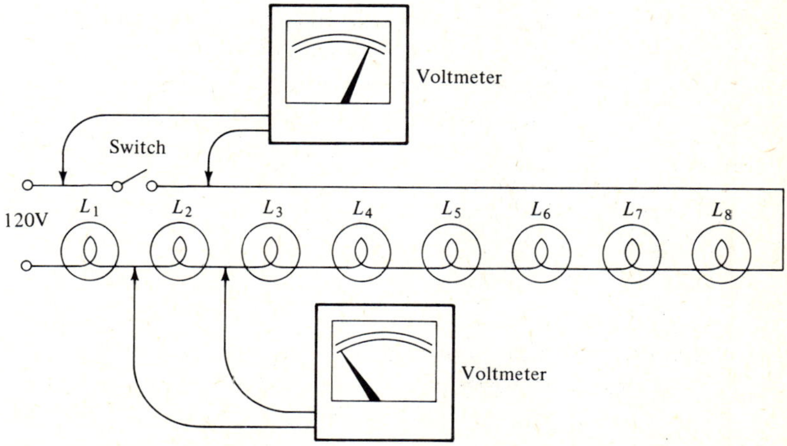

Fig. 6-4
Demonstration to show switch action in a series circuit.

of the lamp filaments. When the switch is open, observe that the first voltmeter reads 120 volts, and the second voltmeter reads zero. On the other hand, when the switch is closed, the first voltmeter reads zero, and the second voltmeter reads 15 volts.

Questions:

1. If the switch is closed in Fig. 6–4 and the filament in lamp L_3 burns out, will the reading of either voltmeter change? Explain your answer.
2. Next, with the filament of lamp L_3 burned out, the switch is opened. What is the reading of each voltmeter under this condition of operation? (Note: remember that each voltmeter has a very high resistance, which allows some small amount of current to flow.)

When a lamp filament burns out in a series string, we say that there is a circuit fault. This is an example of an open-circuit fault. Next, let us consider what happens when a *short-circuit* fault occurs. Figure 6–5 shows a short circuit in a

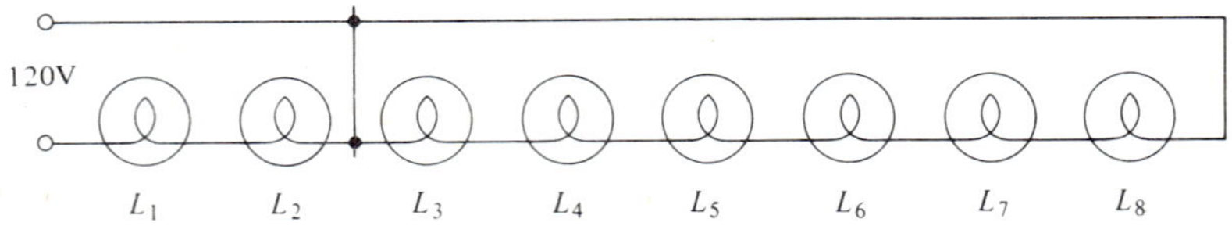

Fig. 6–5
A short circuit in a series string of lamp bulbs.

series string of lamp bulbs. A short circuit is often caused by frayed or worn-out insulation on wires. When the bare wires touch together, a short circuit occurs. In the example of Fig. 6–5, current can flow through lamps L_1 and L_2. However, no current can flow through the other lamps, because a short circuit has a very low resistance compared to a lamp filament. Therefore, the short circuit in this example causes 120 volts to appear across L_1 and L_2.

Let us consider what will happen in Fig. 6–5 when a short circuit occurs. Since 120 volts appear across L_1 and L_2, there will be 60 volts across L_1, and there will be 60 volts across L_2. Since this is four times too much voltage for the filaments, at least one of the filaments will burn out quickly. Sometimes both filaments will burn out. On the other hand, the six lamps L_3-L_4-L_5-L_6-L_7-L_8 will not be damaged.

6.2 Operation of Parallel Circuits

Parallel circuits are widely used in many types of electrical and electronic equipment. For example, Christmas-tree lights may be operated in parallel, as shown in Fig. 6–6a. This string of lights has eight bulbs; all the bulbs are the same. A schematic diagram for the parallel string is shown in Fig. 6–6b. There are 120 volts across each bulb. Also, each bulb provides a complete path for current flow. An equivalent circuit for the parallel string is shown in Fig. 6–6c. In this equivalent circuit, each lamp filament is represented by a resistor. Since all the bulbs are the same, all the resistances in the equivalent circuit are the same.

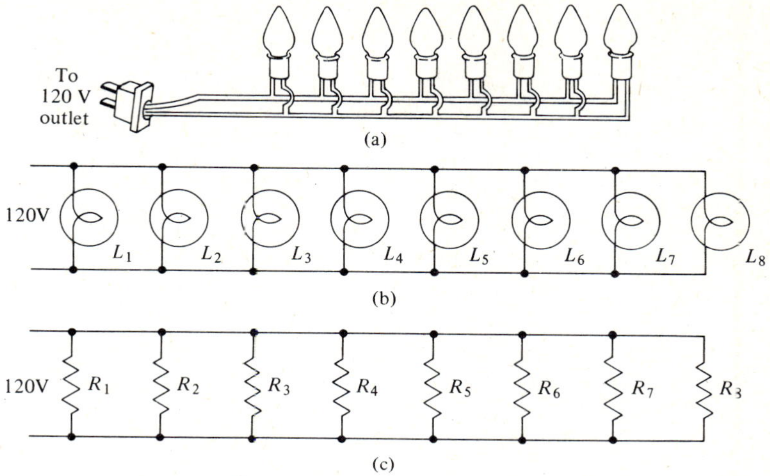

Fig. 6-6
Parallel string of Christmas-tree lights. (a) Appearance of string. (b) Schematic diagram. (c) Equivalent circuit.

Instructor Demonstration or Student Experiment:

Plug a parallel string of Christmas-tree lights into a power outlet. Remove one of the bulbs from its socket, and note that all the other bulbs remain lighted. Explain why this happens.

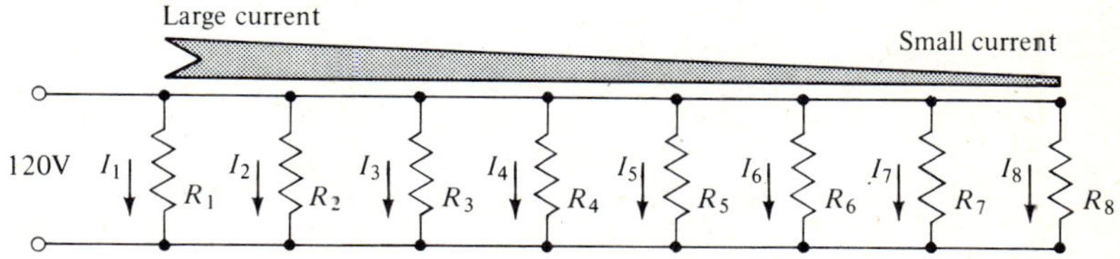

Fig. 6-7
Current flow paths in a parallel circuit.

Figure 6-7 shows the current paths for the parallel circuit in this example. Notice that there are eight separate paths, or *branches*. Since each of the resistances is the same, each of the branch currents is the same. We see that the total current drawn by the parallel string is eight times the amount of current drawn by a single branch. Also, the current at the beginning of the circuit is eight times as great as the current at the end of the circuit. Let us observe a demonstration which proves this fact.

Instructor Demonstration or Student Experiment:

Connect three ammeters into the parallel circuit to show the current which flows at three different points in the circuit (Figure 6-8). Observe that A_2 and A_3 read the same, and that A_1 reads eight times as much current as either A_2 or A_3. Disconnect R_5 and notice that A_2 and A_3 read the same as before,

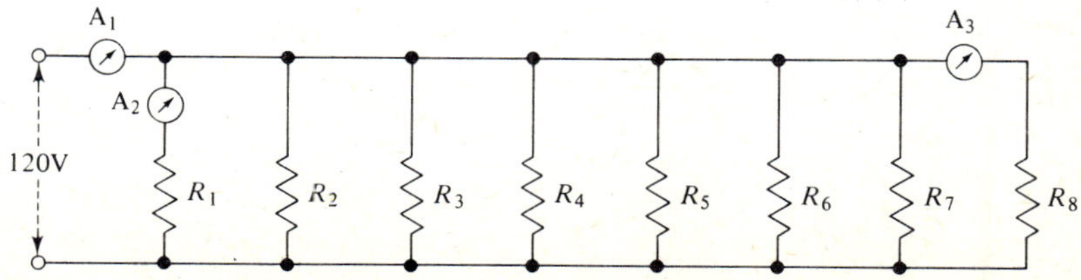

Fig. 6-8
Demonstration of currents in a parallel circuit.

but A_1 now reads seven times as much current as either A_2 or A_3.

Questions:

1. What current does ammeter A_2 read?
2. What current does ammeter A_3 read?
3. How many branch currents are there flowing through A_1 when all the re-resistors are connected into circuit?
4. Would the current through A_1 be reduced more if R_2 were disconnected instead of R_5?

Now let us consider the operation of switches in parallel circuits. The simplest switch arrangement is shown in Fig. 6-9. When the switch is open, all the lamps

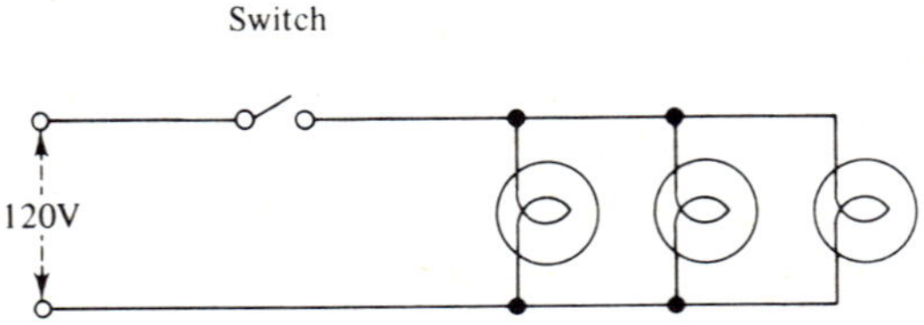

Fig. 6-9
Basic switch connections in a parallel circuit.

are dark. On the other hand, when the switch is closed, all the lamps are lighted. This arrangement is often used to turn the lights on or off in a kitchen or living room.

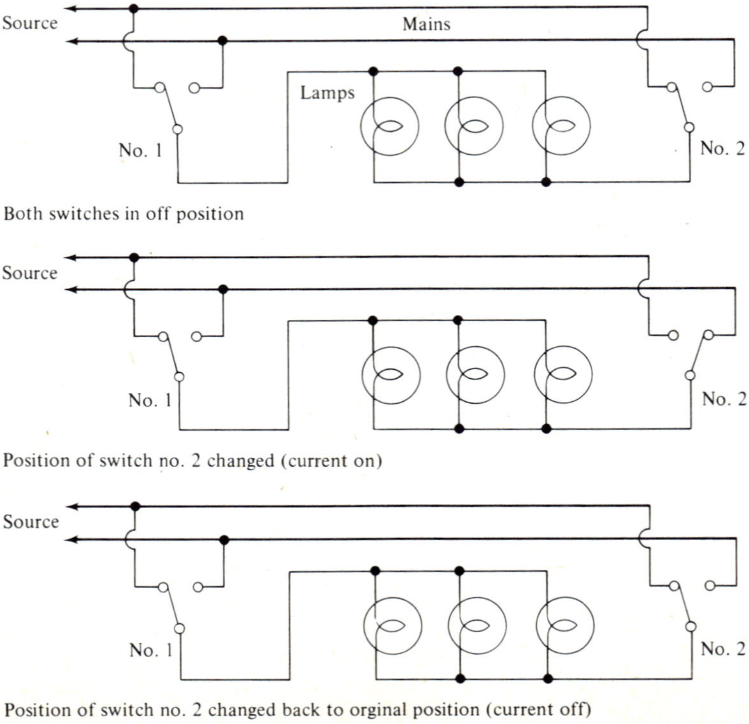

Fig. 6-10
A 3-way switch circuit

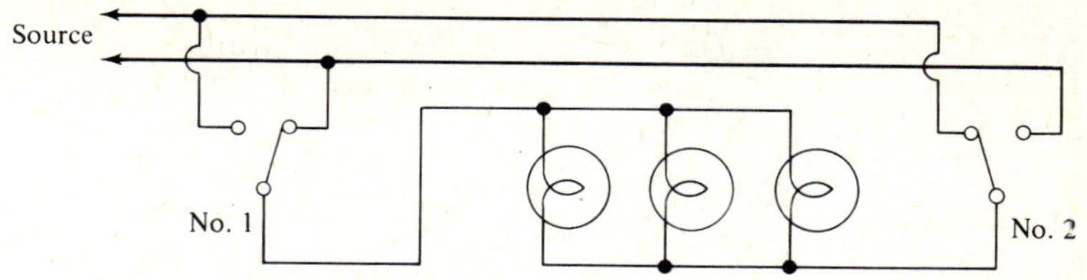

Position of switch no. 1 changed (current on)

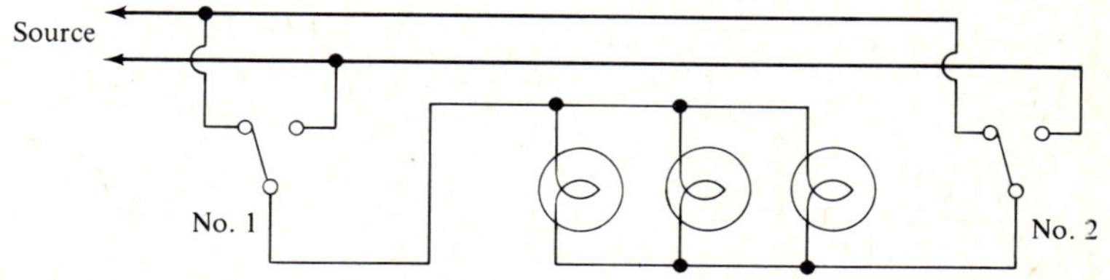

Fig. 6-10 (cont'd) Position of switch no. 1 changed back to original position (current off)

Another widely used switch arrangement has a pair of switches to control the same circuit, as shown in Fig. 6-10. This is called a three-way switch circuit. We find this arrangement used in stairways or hallways, for example. A switch is provided at each end of the circuit, and a person at either end can turn the lights on or off. We see in Fig. 6-10 that the lamps will be off when both switches are thrown to the left. However, if either of the switches is thrown to the right, the lamps will be turned on. Next, if the position of either switch is changed, the lamps will be turned off. If both switches happen to be thrown to the right, for example, the lamps will be turned off. Then, if either switch is thrown to the left, the lamps will be turned on.

6.3 Operation of Series-Parallel Circuits

Series-parallel circuits consist of one or more series circuits combined with one or more parallel circuits. A simple example of a series-parallel circuit is shown in Fig. 6-11. This is an arrangement for dimming lights. An adjustable resistor, or

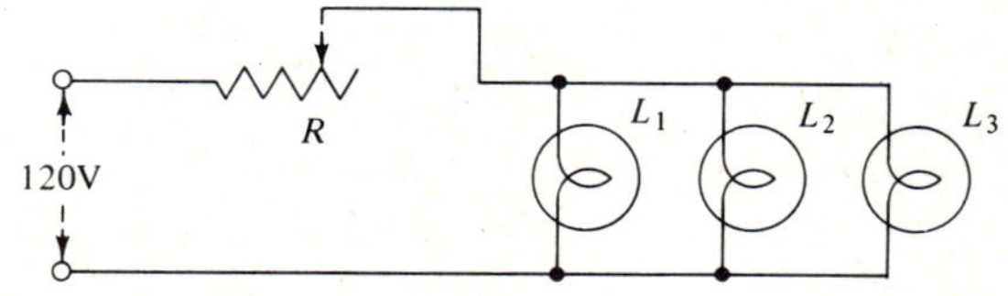

Fig. 6-11 A light-dimming arrangement that uses a series-parallel circuit.

rheostat, is connected in series with the group of parallel lights. As the contactor is moved to the right on the resistor R, more resistance is placed in series with the circuit. In turn, the lights become dimmer. Let us observe a demonstration that shows the operation of this series-parallel circuit.

Instructor Demonstration or Student Experiment:

Connect the experimental arrangement shown in Fig. 6-12 to show the volt-

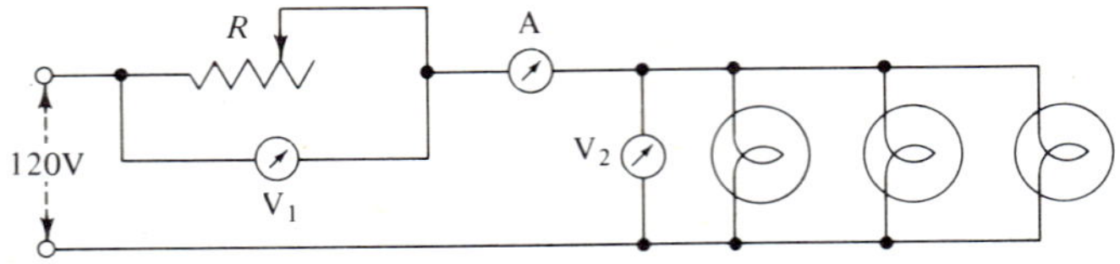

Fig. 6–12
Demonstration of voltages and current in a series–parallel circuit.

ages and current in a series–parallel circuit. As the contactor is moved along resistor R, we observe that the lights become brighter or dimmer. Also, as the lights become dimmer, the reading of ammeter A becomes less. As the current becomes less, we see that the voltmeter V_2 shows that there is less voltage across the lamps. Finally, we notice that the reading on V_1 plus the reading on V_2 is always 120 volts, no matter where the contactor is moved along resistor R.

Questions:

1. Why does the current demand become less when more resistance is inserted in series with the parallel circuit?
2. If the reading of voltmeter V_1 becomes greater, does the reading of voltmeter V_2 then become greater or smaller?
3. Can you explain why the sum of the voltage readings on V_1 and V_2 is always equal to 120 volts? (Hint: Ohm's law states how much current flows through a resistance when voltage is applied.)

Questions Chapter 6

1. Give two reasons why series circuits are not used to connect the electrical appliances in a home.
2. Suppose that two lamps of equal resistance are connected across 115 volts. What is the voltage across each lamp?
3. If the resistance of one of the lamps in question two is twice the resistance of the other, what is the voltage across each lamp?
4. What are the effects when one lamp burns open in a series connection?
5. Define a "short" circuit.
6. The total current into three identical parallel-connected lamps is twelve amperes. What is the current in each lamp?
7. What is the purpose of a switch in an electrical circuit?
8. Suppose the total current in the circuit, shown in Fig. 6–12, is twelve amperes. What is the current in each component if the three lamps are identical?
9. What happens to the total current in the circuit shown in Fig. 6–12 when the resistance is increased?
10. How are voltmeters and ammeters connected in electrical circuits?
11. What is the usual cause of a short circuit?

True–False:

1. The current is the same in all parts of a series circuit.
2. The voltage is the same across each component in a series circuit.
3. The current is the same in each component in a parallel circuit.
4. The voltage is the same across each resistor in a parallel circuit.
5. Removing one lamp from a series circuit will stop the current in the other lamps.
6. Removing one lamp in a parallel circuit consisting of six lamps will stop the current in all the lamps.
7. The current is the same in each appliance in an ordinary kitchen.
8. A voltmeter is always connected in parallel across the component for which the voltage is being measured.
9. An ammeter is always connected in series with the components in a circuit.
10. A switch is used to short a series circuit.

Multiple choice: choose the best answer.

1. In an electrical circuit,
   a. voltmeters are connected in series
   b. ammeters are connected in series
   c. voltmeters are connected in series or parallel
   d. ammeters are connected in series or parallel
2. The resistance value of a/an
   a. ammeter is very low
   b. voltmeter is very low
   c. voltmeter is zero ohms
   d. ammeter is very high
3. A component that is used to turn on or off the current to a circuit is called a
   a. potentiometer
   b. lamp
   c. short circuit
   d. switch
4. Accidentally touching two bare electrical wires in a circuit causes a/an
   a. open circuit
   b. short circuit
   c. potentiometer
   d. switch
5. The head lamps and tail lights of an automobile are connected in
   a. series
   b. parallel
   c. series parallel
   d. opposition

6. A switch in an electrical circuit is usually connected in
   a. parallel
   b. series parallel
   c. series
   d. either series or parallel
7. The resistance of a short circuit is usually
   a. short
   b. high
   c. long
   d. low
8. The resistance of an open circuit is
   a. high
   b. low
   c. either high or low
   d. neither high nor low
9. An example of a series circuit is
   a. the lights in a house
   b. the appliances in a house
   c. the lights of an automobile and the horn of the automobile
   d. a switch and another electrical component

7

PRINCIPLES OF MAGNETISM

7.1 The Magnetic Field

Magnetism has been known since ancient times. Chinese sailors are said to have first used lodestones as compasses. Lodestone is a magnetic iron ore which has the appearance of a black stone. When a lodestone is suspended by a thread, as shown in Fig. 7-1, it points north and south. This compass action was a mystery

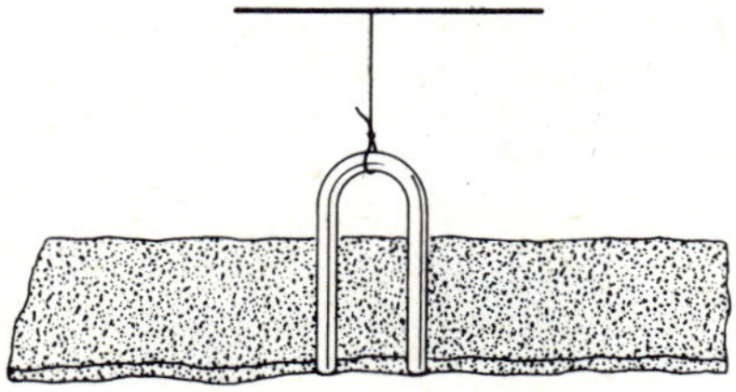

Fig. 7-1
A lodestone points north and south when suspended by a thread.

for many centuries. Today, however, we know that a lodestone acts as a compass because the earth is a huge magnet, although its magnetism is rather weak. Figure 7-2 shows how the earth acts as a magnet.

Any magnet is surrounded by a *magnetic field.* The earth's magnetic field is indicated by the curved lines in Fig. 7-2. A magnetic field is simply an area in which the force of magnetism is exerted. This force may be exerted on a piece of iron or it may be exerted on a magnet, such as lodestone. A compass contains a magnet, although lodestone is no longer used for this purpose. Modern com-

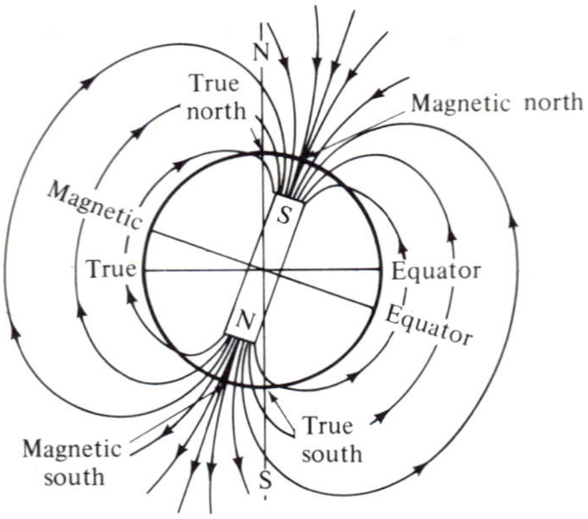

Fig. 7-2
The earth acts as a weak magnet.

passes, such as the one shown in Fig. 7-3, contain a magnetized steel needle which is mounted on a pivoted bearing. Let us observe a demonstration which shows

Fig. 7-3
A magnetic compass.
Courtesy: Sargent-Welch Scientific Co.

the behavior of a magnetic compass in a magnetic field.

Instructor Demonstration:

As shown in Fig. 7-4, a compass can be used to explore the field in the space around a magnet. We observe that the north pole of the compass needle always points in the direction of the arrowheads. The important facts for us to notice in this demonstration are these:

1. The north pole of the compass needle is attracted by the south pole of the magnet. Also, the south pole of the compass needle is attracted by the north pole of the magnet.
2. The north pole of the compass needle is repelled by the north pole of the magnet. Also, the south pole of the compass needle is repelled by the

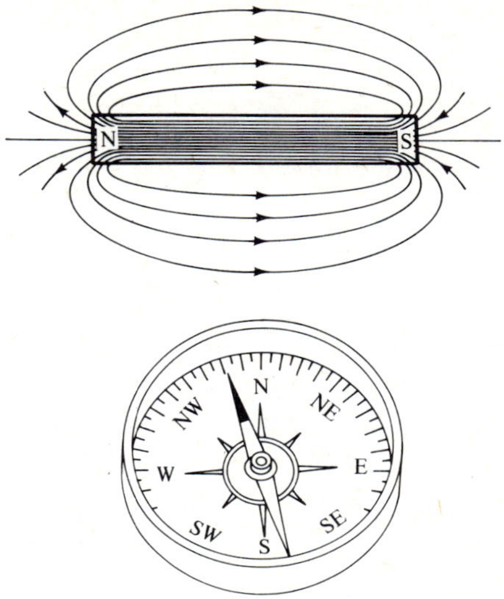

Fig. 7–4
The field of a bar magnet can be explored with a compass.

south pole of the magnet.

3. Notice in Fig. 7-2 that the *magnetic* north pole of the earth has south magnetic polarity. The reason for this fact is seen from (1) in the above demonstration.

It is also interesting to notice in Fig. 7-2 that the magnetic north pole is not located at quite the same place as the true north pole of the earth. The reason for this separation has not been discovered so far. Sailors call the angle between true north and magnetic north the angle of *declination,* or simply the *deviation of the compass.*

When we draw a representation of a magnetic field, the number of *lines of magnetic force* represent the strength of the magnetic field. That is, if we draw twice as many lines of force at the pole of a magnet, we indicate double strength of the magnetic field. Several important facts about magnetic lines of force are the following:

1. Lines of force are always considered to come out of the north pole and to go into the south pole of a magnet.
2. A line of force is continuous both inside and outside a magnet.
3. Lines of force never cross each other.
4. The magnetic field becomes weaker as we go farther away from a magnet, but there is no point at which the field strength disappears completely.

A magnet must always have two poles, with north and south polarity. That is, there is no such thing as a north pole by itself, or a south pole by itself. As shown in Fig. 7-5, we cannot obtain separate north and south poles by breaking a magnet in half. In other words, if we break a magnet in half, we find a south pole opposite the north pole in each half. Then, if we place the broken magnet back together again, we find the original north and south poles, and the new poles which were formed by breaking the magnet disappear.

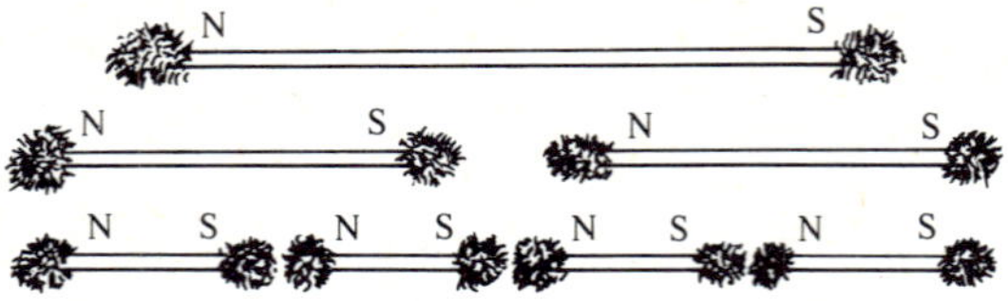

Fig. 7-5
Showing the effects of breaking a magnet into several parts.

7.2 Types of Permanent Magnets

The types of magnets that we have considered are called *permanent magnets.* This means that their magnetism is self-contained, and does not require electricity to produce their magnetic fields. A lodestone is called a *natural magnet,* because it is a type of iron ore found in mines. A compass needle is an example of an *artificial magnet,* because it must be made with the aid of another magnet or with the aid of electricity. The earth is an example of a natural magnet.

Figure 7-4 shows an example of a bar magnet. This is a type of artificial permanent magnet. If a bar magnet is bent into a "U" shape, it is called a horseshoe magnet. An example of a horseshoe magnet is shown in Fig. 7-6. This particular

Fig. 7-6
A horseshoe magnet that can lift 350 pounds of iron. Courtesy: Edmund Scientific Co.

artificial permanent magnet will lift 350 pounds of iron. We find various other shapes of magnets used in particular electrical and electronic equipment. However, the basic magnetic principles are always the same.

If a permanent magnet is handled with care, it will retain its magnetism indefinitely. However, rough treatment will cause a magnet to lose some or all of its magnetism. For example, pounding or jarring a permanent magnet will weaken its field. A permanent magnet should not be subjected to vibration. Heat will also weaken a permanent magnet; for example, a magnet may lose part of its field strength if it is placed in boiling water. Remember also that a weak permanent magnet may lose part or all of its field strength if it is placed in the field of another very strong permanent magnet.

When permanent magnets are stored or shipped, they are less likely to lose field strength if they are placed in pairs with their north and south poles together, as shown in Fig. 7-7. A single permanent magnet is less likely to lose field strength if it is stored or shipped with a *keeper,* as shown in Fig.7-8. A keeper is

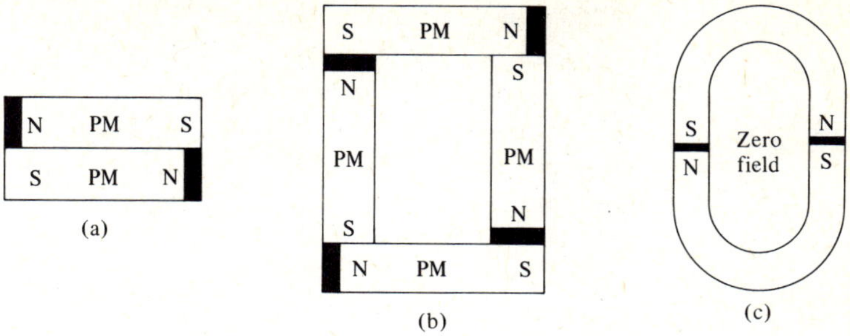

Fig. 7-7
How to store or ship permanent magnets.
(a) Two bar magnets.
(b) Four bar magnets.
(c) Two horseshoe magnets.

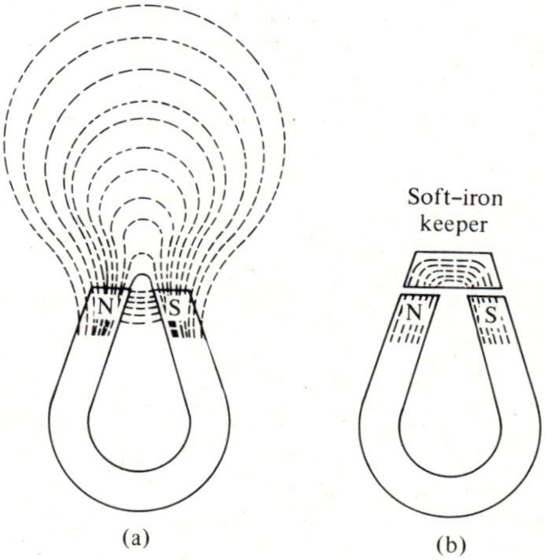

Fig. 7-8
Action of a keeper.
(a) Magnetic field of a horseshoe magnet.
(b) Magnetic field with keeper.

simply a piece of soft iron that is placed across the poles of the permanent magnet. In any case, the principle is to keep the magnetic lines of force confined within the metal.

7.3 Formation of Permanent Magnets

Permanent magnets can be formed from steel, but not from soft iron. When a permanent magnet is formed, the molecules in the steel line up in the same direction, as shown in Fig. 7-9. That is, each molecule in the steel bar acts as a tiny

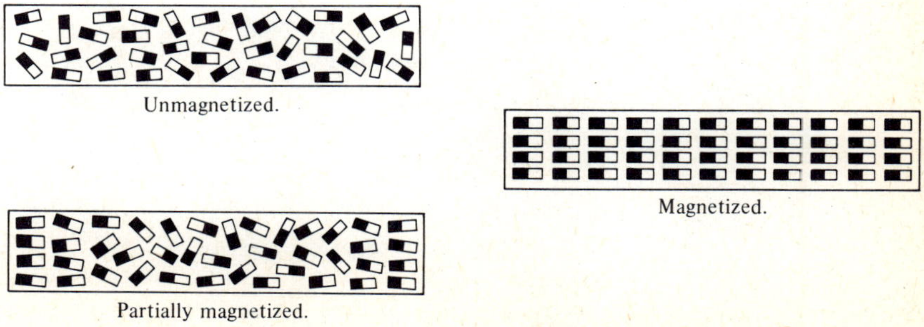

Fig. 7-9
Representation of molecular magnets in a steel bar.

magnet. Because steel is very hard, the molecules stay lined up, thereby forming a permanent magnet. On the other hand, if the molecules are forced to line up in a soft-steel bar, they will fall back into the unmagnetized condition after the magnetizing force is removed.

Figure 7-10 shows a common method of applying magnetizing force to a steel bar in order to form a permanent magnet. A strong permanent magnet is used,

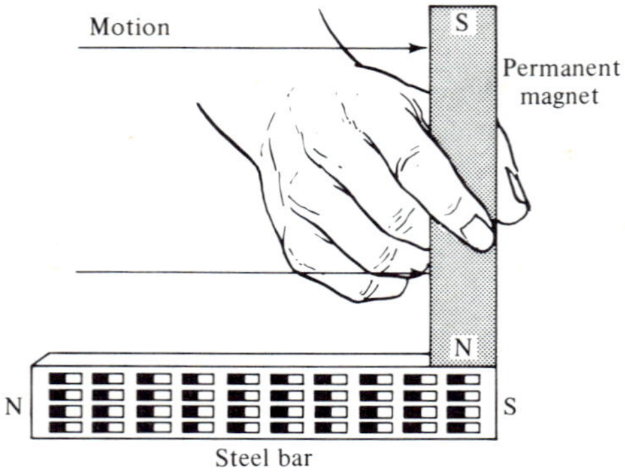

Fig. 7-10
Magnetization of a steel bar.

and the steel bar is stroked in one direction. After the bar has been stroked once, it will be found to have a magnetic field. If the steel bar is stroked again in the same direction and with the same pole of the permanent magnet, it will be found to have a stronger magnetic field. After the bar has been stroked many times in the same direction (with the same pole of the permanent magnet), it will have the strongest magnetic field that can be obtained by this method of magnetization.

Questions

1. If a steel bar is stroked in a certain direction with the north pole of a permanent magnet, and is then stroked again in the same direction with the south pole of the magnet, will the field strength of the steel bar be increased?
2. If a steel bar is stroked in a certain direction with the north pole of a permanent magnet, and is then stroked again in the opposite direction with the south pole of the magnet, will the field strength of the steel bar be increased?

7.4 Action of Two Magnetic Fields on Each Other

Now let us consider how two magnetic fields act upon each other. The two basic conditions are shown in Fig. 7-11. If unlike poles are facing each other, the magnetic lines of force crowd together between the poles, and the permanent magnets attract each other. On the other hand, if like poles are facing each other, the magnetic lines of force are pushed out from between the poles, and the magnets repel

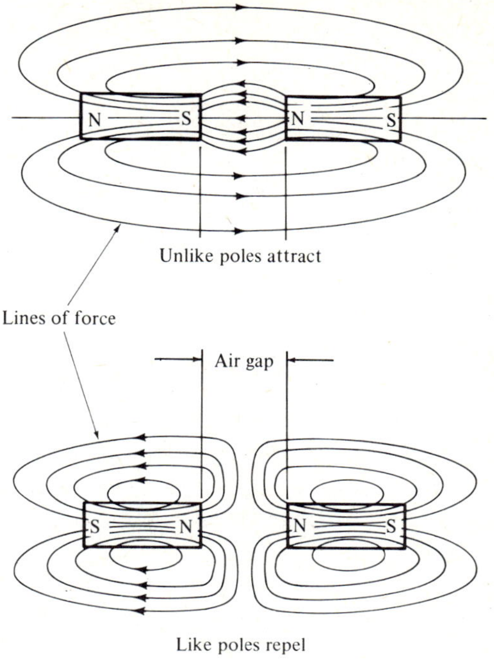

Fig. 7-11
Lines of force between unlike and like poles.

each other. When a pair of very strong bar magnets are placed as shown in Fig. 7-12, the upper magnet will float in the air above the lower magnet. This type of demonstration equipment is called "Mohammed's Coffin."

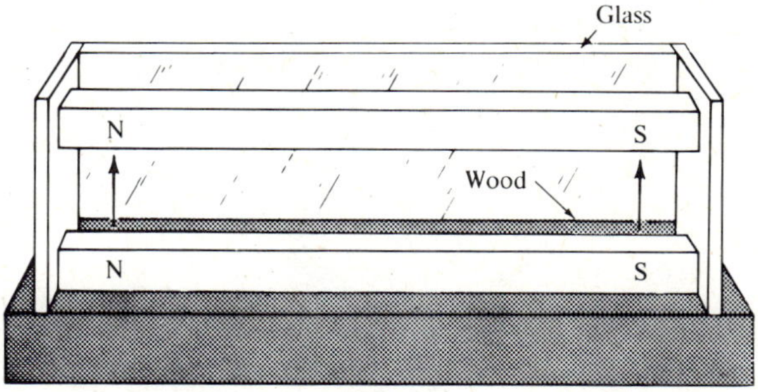

Fig. 7-12
Magnet characteristics.

Whenever we are making experiments in the repulsion of magnetic fields, the two magnets should have about the same field strength. If one of the magnets is very strong and the other magnet is very weak, we will not find any repelling action. The reason for this is that the strong magnet overcomes the magnetic field of the weak magnet, and reverses the poles of the weak magnet. This causes the strong magnet to attract the weak magnet, instead of repelling it. In the following experiment we can easily "map" the shape of a magnetic field.

Instructor Demonstration or Student Experiment:

As shown in Fig. 7-13, iron filings are sprinkled on a sheet of cardboard which

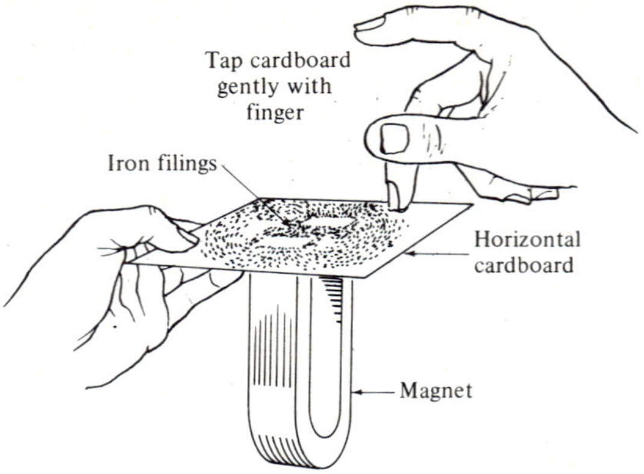

Fig. 7-13
Mapping a magnetic field with iron filings.

is held over a magnet. When the cardboard is tapped, the filings fall into a pattern that maps the magnetic field. This happens because each of the iron filings becomes a temporary magnet in the field of magnetizing force. The result is that the filings turn and line up end to end in the direction of the lines of magnetic force.

Questions:

1. What will happen to the iron filings if the ends of the magnet are reversed under the cardboard?
2. What would be the results if plastic filings were sprinkled on the cardboard?

7.5 Magnetic Shielding

We find that there is no insulator for magnetism. In other words, lines of magnetic force pass through all materials. For example, the sheet of cardboard in Fig. 7-13 does not stop the magnetic force lines. Magnetism is not affected by glass, wood, plastics, vacuum, plants, or animal tissues. In spite of the fact that there is no insulator for magnetism, we can make magnetic shields from soft iron, as shown in Fig. 7-14. A magnetic shield does not stop the magnetic force lines, but it concen-

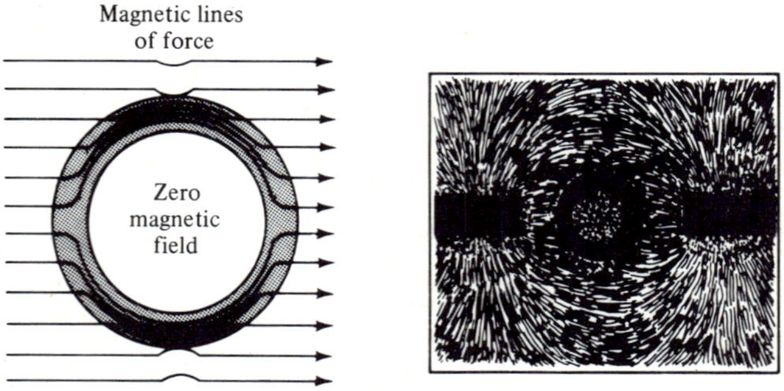

(a) (b)

Fig. 7-14
Principles of magnetic shielding. (a) A soft iron ring carries magnetic lines of force around the enclosed space. (b) Mapping the magnetic field with iron filings.

trates the lines in the soft iron. Therefore, very few lines of magnetic force stray into the space inside the magnetic shield. Such shields are widely used in electronic equipment, as seen in Fig. 7-15.

7.6 Principles of Electromagnetism

There is a close relation between electricity and magnetism. When a magnetic compass is brought near a current-carrying wire, the compass needle is deflected, as shown in Fig. 7-16. This experiment demonstrates the fact that a current-carry-

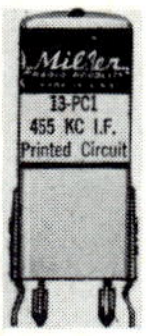

Fig. 7-15
Shield can encloses radio coils. Courtesy: Miller Radio Products

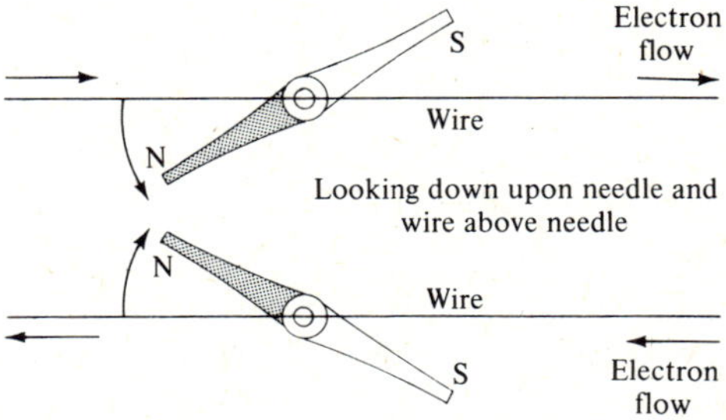

Fig. 7-16
Deflection of a compass needle by electric current in a wire.

ing wire is surrounded by magnetic lines of force. If we map the lines of force, we find that they encircle the wire, as seen in Fig. 7-17; The left-hand rule is used

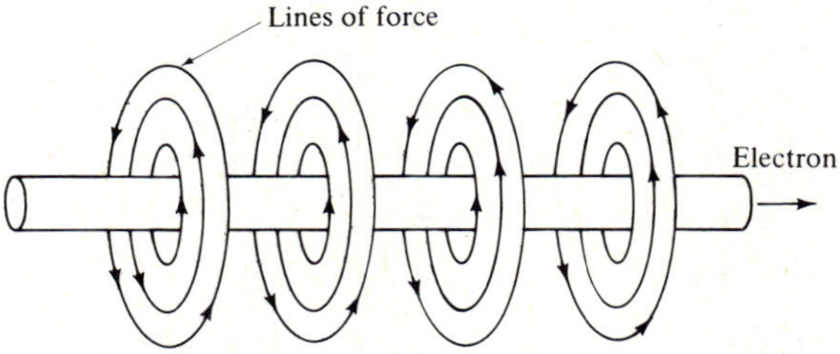

Fig. 7-17
Magnetic lines of force surround a current-carrying wire.

to remember the directions of the electron flow and the magnetic force lines, as shown in Fig. 7-18. That is, if you grasp a current-carrying wire in your left hand,

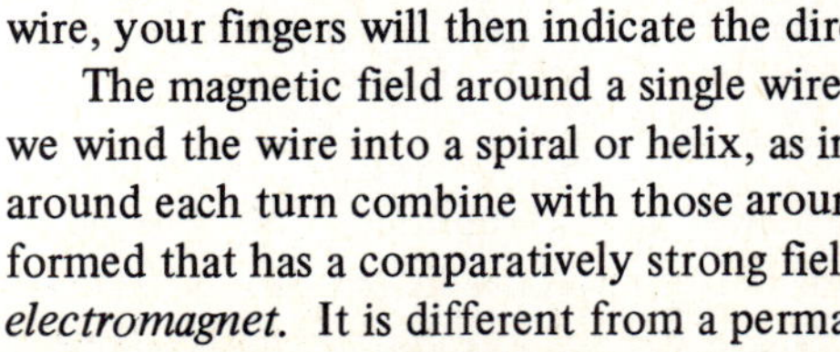

Fig. 7-18
Left-hand rule used to determine direction of magnetic force lines around a current-carrying wire.

with your thumb pointing from the negative end toward the positive end of the wire, your fingers will then indicate the direction of the magnetic lines of force.

The magnetic field around a single wire is comparatively weak. However, if we wind the wire into a spiral or helix, as in Fig. 7-19, the magnetic force lines around each turn combine with those around the next turn, and a magnet is formed that has a comparatively strong field. This kind of magnet is called an *electromagnet.* It is different from a permanent magnet, because the magnetic field is produced only when turns of wire are carrying electric current. The left-hand rule can be used to find the polarity of an electromagnet, as shown in Fig.

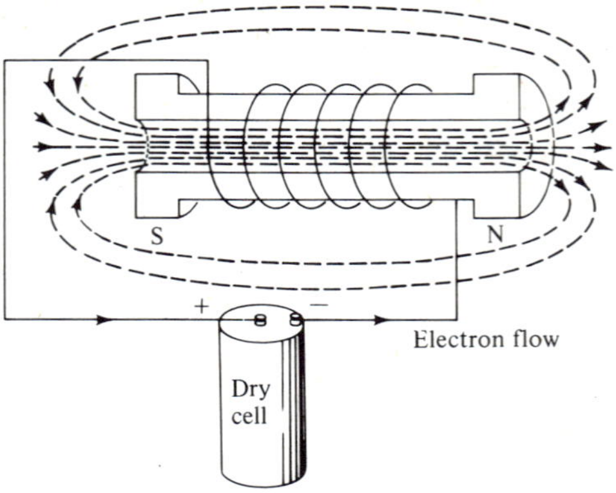

Fig. 7-19
Magnetic field around an air-core electromagnet.

7-20. If you grasp the coil with your left hand so that your fingers are in the direction of electron flow, your thumb will point toward the north pole of the electromagnet.

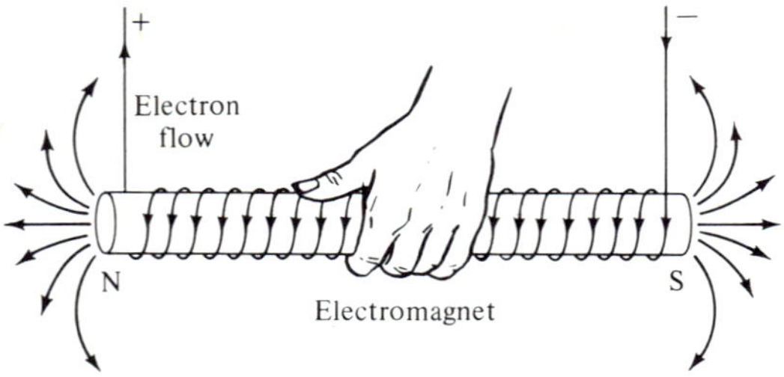

Fig. 7-20
Showing method of finding the magnetic polarity of a coil by means of the left-hand rule.

Instructor Demonstration or Student Experiment:

Connect a coil of wire to a dry cell, as shown in Fig. 7-21. Observe that when

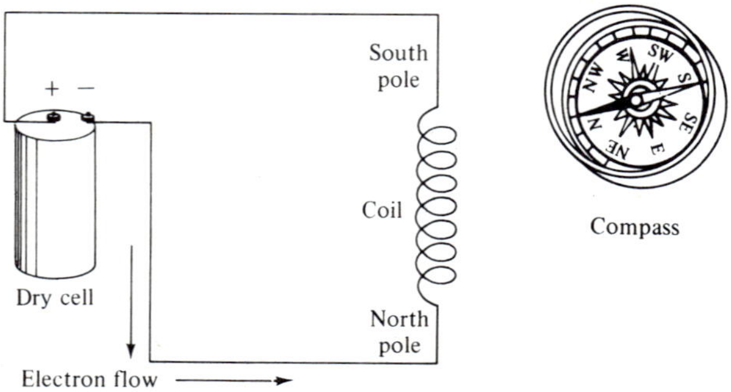

Fig. 7-21
Demonstration of electromagnetism.

a magnetic compass is brought near one end of the coil, the north-seeking end of the compass needle points toward that end of the coil. Therefore, that end of the coil has south polarity. We can observe that the left-hand rule indicates the same fact. Next, when the direction of current is reversed in the coil, we find that the north and south poles of the coil are also reversed.

Questions:

1. What are the effects on the compass as it is moved around the coil?
2. What are the effects on the compass when the direction of current in the coil is reversed?

There are four ways of increasing the magnetic field from an electromagnet:

1. More voltage can be applied across the coil.
2. More turns can be used in the spiral of wire.
3. More layers can be wound on the coil.
4. A soft-iron core can be inserted in the coil.

When a coil of wire is wound on a nonmagnetic substance, such as a plastic tube, it is called a *solenoid,* or an *air-core* electromagnet. However, if a soft-iron core is inserted into the coil, as shown in Fig. 7-22, an *iron-core* electromagnet is formed. An iron core greatly increases the strength of an electromagnet.

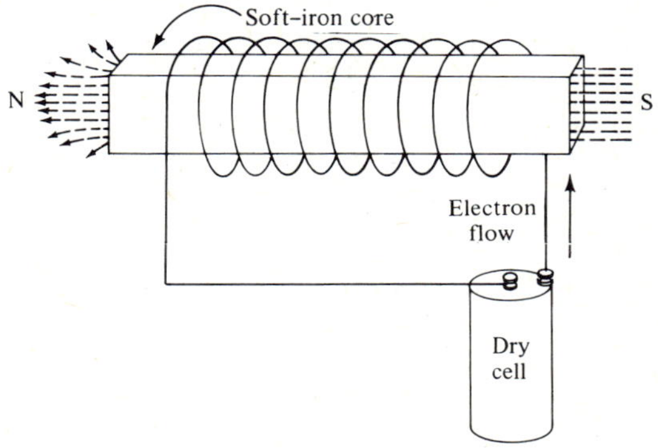

Fig. 7-22
A soft-iron core greatly increases the strength of an electromagnet.

7.7 Electromagnetic Devices

Electromagnets are used in many familiar electrical devices. For example, a doorbell has the construction shown in Fig. 7-23. It consists of an electromagnet and

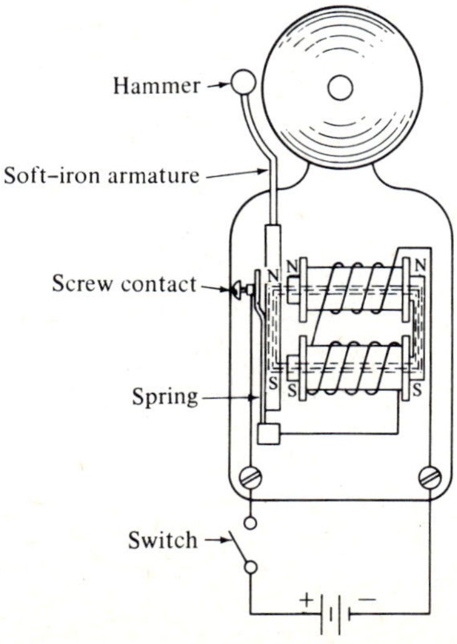

Fig. 7-23
Construction of an electric bell.

a soft-iron armature which is mounted on a flat spring. The armature also has a small hammer arrangement which can strike the bell. Notice that the circuit of the electromagnet contains a battery, a switch, and a contact which makes connection to the armature. This contact operates as an interrupter, which opens and closes the circuit to keep the armature vibrating. The circuit action is as follows:

1. When the switch is closed in Fig. 7-23, current from the battery flows through the coils on the electromagnet.
2. The electromagnet attracts the armature, and the hammer strikes the bell.
3. As the hammer strikes the bell, the contact to the armature is opened and current stops flowing through the coils of the electromagnet.
4. Since the armature is not being attracted by the electromagnet, the spring returns the armature to its original position, thereby closing the contact to the armature.
5. The cycle of operation is then repeated.

Another familiar electromagnetic device is the *circuit breaker,* which is often used instead of a fuse in residential electrical systems. A basic type of circuit breaker is shown in Fig. 7-24. A circuit breaker automatically stops the current

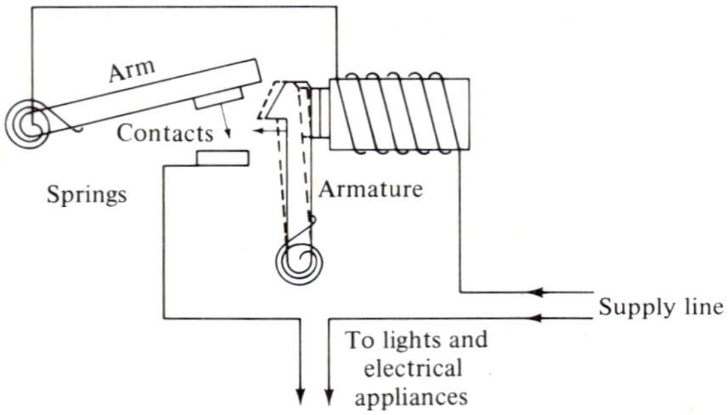

Fig. 7-24
Principle of the magnetic circuit breaker.

flow in case a short circuit occurs in an electrical appliance, for example. This action prevents the possibility of fire. In operation, the arm is pressed down to close the contacts, and is latched in place by the armature. The circuit is then closed, and current is drawn through the electromagnet. Although the electromagnet attracts the armature, ordinary current drains do not produce enough magnetic force to pull the armature back against the spring pressure. However, if a short-circuit occurs, the excessive current drain causes the armature to pull back. The arm is then unlatched and springs up to open the circuit. The circuit then remains open until the arm is pressed down manually.

The *electric telegraph* (Fig. 7-25) was formerly used in every city and town in the country, and is still used in outlying areas for communication over wire lines by means of the Morse code. A telegraph circuit includes a key, a sounder, and a battery. The key is a manually operated switch whereby the telegraph operator forms the dots and dashes of the Morse code. The sounder consists of an electromagnet which attracts a soft-iron armature when current flows through the coils. Notice that the armature is held up by spring pressure. When the armature is attracted by the electromagnet, it is pulled down and makes a

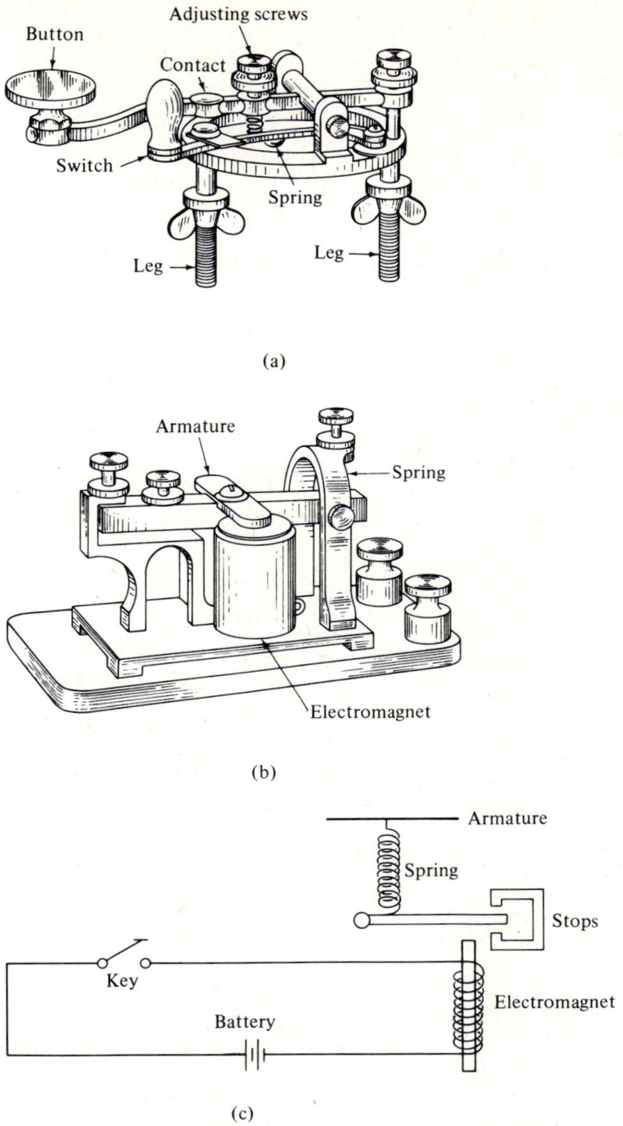

Fig. 7–25
The electric telegraph.
(a) Key. (b) Sounder.
(c) Circuit.

click against the lower stop. When the armature is released by the electromagnet, it is pulled up by spring pressure and makes a click against the upper stop. The operator reads these clicks as dots and dashes in the telegraph code, and so receives the message that is sent from the other end of the telegraph line.

Another important electromagnetic device is called a *relay*. As seen in Fig. 7–26, a relay is a switch which is operated electromagnetically. It opens or closes a circuit when current through the coil is started or stopped. When the operating switch SW is closed, current flows through the coil. In turn, the armature is attracted by the electromagnet and the contacts are closed. That is, how power is ap-

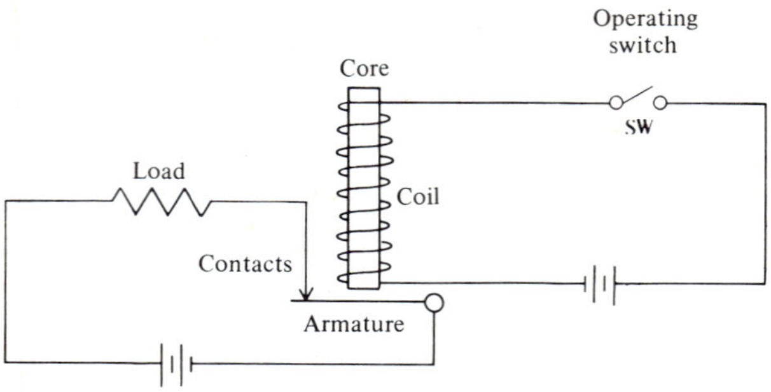

Fig. 7-26
Simple relay circuit.

plied to the load. When operating switch SW is opened, the contacts also open, and power is removed from the load. Relays are used in many applications, such as automobile electrical systems. Telephone systems use relays with many contacts, as shown in Fig. 7-27.

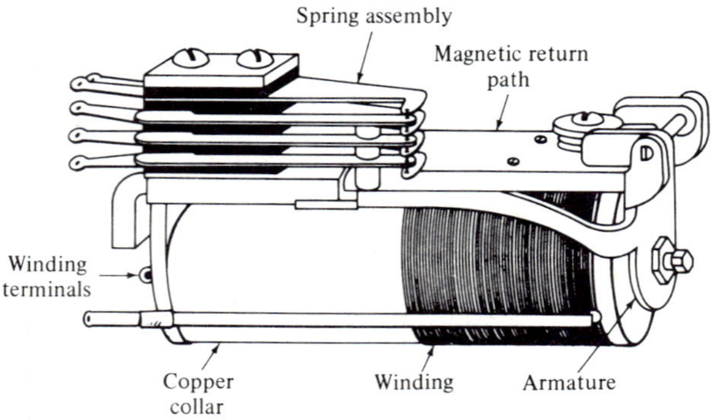

Fig. 7-27
A telephone system relay.

Another widely used type of relay is shown in Fig. 7-28. This is called a *sole-*

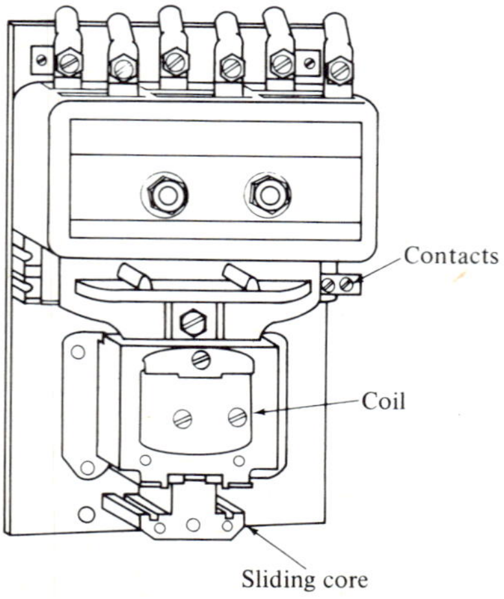

Fig. 7-28
Solenoid type of relay.

noid relay. It consists of a coil, and a sliding core which is drawn up into the coil when current flows through the turns. This arrangement is sometimes called a *solenoid sucking coil.* The iron core is drawn up against the force of gravity. In turn, the sliding core presses electrical contacts together and thus applies power to the load. We find that solenoids with sliding cores are used also to open latches in automobiles, to open and close valves in washing machines, and for various other applications.

Questions Chapter 7

1. A piece of iron is stroked with a permanent magnet, thereby producing another permanent magnet. Where does the force come from that aligns the molecules? MAGNETIC LINES OF FLUX
2. Why does a magnet act as a compass when it is suspended?
3. What are the two basic rules about like and unlike poles of magnets?
4. How should permanent magnets be handled when shipped?
5. Explain how you would map the magnetic field of a permanent magnet.
6. Explain how you would shield an object from magnetic lines of force.
7. Explain the operation of a doorbell relay.
8. Explain the operation of a circuit breaker.
9. Explain the operation of a solenoid relay.

True–False:

T 1. Lines of force are always considered to come out of the north pole and enter the south pole of a magnet.
F 2. The magnetic lines do not exist inside a magnet.
F 3. Magnetic lines of force are free to cross each other.
T 4. A magnet must have both a north and a south pole.
T 5. A lodestone is a natural magnet.
F 6. Vibration and heat will not damage a permanent magnet.
F 7. A keeper is a storage box for a magnet that prevents vibrations from affecting the magnet.
F 8. Permanent magnets are formed from soft iron.
T 9. Stroking a bar of steel with a permanent magnet will magnetize the steel.
T 10. Iron filings may be used to map the magnetic field of a magnet.
F 11. There is no insulator for magnetism.
? 12. Magnetic shields are made from soft iron.
T 13. A magnetic field surrounds a current-carrying conductor.
T 14. An electromagnet requires a coil of wire and an electric current.
F 15. If you grasp a coil with your left hand so that your fingers are in the direction of electron flow, your thumb will point in the direction of the south pole of the electromagnet.
T 16. Reversing the direction of current flow in a coil will cause the magnetic field of the coil to reverse.

17. A solenoid is an electromagnet with an iron core.
18. A circuit breaker is an electromagnetic device.
19. A relay is an electromagnetic switch.

Multiple choice: choose the best answer.

1. Man-made magnets are called
   a. lodestones
   b. artificial magnets
   c. domains
   d. molecules
2. The field of a bar magnet is concentrated near the
   a. north pole
   b. south pole
   c. ends
   d. center
3. When a piece of steel is stroked with a magnet to produce another magnet, the energy that produces the magnet comes form the
   a. the first magnet
   b. the motion
   c. the heat
   d. the air
4. The best material to produce magnetic shielding is
   a. steel
   b. soft iron
   c. paper
   d. copper
5. At the north pole a compass would point
   a. north
   b. south
   c. down
   d. in any direction
6. The magnetic field of the earth is
   a. very strong
   b. weak
   c. inconsistent
   d. not a true magnetic field
7. One device that does not use electromagnetism in its operation is a
   a. relay
   b. solenoid
   c. telegraph
   d. compass
8. A keeper is placed on a magnet to
   a. change the magnet
   b. form another magnet
   c. prevent loss of magnetism
   d. protect nearby objects from magnetic fields

8

PRINCIPLES OF ELECTROMAGNETIC INDUCTION

8.1 Induced Voltages

We know that electricity produces magnetism in an electromagnet. Now we find that magnetism can produce electricity. Figure 8-1 shows how a voltage is *induced* in a coil when a permanent magnet is moved into or out of the coil. This

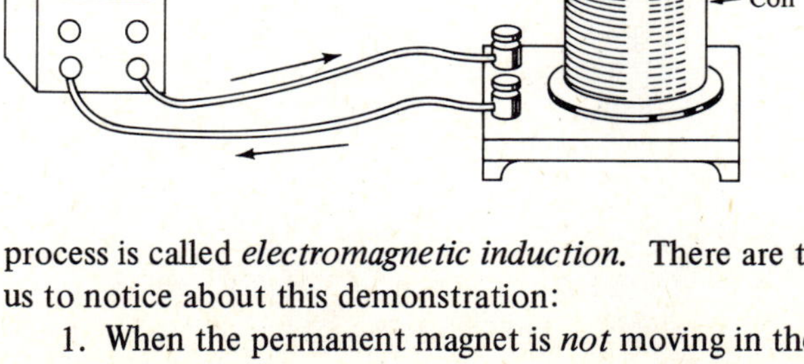

Fig. 8-1
An example of electromagnetic induction.

process is called *electromagnetic induction.* There are three important facts for us to notice about this demonstration:

1. When the permanent magnet is *not* moving in the coil, there is no voltage produced.

2. If the permanent magnet is moved out of the coil, the voltmeter shows a voltage in one polarity (current flowing in one direction).
3. If the permanent magnet is moved into the coil, the voltmeter shows a voltage in the other polarity (current flowing in the other direction).

A voltage is induced in a coil if the strength of the magnetic field changes. When we move the permanent magnet into the coil, the field becomes stronger. On the other hand, when we move the permanent magnet out of the coil, the field becomes weaker. Of course, if the permanent magnet is not moving in the coil, there is no change in the magnetic field strength, and no voltage is induced in the coil. This is one of the basic laws of electricity.

Another important form of electromagnetic induction is called the *self-induction* of a coil. We can observe that when a switch is opened in the circuit of a strong electromagnet, a large flaming spark, or arc, jumps between the switch contacts. This large electrical flash is produced by the voltage of self-induction. Figure 8-2 shows an experimental electromagnet which can be used to demonstrate self-induction of a coil. Let us consider this electromagnetic action.

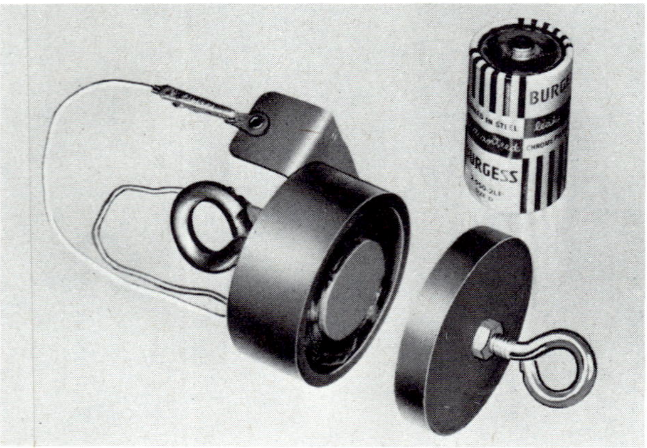

Fig. 8-2
An electromagnet used to demonstrate self-induction. Courtesy: Edmund Scientific Co.

In Fig. 8-3, the switch has been closed and the coil is surrounded by a magnetic field. At the instant that the switch is being opened, the current stops flowing in the coil, and the magnetic field starts to weaken. This change in the field strength induces a voltage in the coil, and the induced voltage causes a spark to flash between the switch contacts. We can observe that self-induction produces a larger spark if a larger battery is placed in the circuit. That is, a larger battery causes more current to flow, which produces a stronger magnetic field. In turn, the collapse of the stronger field produces a greater voltage of self-induction.

We also find that a greater voltage of self-induction is produced if a soft-iron core is placed in the coil. When the core is in place, the strength of the magnetic field is increased. Therefore, when this magnetic field collapses, it produces a greater voltage of self-induction. Increasing the number of turns in the coil also increases the induced voltage. Let us now observe a demonstration which shows that the self-induced voltage of a coil is much higher than the battery voltage.

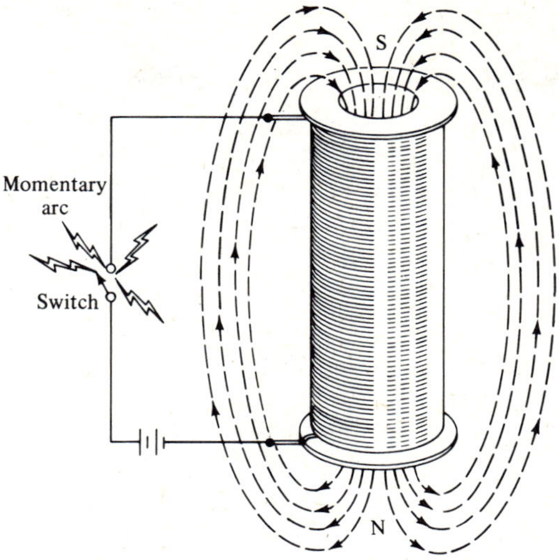

Fig. 8-3
Self-induction causes arcing at the switch contacts.

Instructor Demonstration or Student Experiment:

Connect a neon bulb across a large electromagnet, as shown in Fig. 8-4. A battery and switch are used to apply voltage across the bulb and the coil. Notice the following facts:

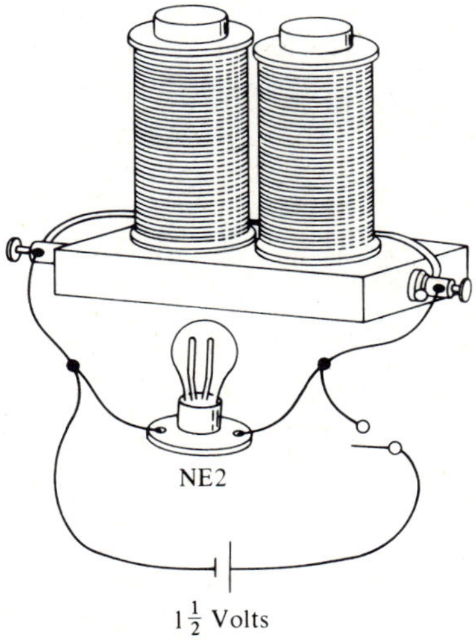

Fig. 8-4
Self-induction of an electromagnet.

1. When the switch is closed, the lamp neon bulb remains dark. That is, the battery does not have enough voltage to make the neon bulb light up.
2. When the switch is opened, the bulb flashes brightly for an instant. This shows that the voltage of self-induction is much greater than the battery voltage (probably about 50 volts.)
3. If a soft-iron armature is placed across the poles of the electromagnet, the bulb flashes still brighter, and may explode when the switch is opened.

This shows the manner in which an armature increases the strength of a magnetic field.

We should note here that the voltage induced as the switch is opened is much greater than the battery voltage because the magnetic field collapses in such a short period of time. The greater the number of flux lines that cut the coils of wire, the greater the induced voltage. This is the reason that either a larger current or a greater number of turns on the coil will cause a larger induced voltage.

Questions:

1. Can you explain why an iron armature increases the strength of an electromagnet?
2. If you touch your fingers between the switch contacts, you will feel a sharp shock when the circuit to the electromagnet is opened. Why does this happen?
3. Why does an electromagnet produce its voltage of self-induction for only a brief instant after the switch is opened?

8.2 Induction Coils and Ignition Coils

We know that a voltage is induced in a coil by means of a changing magnetic field. This magnetic field might be obtained from a permanent magnet or it might be obtained from an electromagnet. For example, Fig. 8-5 shows a demonstration arrangement to show the production of induced voltage in a coil by the field from an electromagnet. When the electromagnet *P* is moved into the coil *S,* the galva-

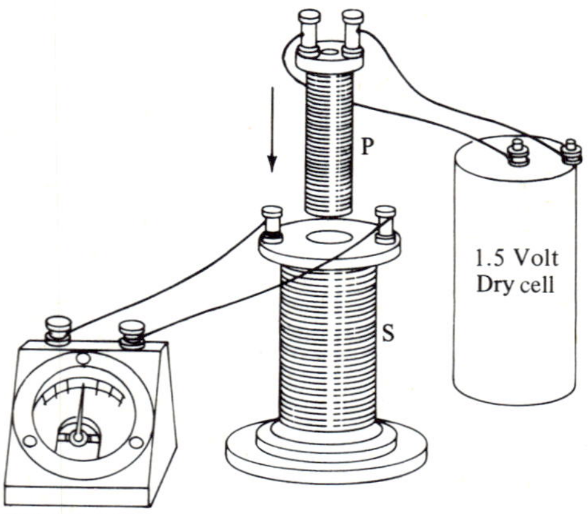

Fig. 8-5
A moving electromagnet induces current.

nometer deflects in one direction. Next, when the electromagnet is moved out of the coil, the galvanometer deflects in the other direction. However, if the electromagnet is not moving in the coil, the galvanometer does not deflect.

In the arrangement of Fig. 8-5, we call the electromagnet *P* the *primary* winding, and we call the coil *S* the *secondary* winding. This is a basic form of *induc-*

tion coil. Now, let us notice an important fact about this induction coil. If we place the primary *P* inside of the secondary *S,* and then disconnect the dry cell from the primary, the galvanometer gives a "kick" in one direction. Next, if we reconnect the dry cell to the primary, the galvanometer gives a "kick" in the other direction. This experiment shows that a changing current in the primary will induce a voltage in the secondary of an induction coil. This principle is used to operate the ignition coil of an automobile, as shown in Fig. 8-6. A pair of contacts

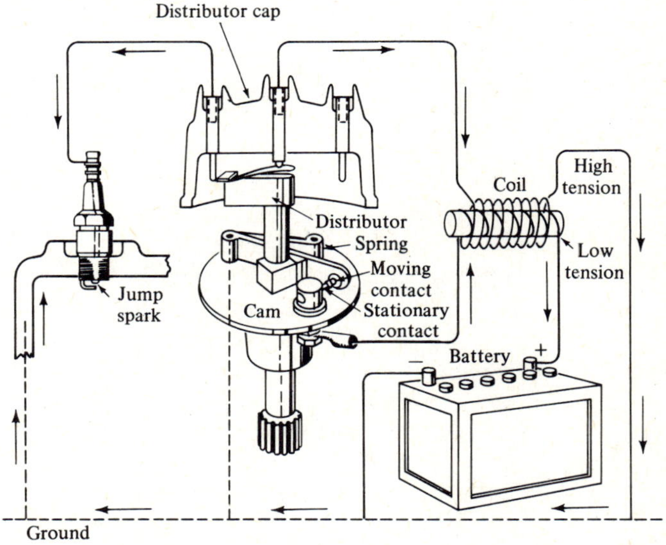

Fig. 8-6
Diagram of automobile ignition system using battery, ignition coil and a distributor.

in the *distributor* are opened and closed by a revolving cam. These contacts operate as a switch to open and close the primary circuit. The primary circuit obtains its current from a storage battery. When the primary circuit is opened, a very high voltage is induced in the secondary of the ignition coil. To obtain this very high voltage, the secondary coil is wound with many turns of wire. The induced secondary voltage is sufficient to jump across the space between the points of the spark plug. If you accidentally touch a spark plug, you will get a violent shock. The *ground circuit* is shown by the dotted lines in Fig. 8-6. The dotted lines represent the engine block and car frame that are used as part of the ignition circuit. Both the primary and the secondary currents travel through parts of this ground circuit. Such an arrangement simplifies the ignition wiring. We might also note that the primary circuit is often called the low-tension circuit, and the secondary circuit is called the high-tension circuit.

Spark coils, used in specialized military and industrial applications, are similar to ignition coils. Figure 8-7 shows the construction of a spark coil, or induction coil. Notice that the contacts in the primary circuit are opened and closed by electromagnetic action. That is, the primary circuit operates in the basic manner of an electric bell. Starting and stopping the current in the primary winding produces a magnetic field that rises and falls. This change in magnetic field strength

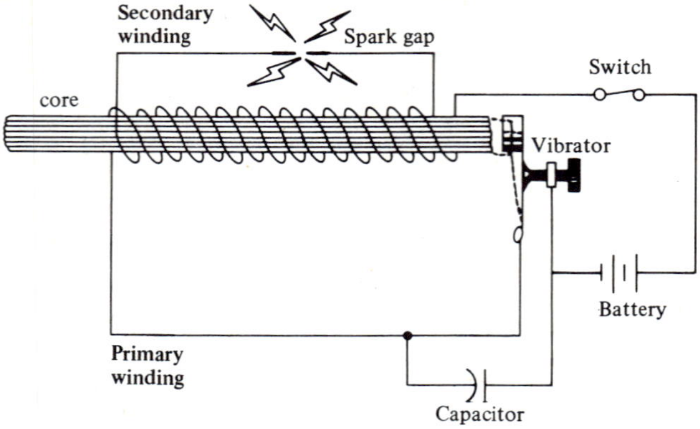

Fig. 8-7
Arrangement of a spark coil.

induces a voltage in the secondary winding. Since there are many turns in the secondary winding, the voltages induced in these numerous turns add up to a very high secondary voltage that can jump across a spark gap.

Also notice in Fig. 8-7 that a *capacitor* is connected across the vibrator contacts. This important device consists basically of a pair of metal plates with a sheet of insulating material between the plates. Waxed paper, mica, or plastic are commonly used as the insulating material. A capacitor stores electrical charges in the same manner as a Leyden jar. As an example of the function of the capacitor, consider the simple circuit shown in Fig. 8-8. When the switch is closed,

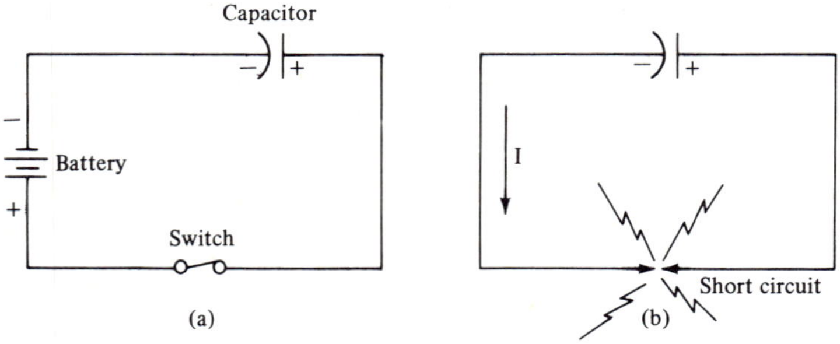

Fig. 8-8
Capacitor charge and discharge.
(a) Capacitor charges when switch is closed.
(b) Stored electricity in capacitor produces a momentary current.

current flows in the circuit for an instant. That is, the battery forces electrons into the negative plate and attracts electrons out the positive plate. The capacitor quickly charges up to the voltage of the battery, and the current flow then stops.

After a capacitor has been charged, it has a terminal voltage in the same way that a battery has a terminal voltage. The capacitor contains electrons under electrical pressure, and these electrons can be returned to a circuit. When the switch in Fig.8-8 is opened, the capacitor remains charged because there is no closed circuit between the terminals of the capacitor. When a wire is connected across the capacitor terminals, however, a current flows for an instant, and we may see a flashing spark at the point where the short circuit is completed.

Now let us again consider the circuit in Fig. 8-7. The capacitor is used to stop sparking at the contacts when the circuit is opened. It is necessary to prevent contact sparking, because the spark is very hot and will burn out the contacts quickly. Also, a spark means wasted electrical energy, and the secondary voltage will be much greater if sparking is eliminated at the primary contacts. The capacitor in Fig. 8-7 prevents contact sparking when the circuit is being opened, because the self-induced voltage from the primary winding is absorbed by charging the capacitor. In other words, the voltage of self-induction is used to charge the capacitor instead of being wasted as a spark discharge. As soon as the charging action is complete, the capacitor then starts to discharge back through the primary winding. This discharge current induces additional voltage in the secondary, and produces a greater output. A typical type of capacitor that is used in electronics is shown in Fig. 8-9.

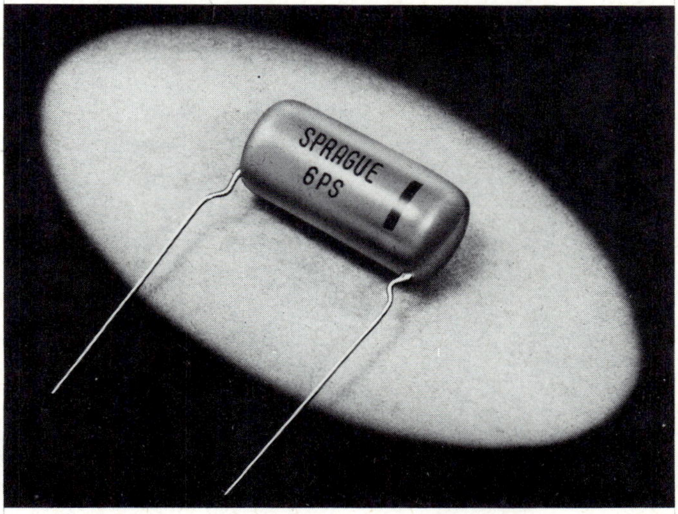

Fig. 8-9
A typical capacitor.
Courtesy: Sprague Products Co.

8.3 Alternators, or AC Generators

Most automobiles have alternators, also called *alternating-current (AC) generators,* that convert mechanical energy to electrical energy to charge the battery. An alternator operates on the principle of electromagnetic induction. In other words, coils are turned in a magnetic field, and a voltage is induced in the coils. Figure 8-10 shows the plan of an alternator. A one-turn coil is seen in four positions as it turns in a magnetic field. In (a), the conductors are moving in the same direction as the magnetic force lines, and there is no change in field strength at this moment. Therefore, no voltage is being induced in the coil. In (b), the conductors are moving across and through the magnetic force lines, and the induced voltage has risen to its greatest amount. We say that the conductors are *cutting* the magnetic force lines. At (C) in Fig. 8-10, the conductors are again moving in the same direction as the magnetic force lines. In turn, there are no magnetic force

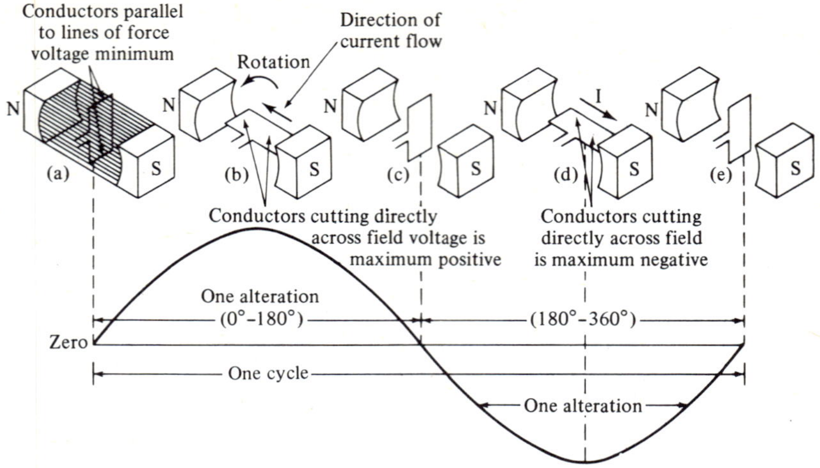

Fig. 8-10
Alternator action.

lines being cut at this moment, and therefore, no voltage induced in the coil. Next at (D), the induced voltage once more has risen to its greatest amount, but we notice that polarity of the induced voltage has now been reversed. Finally at (E), the conductors have returned to their starting point, and there is no voltage induced in the coil at this moment. Over a complete revolution of the coil (armature), the induced voltage rises and falls as shown by the graph in Fig. 8-10. This waveform of voltage is called an *alternating voltage,* because its polarity changes back and forth from positive to negative. If an alternating (AC) voltage is connected across a resistor or a light bulb, an alternating current flows through the load. This ac current has the same kind of graph as the ac voltage.

Fig. 8-11
A telephone magneto.
Courtesy: Edmund Scientific Co.

Another application of alternators is in field telephone systems to ring telephone bells. This type of alternator is called a *telephone magneto.* Figure 8-11 shows a typical hand-cranked telephone magneto. Some form of sliding contacts for connection of the rotating coil (armature winding) to a circuit outside the generator is necessary in all AC generators. The usual method is to use a pair of slip rings and brushes, as shown in Fig. 8-12. Since the slip rings are attached to

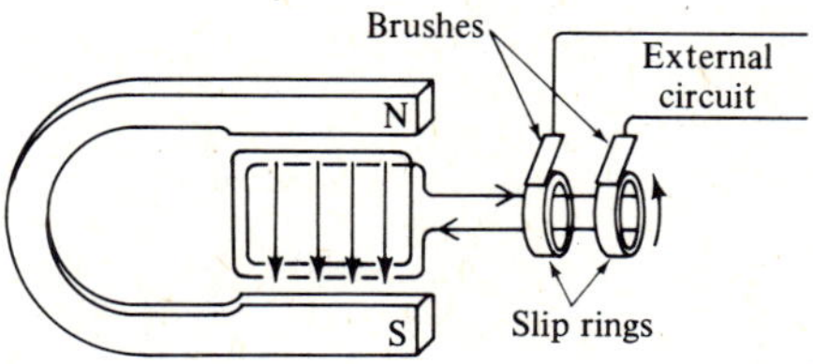

Fig. 8-12
Slip-rings take alternating current from the armature.

the armature shaft, they rotate with the armature. The brushes are held in a fixed position, and make a pressure contact against the slip rings. In this manner, the ac current generated by the alternator is conducted to the external circuit. The alternator may be rotated by a gasoline engine, as in an automobile, or by any other source of rotating mechanical power.

Instructor Demonstration or Student Experiment:

Connect a coil with several turns of wire to a galvanometer, as shown in Fig. 8-13. Observe that, when the coil is moved down over the pole of the per-

Fig. 8-13
Induction of electric currents.

manent magnet, the galvanometer needle deflects in a certain direction. Then, when the coil is moved up from the pole of the magnet, the galvanometer needle deflects in the other direction. If the coil is moved faster, the galvanometer needle deflects a longer distance on the scale.

Questions:

1. Is there any basic difference between the experiment shown in Figs. 8-13 and 8-1? Explain your answer.
2. Why does the galvanometer deflect only when the coil is moving in Fig. 8-13?

3. Why does the galvanometer show a greater deflection when the coil is moved faster?

8.4 DC Generators

In some automobiles, direct current (DC) generators are used to charge storage batteries. We find dc generators used to develop power for many industrial electrical and electronic devices. A DC generator produces direct current. That is, the current flows in one direction only. Although a DC generator is quite similar to an alternator, there is however, one important difference, as seen in Fig. 8-14.

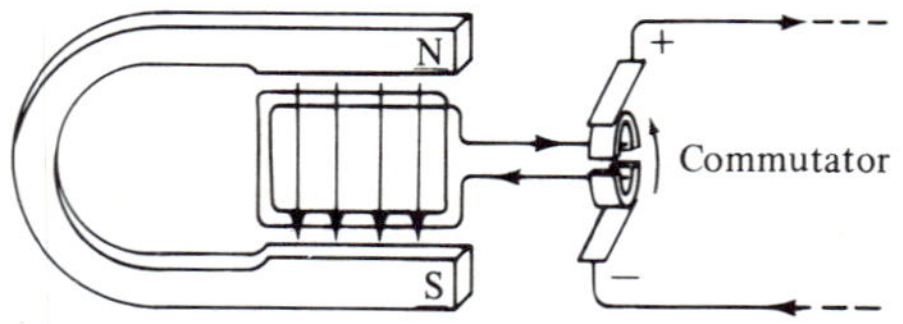

Fig. 8-14
The commutator changes alternating current to direct current.

That is, a *commutator* is used instead of a pair of slip rings. The commutator, consisting of a split ring and a pair of brushes, operates as an automatic switch to keep the polarity of the output voltage the same at all times. Each half of the split ring is called a *segment* and each end of the armature winding (rotating coil) is connected to a segment. When the armature completes one half of a revolution, the commutator segments reverse the coil connections to the brushes. This switching action changes the alternating current into direct current.

This process of changing alternating current into direct current with a commutator is shown in Fig. 8-15. Notice that the output from a commutator always has the same polarity, although in this example the voltage rises and falls between

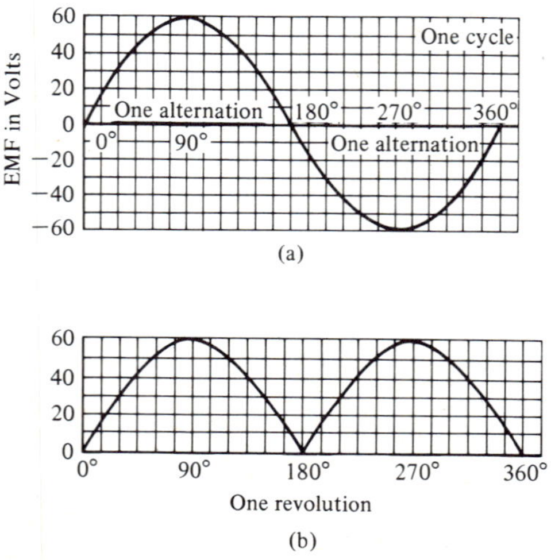

Fig. 8-15
Graph of DC generator output, compared with graph of alternator output.
(a) A curve of the type produced by an alternating EMF.
(b) Direct EMF curve as produced by use of a commutator.

zero and sixty volts. We see that, although a DC generator produces direct current, the output is not a steady direct current such as is produced by a battery. Therefore, we state that a DC generator produces a *pulsating DC* output, whereas a battery produces a *pure DC* output. The word "pulsating" means rising and falling. It is important to remember that a commutator produces *rectification* of an AC voltage to a DC voltage. In other words, a commutator is a basic example of a *mechanical rectifier.* As we continue in our study of electronics, we will learn about other types of rectifiers.

A picture of a small DC generator is shown in Fig. 8-16. It is operated at a

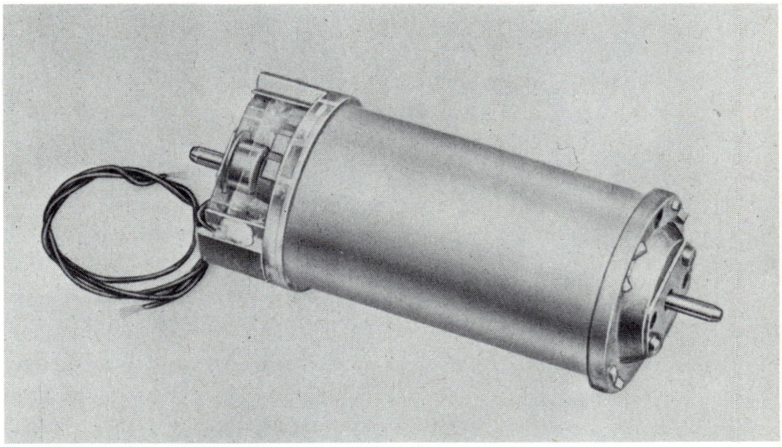

Fig. 8-16
A small DC generator.
Courtesy: Edmund Scientific Co.

speed of 2,000 to 3,000 revolutions per minute, and generates up to three volts, depending upon its speed. If a friction wheel is attached to the armature shaft, the generator can be operated by pressure of the friction wheel against a bicycle tire. The electricity that is generated can be used to operate a horn or a light on the bicycle. The generator can also be used as a speedometer. That is, if the output from the generator is connected to a DC voltmeter, the scale of the meter can be marked in miles per hour.

Questions: Chapter 8

1. Explain the requirements for a voltage to be induced into a coil of wire by a magnetic field. COIL MUST CUT FEILD
2. What causes the arc across a switch when an inductive circuit is opened, in which there is a current through the coil?
3. Why does the insertion of an iron core into a coil increase the induced voltage in the coil? CONCEN TRATES LINES OF FORCE
4. How can one coil be used to induce a voltage in another coil?
5. Explain the operation of the induction coil and the distributor in an automobile ignition system.

6. What is the purpose of the ground circuit in an automobile ignition system?
7. Why is a capacitor connected across the points in an ignition system?
8. What is the basic difference between an alternator and a DC generator?
9. Explain briefly the operation of an alternator.
10. Discuss the purpose of the commutator rings on the armature of a DC generator.
11. What is the name given to the process that is used to change an AC voltage into a pulsating DC voltage?
12. Why could a DC generator be used as a speedometer on a bicycle?

True–False:

1. To induce a voltage in a coil with a magnetic field, the field must cut across the coil.
2. Moving a loop of wire across a magnetic field will not induce a voltage in the wire.
3. The voltage induced in a coil of wire, when current through the wire is decreased, is caused by self-induction.
4. An iron core increases the strength of an electromagnet.
5. The polarity of the voltage induced in a coil is not affected by changing the pole of the magnet that is used to induce the voltage.
6. The spark across a spark plug would be normal if the coils of the ignition coil were interchanged.
7. Using the engine as a ground results in a greater induced voltage than if a separate conductor were used.
8. Placing a capacitor across the points in an ignition system causes a greater arc at the points.
9. A capacitor can store a charge.
10. Slip rings are used in an alternator to change the AC current to DC current.
11. An alternating voltage changes polarity back and forth each cycle.
12. The output of a DC generator is pure DC.

9

INDUCTANCE AND BASIC INDUCTIVE CIRCUITS

9.1 Inductors and Inductance

Coils are also called *inductors.* Inductors are widely used in electronic equipment to control the flow of alternating current. Any coil or inductor has a certain amount of *inductance.* A large inductance cuts down the flow of alternating current more than a small inductance. Let us observe a simple demonstration which shows the effect of inductance in an AC circuit. This is a light-dimmer arrangement, such as is used in theaters to control stage lighting.

Instructor Demonstration or Student Experiment:

Connect a 117-volt lamp in series with a coil which has a movable iron core, as in a solenoid (Fig. 9-1). Plug the circuit into a 117-volt outlet. Notice

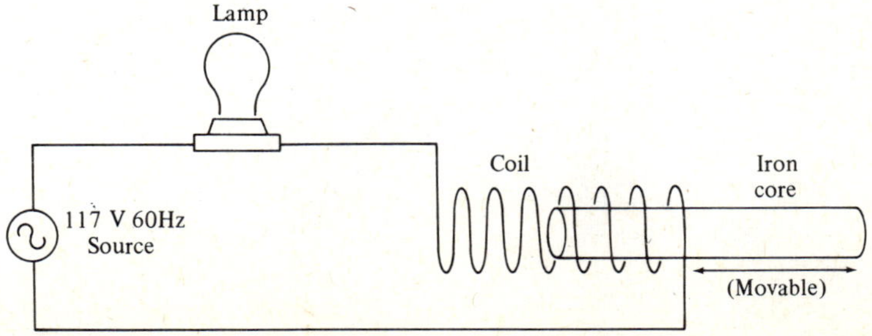

Fig. 9-1
Basic lighting-control circuit.

that the lamp glows brightest when the iron core is out of the coil. On the other hand, the lamp is dimmest when the iron core is slid completely into the coil. This experiment is based on the fact that an inductor with an iron core is capable of greater inductance than an inductor with an air core.

A coil with a movable iron core is called a *variable inductor,* or a *variable reactor.* Any coil which operates in an AC circuit has a certain amount of *reactance.* Reactance is the opposition to AC current flow, and is measured in ohms. Although reactance can be compared with resistance, we find that reactance is not the same thing as resistance. For example, reactance does not cut down DC current flow, as we will observe in the next demonstration.

Instructor Demonstration or Student Experiment:

Connect the circuit in Fig. 9-2 in the same arrangement as in Fig. 9-1, except that a 117-volt DC supply is used. Notice that the lamp continues to glow at

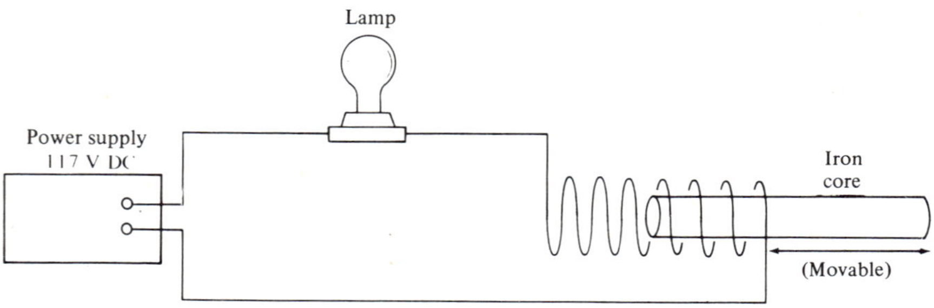

Fig. 9-2
The core has no effect on DC current flow.

the same brightness, no matter where the iron core is placed in the coil. This experiment shows that the reactance of the coil has no effect on the flow of DC current.

The principle of reactance is based on changing magnetic fields. Since an alternating current rises and falls, the magnetic-field that it produces in an inductor also rises and falls. The changing magnetic field in a coil produces self-induction. This voltage of self-induction is called *counter-voltage* or *counter electromotive force,* and it opposes a change in magnetic-field strength. Therefore, this counter emf also opposes AC current flow through the inductor.

Inductance is measured in *Henrys.* Figure 9-3 shows a ten-Henry inductor.

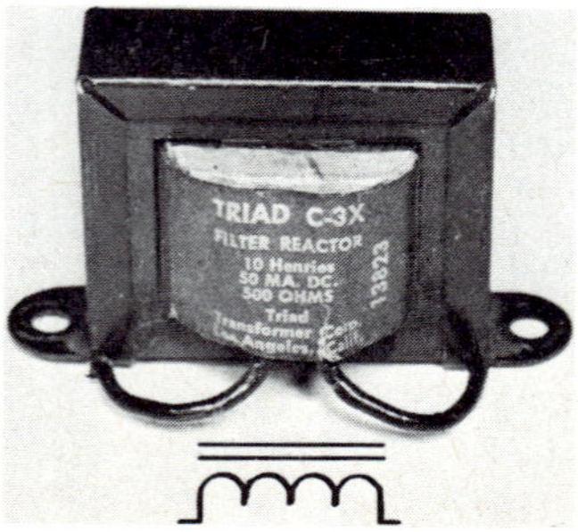

Fig. 9-3
A 10-Henry Inductor.
Courtesy: Triad Transformer Corp.

A coil has an inductance of one Henry if it allows a current change of one ampere per second when a DC voltage of one volt is connected across the coil. This principle is shown in the graph in Fig. 9-4. If we close the switch for five seconds, the

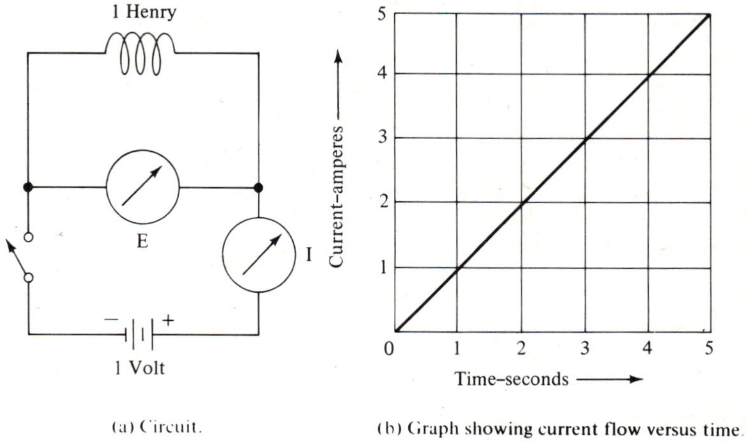

Fig. 9-4
A voltage of one Volt applied to one Henry of pure inductance.
(a) Circuit.
(b) Graph showing current flow versus time.

current through the one-Henry coil will build up at the rate of one ampere per second. At the end of five seconds the circuit current will be five amperes. It is important to remember that this experiment requires the use of a coil which is wound with large wire, so that the winding resistance of the coil is very small. If the coil has quite a bit of winding resistance, the current build-up will be slower than if the coil has very little resistance. A coil that has no winding resistance is called a *pure inductance.*

Inductive reactance is measured in ohms. We can find how many ohms of reactance an inductor has by measuring the current that flows when a certain amount of voltage is applied, and then using Ohm's law. For example, Fig. 9-5 shows a one-Henry inductor connected to a 117-volt 60-Hz outlet, with an AC voltmeter and an AC ammeter. The ammeter indicates that 0.5 ampere is flowing

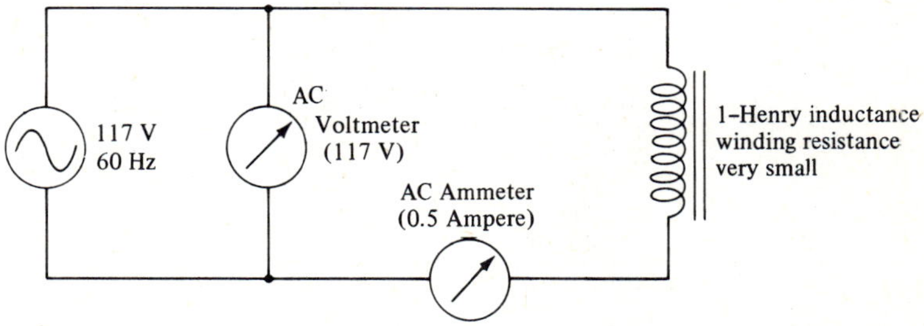

Fig. 9-5
Illustration of Ohm's law for an inductive circuit with little or no resistance.

through the one-Henry inductor. In turn, we write Ohm's law:

$$I = \frac{E}{R_{AC}}$$

where R_{AC} represents the ohms of reactance in the circuit, or

$$0.5 = \frac{117}{R_{AC}}$$

or

$$R_{AC} = 234 \text{ ohms}$$

The 117-volt supply in Fig. 9-5 is labeled as having a *frequency* of 60 Hertz. This means that the AC voltage goes through a complete *cycle* each $\frac{1}{60}$ second. To clearly understand what is meant by frequency, let us observe the sine graph for an alternator, shown in Fig. 9-6. In this example, the armature makes one

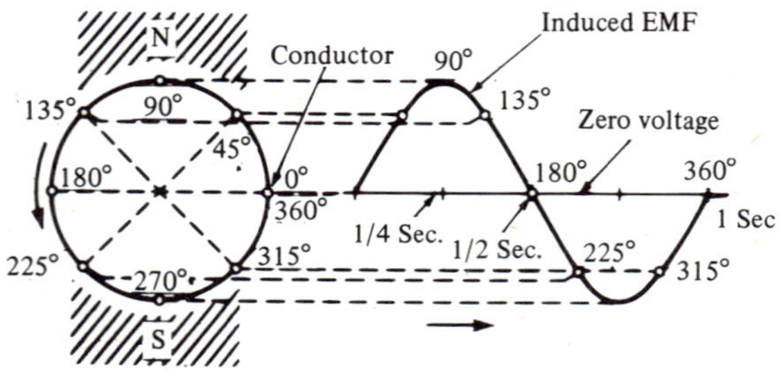

Fig. 9-6
Curve showing relationship between time and induced voltage in an alternating current circuit.

revolution per second. Therefore, the output voltage goes through one cycle in one second. The frequency of the output voltage in this example is one Hertz. *One Hertz means one cycle per second.* We usually abbreviate the word Hertz, and write Hz.

Ordinary house current has a frequency of 60 Hz. This is a comparatively low frequency. As an example of a comparatively high frequency, we may consider a broadcast radio receiver. The radio receiver operates at a frequency near

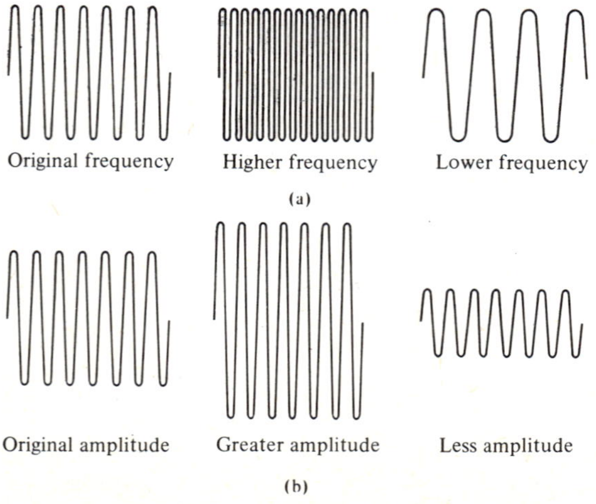

Fig. 9-7
Distinction between amplitude and frequency
(a) Sine waveforms with same amplitude.
(b) Sine waveforms with same frequency.

1,000,000 Hz. This means that the radio signal voltage goes through one cycle in 1/1,000,000 second. As an example of a very high frequency, we may consider a television receiver. The television receiver operates at a frequency near 100,000,000 Hz. This means that the television signal voltage goes through one cycle in 1/100,000,000 second.

When we draw a sine curve, its voltage is represented by the height of the curve, as seen in Fig. 9-7. Frequency is represented by the number of cycles in a given interval, such as ten cycles in one inch to represent ten Hertz. The important facts for us to observe at this time are that AC voltages have various frequencies, various amplitudes, and that the frequency does not change if we change the voltage, nor does the voltage change if we change the frequency.

9.2 Transformer Construction and Operation

A transformer can be compared with an ignition coil, or with a spark coil. However, a transformer is operated by AC voltage, whereas an ignition coil is operated by DC voltage. Transformers are widely used in all branches of electricity and electronics. For example, Fig. 9-8 shows a power transformer mounted on an

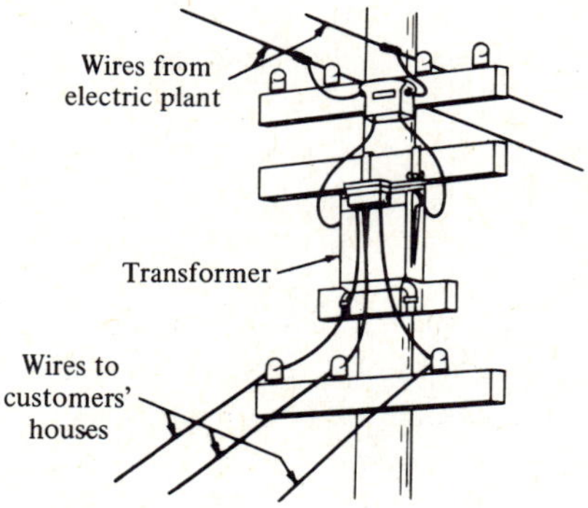

Fig. 9-8
Transformer on electric-light pole.

electric-light pole. The principle of a basic transformer is shown in Fig. 9-9.

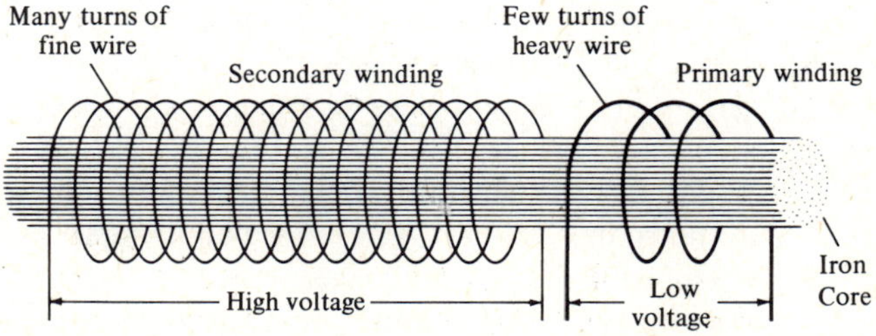

Fig. 9-9
Transformer consisting of an iron core and primary and secondary windings.

This transformer has more turns on the secondary winding than on the primary winding. An iron core is used to increase the magnetic field and make the transformer more efficient. It is called an *open-core* transformer. Notice in Fig. 9-10

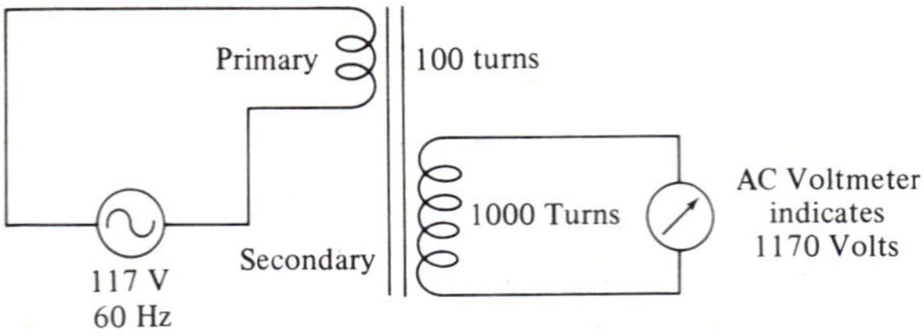

Fig. 9-10
Step-up transformer action.

that, if we connect the primary to a source of AC voltage, the secondary output voltage will be greater than the primary voltage. Therefore, this arrangement is called a *step-up transformer.*

We observe in Fig. 9-10 that the primary winding has 100 turns, whereas the secondary has 1,000 turns. Thus, the secondary has ten times as many turns as the primary. The transformer operates as a step-up transformer and the secondary output voltage is 1,170 volts when the primary input voltage is 117 volts. If the transformer is connected the other way, as shown in Fig. 9-11, it operates as a *step-down transformer.* There are now ten times as many turns on the primary

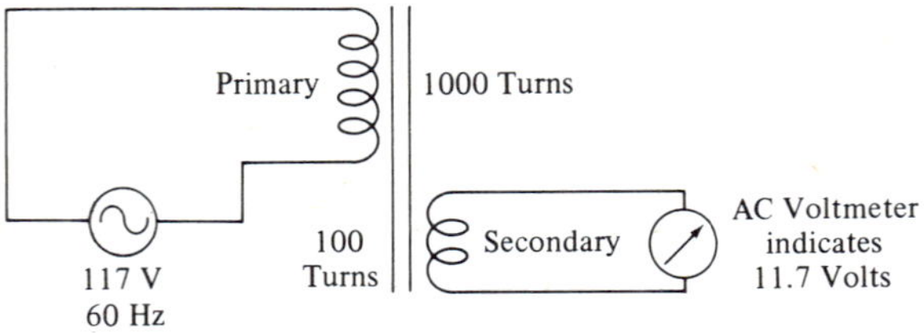

Fig. 9-11
Step-down transformer action.

as there are on the secondary. Therefore, the secondary output voltage is 11.7 volts when the primary input voltage is 117 volts. Step-down transformers are used in most houses to operate door bells. Figure 9-12 shows the appearance of a doorbell-ringing transformer.

There are three basic types of transformer cores, as shown in Fig. 9-13. The *core-type* transformer has its coils wound around two sides of a rectangular iron core; however the *shell-type* transformer has three legs, and the coils are wound on the center leg; and finally, the *H-type* transformer has five legs, with the coils wound on the center leg. We find that the core type is least efficient, and the H type is most efficient. That is, there is less loss between the primary and secondary of the H type than the other types, when we draw more current from the secondary.

Almost all transformer cores are made of thin sheets of iron, or *laminations,* as shown in Fig. 9-14. This construction improves the efficiency of a transformer by reducing the loss due to *eddy currents.* Since iron is a conductor of electricity, the magnetic field from the primary induces a current in the core (eddy current) as well as in the secondary. This induced current in the core represents wasted electrical energy. To reduce the eddy currents induced in the core, the laminations are made from thin sheet iron, and each lamination is insulated from the next. In turn, the induced eddy currents are split up into as many as 100 small currents.

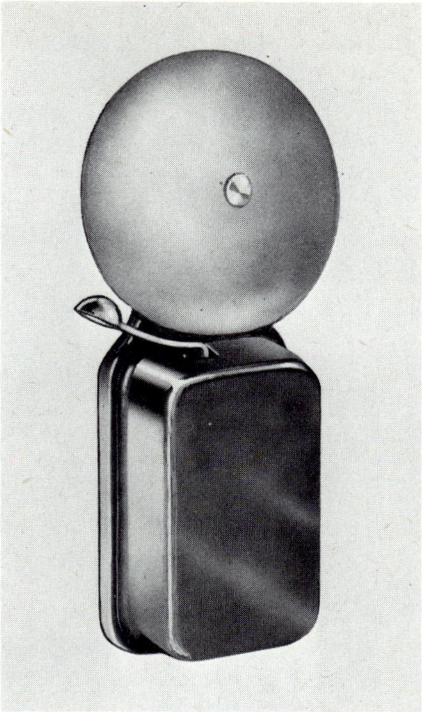

Fig. 9–12
A door-bell ringing transformer.
Courtesy: Sargent-Welch Scientific Co.

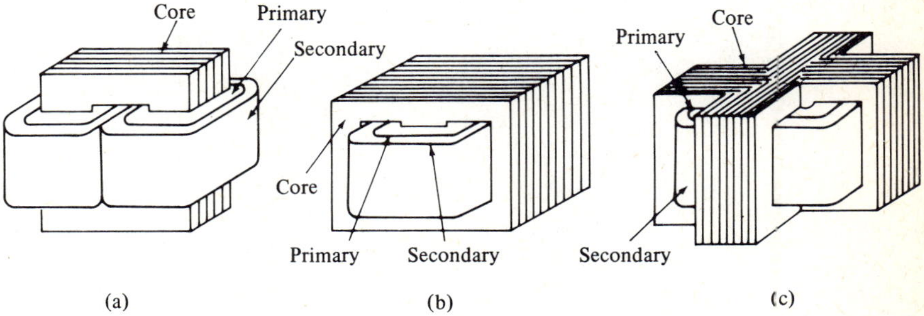

Fig. 9–13
Three types of transformer:
(a) Core type.
(b) Shell type
(c) H-type.

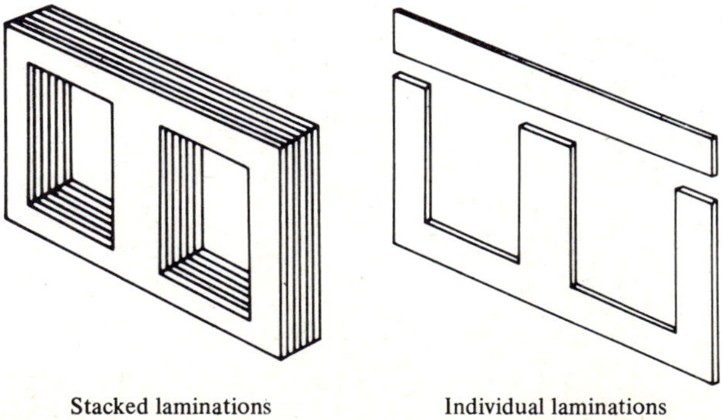

Fig. 9–14
A core formed from stacked laminations.

In effect, the eddy current must travel through 100 times the length of iron represented by a solid core. Therefore, the eddy current loss is cut down to 1/100 of the loss that would occur in a solid core.

Step-up transformers used to operate neon signs may have a secondary output as high as 15,000 volts. High-voltage transformers are dangerous, and a person can be killed by accidental contact with the secondary terminals. Figure 9-15 shows how a transformer for a neon sign is constructed. This type of transformer

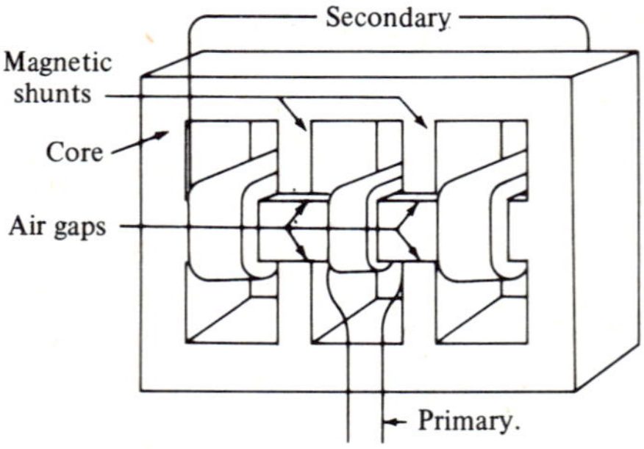

Fig. 9-15
Sign transformer showing the magnetic shunts between winding sections.

is designed to produce only a certain amount of secondary current. This is necessary to prevent burn-out of the neon tube. Notice that a shell-type core is used, to which is added a pair of *magnetic shunts.* These are extensions on the core, with air gaps between their ends and the center leg of the core. In operation, quite a bit of the magnetic field from the primary is taken away by the shunts, and does not pass through to the secondary winding. Therefore, the secondary cannot supply more than 0.03 ampere in a typical example.

Figure 9-16 shows the circuit for a simple neon sign. The glass tube may be filled with neon, helium, argon, or mercury vapor. Sometimes a mixture of gases is used. Mercury vapor produces green light; neon produces an orange-red glow; helium produces a light pink glow; neon and mercury vapor produce a blue glow; neon and argon give a deep lavender glow; helium and argon give a pink glow, while neon, helium, and mercury vapor produce various shades of blue or green. The gas pressure in the tube is low; if an air leak occurs in the glass tube, the gas stops glowing.

Transformers are sometimes designed to supply more than one voltage output. For example, a step-down transformer for operating a toy train has several output voltages available. In turn, the speed of the toy train can be changed by changing the supply voltage. Most transformers of this type have a tap switch, as seen in Fig. 9-17. As shown in Fig. 9-18, the secondary winding is tapped. A typical transformer provides a choice of eight, twelve, sixteen, twenty, and twenty-four volts. We know that a toy-train transformer is much smaller than a power transformer on an electric-light pole. This difference in size is a result of the large current demand on the power transformer, and the small current demand on the toy-train transformer. If a transformer supplies a large amount of current, it must be wound with heavy wire and it must have a large core.

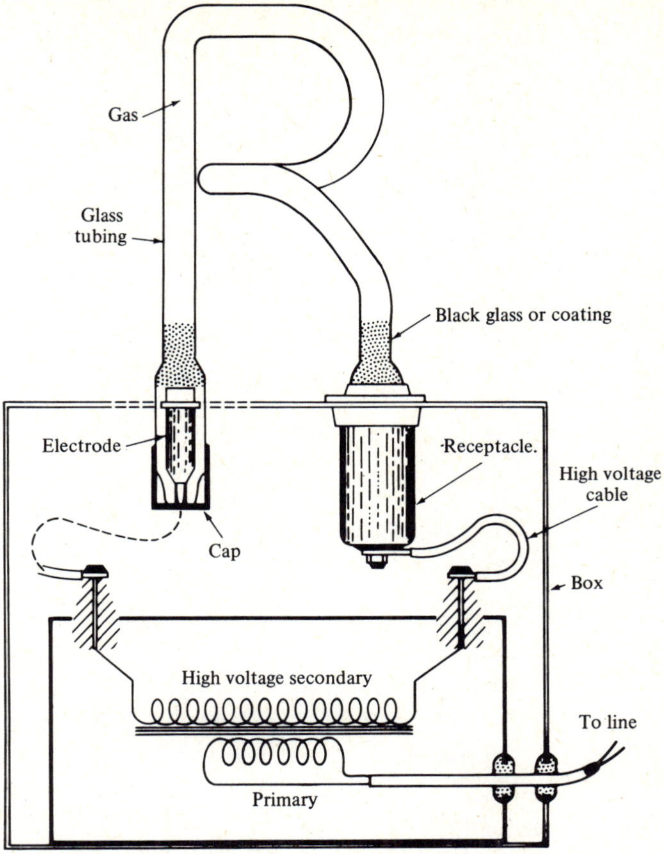

Fig. 9–16
Basic arrangement of a neon sign.

Fig. 9–17
Appearance of an electric train transformer.

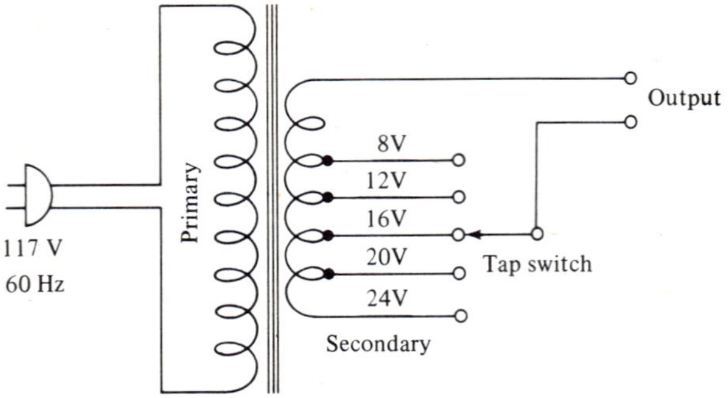

Fig. 9-18
Step-down transformer with tapped secondary.

Questions: Chapter 9

1. Does the counter emf of a coil increase when an iron core is placed in the coil? Explain your answer.
2. Is there any counter emf in a coil when DC current is flowing? Why?
3. Explain why reactance is like resistance in some ways, and in what way it is different.
4. If an AC voltage has a frequency of sixty Hz, how many positive half-cycles are there in one second?
5. If an AC voltage has a frequency of sixty Hz, how many negative half-cycles are there in one second?
6. If an AC voltage has a frequency of sixty Hz, how many half-cycles are there altogether in one second?
7. What precautions should one take when working around a high-voltage transformer?
8. What is the basic difference between a transformer that supplies a small current and a transformer that supplies a large current?

True-False:

1. Coils are called conductors.
2. Any coil has a certain amount of inductance.
3. A coil with a movable iron core is called a variable reactor.
4. Reactance is the same as resistance.
5. The value of DC current in a coil does not change as a core is added to the coil.
6. The voltage of self-induction is called counter electromotive force.
7. The counter electromotive force remains the same in a coil when a core is added.
8. Inductive reactance is measured in ohms.
9. A transformer has at least two windings.
10. In a step-up transformer, the secondary has more windings than the primary.
11. A step-up transformer can be operated as a step-down transformer.

12. Hertz means cycles per second.
13. A sixty-Hz AC waveform has sixty peaks in one second.
14. Losses in the winding of a transformer are called eddy currents.
15. High-voltage transformers can be dangerous.
16. The current demand of a transformer determines the wire size.

Multiple choice: choose the best answer.

1. Coils are also called
   a. resistors
   b. inductors
   c. capacitors
   d. transformers
2. A coil with a movable iron core is called a variable
   a. transformer
   b. controller
   c. reactor
   d. resistor
3. The opposition that an inductor offers to an AC voltage is called
   a. conductance
   b. eddy currents
   c. resistance
   d. reactance
4. Every coil has
   a. a core
   b. a secondary winding
   c. inductance
   d. a primary winding
5. Inductive reactance is measured in
   a. ohms
   b. ergs
   c. volts
   d. amperes
6. A step-up transformer has
   a. more turns on the secondary than the primary
   b. more turns on the primary than the secondary
   c. equal turns on the primary and secondary
   d. two secondaries
7. We usually call cycles per second
   a. cps
   b. reactance
   c. Hertz
   d. ohms

8. Eddy currents are losses in the _____________ of a transformer.
   a. core
   b. windings
   c. load
   d. input

10 CAPACITANCE AND CAPACITIVE CIRCUITS

10.1 Capacitors and Capacitance

Capacitors are also called *condensers,* although capacitor is preferred by most technicians. Capacitors such as the samples shown in Fig. 10-1, are widely used in

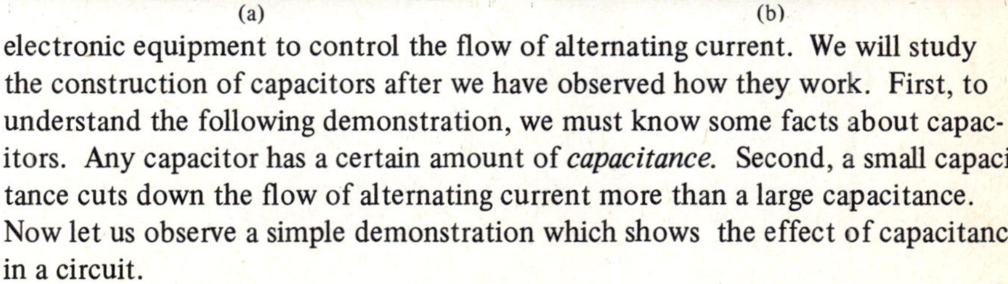

(a) (b)

Fig. 10-1 Typical fixed capacitors. (a) A paper-dielectric capacitor. (b) An electrolytic capacitor. Courtesy: Sprague Products Co.

electronic equipment to control the flow of alternating current. We will study the construction of capacitors after we have observed how they work. First, to understand the following demonstration, we must know some facts about capacitors. Any capacitor has a certain amount of *capacitance.* Second, a small capacitance cuts down the flow of alternating current more than a large capacitance. Now let us observe a simple demonstration which shows the effect of capacitance in a circuit.

Instructor Demonstration or Student Experiment:

Connect an electric light bulb in series with a capacitor and a source of voltage, as shown in Fig. 10-2. Notice that, when the circuit is connected to a

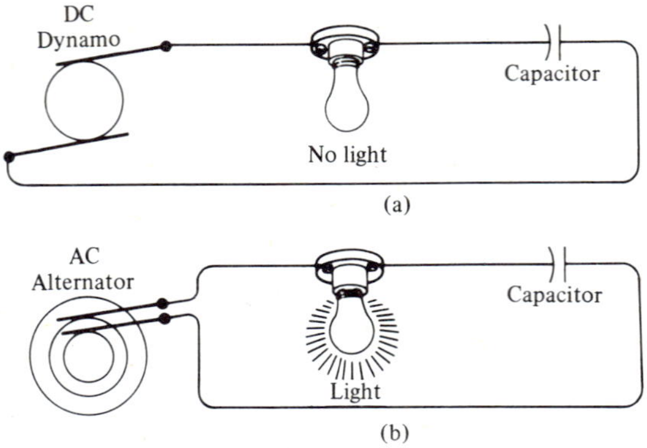

Fig. 10-2
Lamp connected in series with a capacitor.

source of DC voltage, the bulb does not light up; but when the circuit is connected to a source of AC voltage, the bulb glows. This experiment shows that a capacitor blocks the flow of DC current, whereas the capacitor allows the flow of AC current. If a larger capacitor is used in the circuit, the lamp glows more brightly. This shows that a larger capacitance allows more AC current to flow.

The opposition that a capacitor offers to the current flow in an AC circuit is called *reactance*. A capacitor has *capacitive reactance,* in much the same way as an inductor has inductive reactance. Capacitive reactance is measured in ohms, just as inductive reactance is measured in ohms.

Although capacitive reactance can be compared with resistance, we find that capacitive reactance is not the same thing as resistance. For example, if we replace the capacitor in Fig. 10-2(a) with a resistor, the bulb then glows (Fig. 10-2b). Perhaps we can understand capacitor action by the drawings shown in Fig. 10-3. In Fig. 10-3(a) a capacitor is connected in series with a battery switch, and ammeter. Before the switch is closed, there is no voltage across the capacitor, and the capacitor is uncharged. In Fig. 10-3(b), when the switch is closed, there is a momentary surge of electrons out of the upper plate and into the lower plate. This quickly charges the capacitor to the battery voltage, and no current flows after the momentary surge.

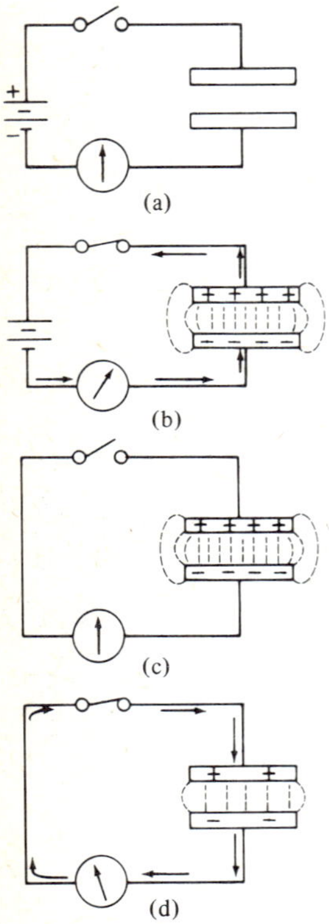

Fig. 10-3
Capacitor action. (a) Charging circuit, switch open. (b) Charging circuit, switch closed. (c) Discharge circuit, switch open. (d) Discharge circuit, switch closed.

Now let us follow the capacitor discharge action. In Fig. 10-3(c), the switch has been opened, and the battery has been replaced by a connecting wire. The capacitor is still charged to the same voltage as the battery that was replaced. However, there is no current flow, because the switch is still open. Finally, the switch is closed in Fig. 10-3(d), and there is a momentary surge of electrons out of the lower plate and into the upper plate. The capacitor is now discharged, and no current flows after the momentary surge. Let us observe a demonstration that shows how AC current flows in a capacitive circuit.

Instructor Demonstration or Student Experiment:

A capacitor is connected in series with a galvanometer and a double-throw single-pole switch to a pair of batteries, as shown in Fig. 10-4. When the

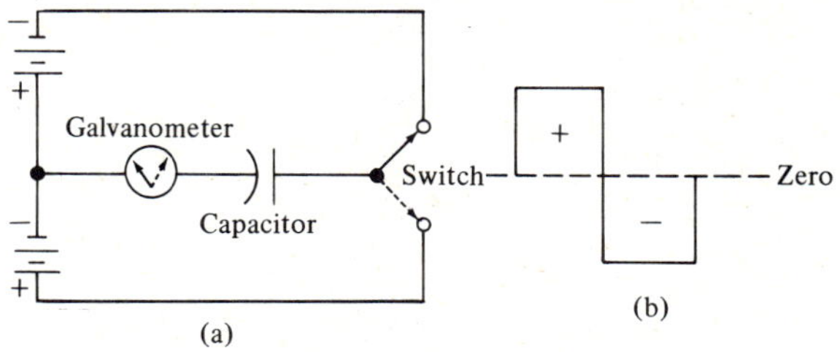

Fig. 10-4
How AC current flows in a capacitive circuit.
(a) Demonstration circuit.
(b) AC current graph.

switch is in the up position, the right-hand plate of the capacitor is charged negatively. On the other hand, when the switch is in the down position, the right-hand plate of the capacitor is charged positively. When the switch is switched up and down rapidly, the galvanometer also deflects rapidly from left to right, as long as the switch action is continued. This experiment proves that AC current flows in a capacitive circuit.

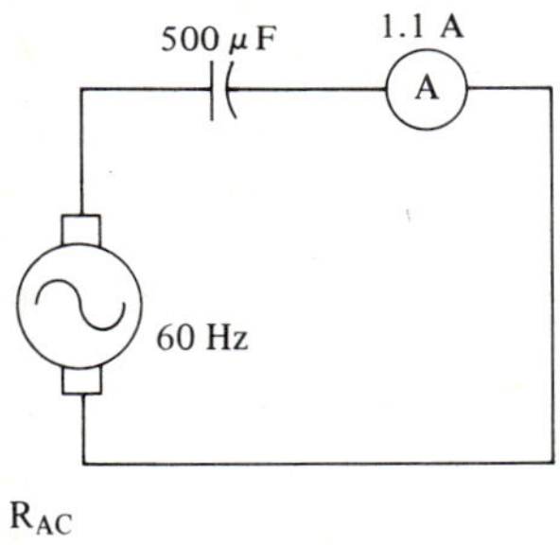

Fig. 10-5
A capacitor connected in a series AC circuit.

The capacitance of a capacitor is measured in *farads.* A capacitor has a capacitance of one farad if it takes one coulomb of charge when connected to a 1-volt source. One coulomb is equal to 6.28×10^{18} electrons. Since a farad is a very large amount of capacitance, we usually work with the *microfarad,* or with the *picofarad.* A microfarad is equal to 1/1,000,000 of a farad, and a picofarad is equal to 1/1,000,000 of a microfarad. Note that microfarad is written μF and picofarad is written pF. The capacitance of capacitors ranges up to 500 microfarads, or more. Capacitors are made with values of less than 1 pF up to over 500 μF. The opposition that a capacitor offers to an AC current is called reactance. The symbol for capacitive reactance is X_c. We can determine the reactance of a capacitor in a circuit by Ohm's law, when we know the conditions of the circuit. For example, in Fig. 10-5, a 500-microfarad capacitor is connected to a 6-volt 60-Hertz source. The AC ammeter shows that there are 1.1 amperes flowing in the circuit. We can find the amount of capacitive reactance from Ohm's law:

$$I = \frac{E}{X_c}$$

where

I is the current in amperes
E is the applied voltage in volts
X_c is the number of ohms of reactance

or

$$1.1 = \frac{6}{X_c}$$

$X_c = 5.4$ ohms, approximately

10.2 Commercial Fixed Capacitors

Many types of fixed capacitors are used in electrical and electronic equipment. *Paper capacitors,* such as shown in Fig. 10-6(a), are constructed from sheets of tin foil separated by waxed paper. Connecting wires, called *pigtail leads,* extend

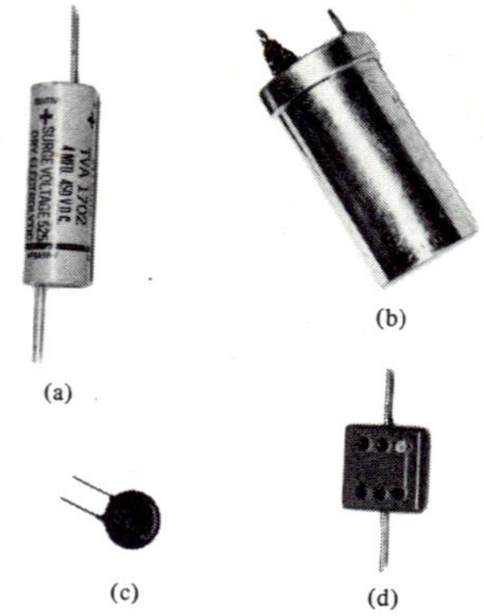

Fig. 10-6
Fixed capacitors. (a) Paper capacitor. (b) Oil-filled capacitor. (c) Ceramic capacitor. (d) Mica capacitor.

from the tin foil sheets through the ends of the plastic case, which is called the *encapsulation*. Although the *oil-filled* type of capacitor shown in Fig. 10-6(b) is constructed in basically the same way as the paper capacitor, it is sealed in an oil-filled metal case. Oil provides increased insulation between the tin foil sheets permitting an oil-filled capacitor to be operated at comparatively high voltages.

Ceramic capacitors, such as shown in Fig. 10-6(c), are in wide use in radio and television receivers, because they have the advantage of small size, compared with paper capacitors. The capacitor plates are made of silver, and insulation (dielectric) between the plates is made from a special ceramic material. A ceramic capacitor is encapsulated in a round case, which looks like a disc with a pair of pigtail leads. The *mica capacitors,* shown in Fig. 10-6(d), are in fairly wide use although they are being replaced by ceramic capacitors in various types of electronic equipment. A mica capacitor is constructed by piling up alternate sheets of tin foil and mica, as shown in Fig. 10-7. The stack of foil and mica sheets is then encapsulated in a plastic case, with a pair of pigtail leads.

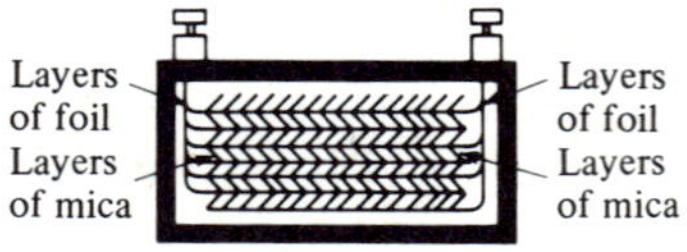

Fig. 10-7
Basic construction of a mica capacitor.

Electrolytic capacitors are manufactured in various shapes and with metal, plastic, or cardboard cases. However, the basic principle is the same in all types.

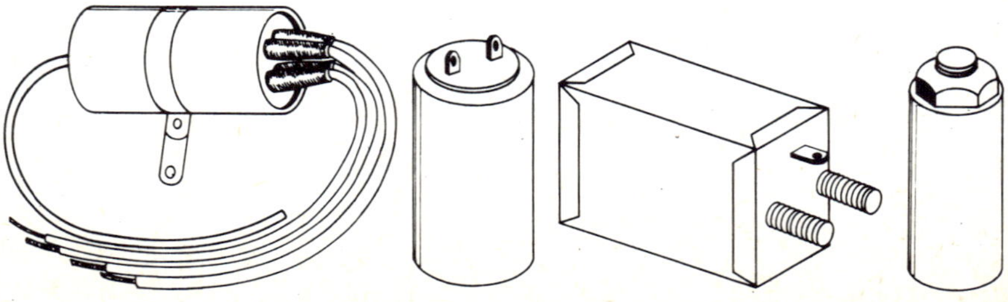

Fig. 10-8
Various types of electrolytic capacitors.

Figure 10-8 shows several kinds of electrolytic capacitors. Two or more separate electrolytic capacitors are often contained in the same case. As shown in Fig. 10-9, one plate is a strip of aluminum foil. A very thin film of insulating substance is

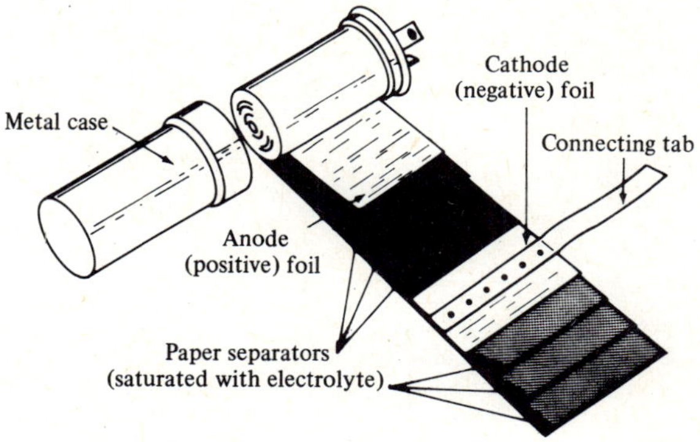

Fig. 10-9
Construction of an electrolytic capacitor.

formed on this plate by electrochemical action. This chemical substance is the dielectric of the capacitor. The other plate is actually a fluid called the *electrolyte.* Porous paper separators are used between the turns of foil to soak up the excess electrolyte. Connection is made to the electrolyte by means of a separate strip of aluminum foil.

The electrolytic capacitor has an advantage in that it can provide a large amount of capacitance in a very small space. There are two basic electrical types, which we should keep in mind. The *nonpolarized* type of electrolytic capacitor has dielectric films on both of the aluminum strips, and can be used in AC circuits. The *polarized* type of electrolytic capacitor has a dielectric film on only one of its aluminum strips, and can be used only in DC circuits. A polarized type of capacitor has positive and negative markings so that it can be connected into a DC circuit correctly.

10.3 Safety Precautions When Working With Capacitors

There are some important safety precautions that we must always remember when working with large capacitors. These precautions are particularly important when working with high-voltage capacitors.

1. A capacitor may hold its charge for many hours. If a high-voltage capacitor is not discharged by short-circuiting its terminals, it may act as a "booby trap" for the unwary technician. Accidental contact with the terminals can result in severe shock, or even death.

2. If an electrolytic capacitor is operated above its correct voltage, it will "leak" excessively. This leakage current will cause the capacitor to heat up, and often to explode. This precaution applies to both the polarized and the nonpolarized types.

3. If a polarized electrolytic capacitor is connected into a DC circuit in the incorrect polarity, it will heat up, and may explode.

4. If a polarized electrolytic capacitor is connected into an AC circuit, it will heat up, and may explode.

5. Technicians who work with radio transmitters or other equipment that contain high-voltage capacitors are required to use a shorting stick, such as that shown in Fig. 10-10, to discharge capacitors before touching any part of the electrical circuit.

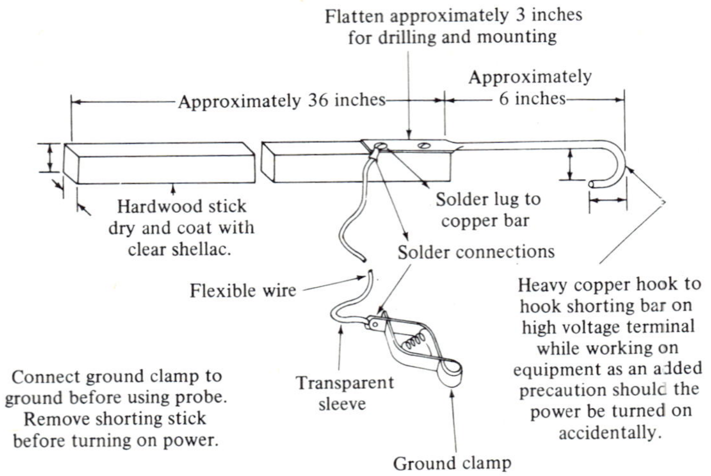

Fig. 10-10
Construction of a shorting stick.

10.4 Commercial Variable Capacitors

Variable capacitors and adjustable capacitors are in wide use in electronic equipment. The appearance of a typical variable capacitor is shown in Fig. 10-11.

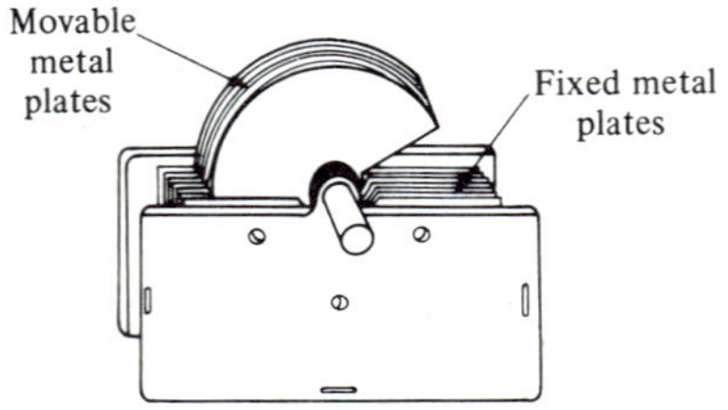

Fig. 10-11
The variable capacitor is used in radio work.

This variable-capacitor design is widely used in radio receivers to select stations, and consists of a number of semicircular metal plates which are mounted on a shaft. When the shaft is turned, these *rotor* plates mesh with the *stator* plates, which are attached to the frame of the capacitor. A variable capacitor has its greatest capacitance value when the plates are fully meshed.

Another type of variable capacitor has no movable parts, but consists of a semiconductor diode which is operated as a voltage variable capacitor. The capacitance of the diode varies with a change of the DC voltage. Such a variable capacitor, often called a *varicap,* is shown in Fig. 10-12. Details of semiconductor action are explained in Chapter 14.

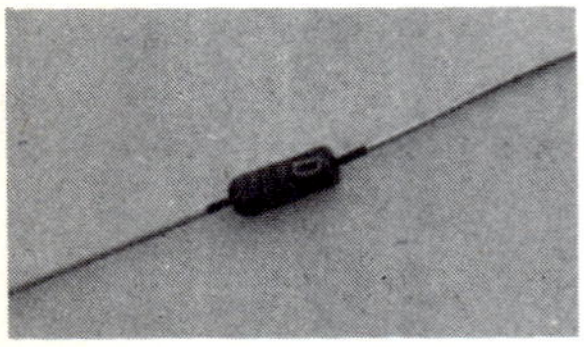

Fig. 10-12
Voltage variable capacitor diode.

Trimmer capacitors, such as shown in Fig. 10-13, are another type of variable capacitors. Most trimmer capacitors have one rotor plate and one stator plate,

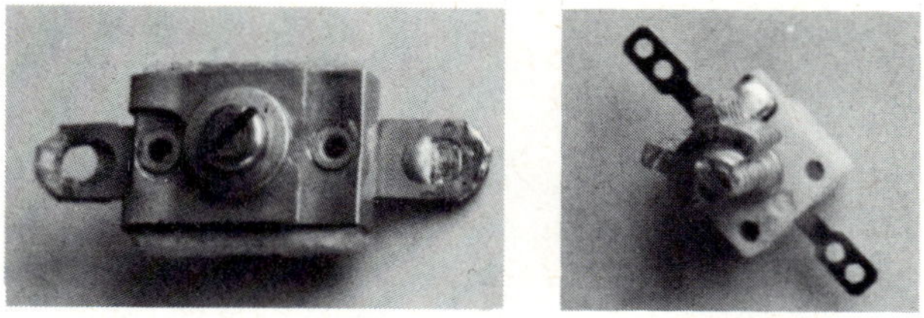

Fig. 10-13
Trimmer capacitors.

and their capacitance is comparatively small. Mica is generally used as a dielectric. The rotor plate is not mounted on a shaft, but on the shank of an adjusting screw. Therefore, adjustment of a trimmer capacitor is made with the aid of a screwdriver. We find that trimmer capacitors are used for maintenance adjustments, rather than for tuning in radio receivers.

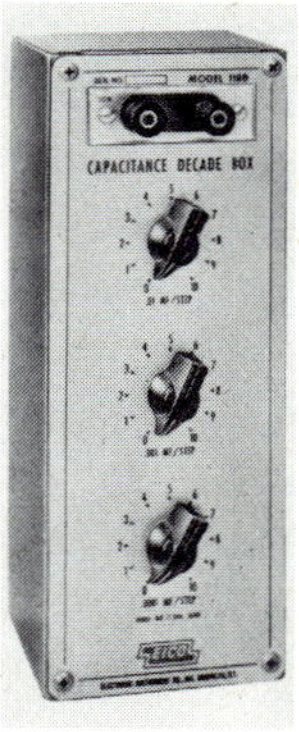

Fig. 10-14
A capacitance decade box.
Courtesy: EICO

Adjustable capacitors are used in many areas of electronics. Figure 10-14 shows a *capacitance decade box.* This arrangement consists of a number of paper

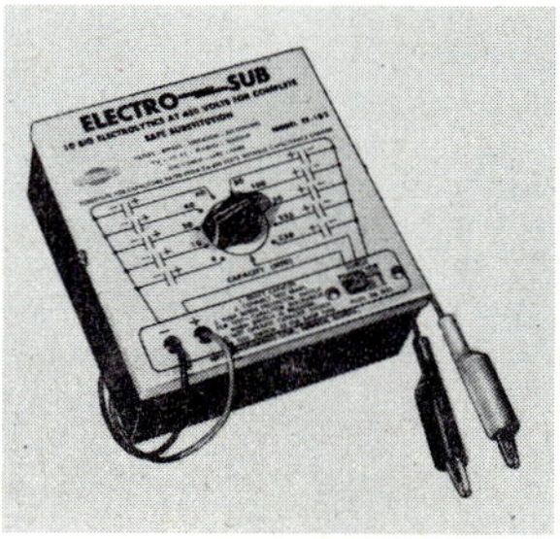

Fig. 10-15
An electrolytic capacitor substitution box.
Courtesy: Sencore Instruments Corp.

capacitors, with a rotary switching circuit which provides capacitance values from 0.00001 microfarad to 0.111 microfarad, in steps of 0.0001 microfarad. When large amounts of adjustable capacitance are required, an *electrolytic capacitor substitution box* is used. Figure 10-15 shows the appearance of this type of device. It provides capacitances from 4 to 350 microfarads in ten steps. The electrolytic capacitors in this example are polarized, and can be used only in DC circuits. Let us now examine an interesting application of capacitors.

10.5 Light Flashers

Neon light flashers are used in various types of electronic equipment, as well as in toys and novelties. A flasher circuit consists of one or more capacitors, resistors, and neon bulbs, with a source of dc electricity. Figure 10-16 shows three types of flasher circuits. In Fig. 10-16(a), the neon bulb flashes on and off, because the capacitor charges slowly through the resistor. When the capacitor charges to about 65 volts, the neon bulb suddenly starts to conduct and glows with an orange-colored light. However, conduction of the neon bulb soon discharges the capacitor, and the bulb becomes dark. Then, the capacitor starts to charge again, and the flasher action is repeated.

In Fig. 10-16(b), we see the circuit for a seesaw type of flasher. It consists of two neon bulbs, a capacitor, two resistors, and a battery. One of the bulbs flashes first, and then becomes dark. As soon as the first bulb becomes dark, the second bulb flashes. Then, the flasher action is repeated back and forth. Notice that, when the upper bulb is conducting, the capacitor starts to charge up through the lower resistor. However, as the capacitor becomes almost fully charged, the current flow becomes very small, and the upper neon bulb stops glowing. But now, the capacitor can be charged in the opposite polarity through the lower neon bulb. Therefore, the lower neon bulb conducts. This seesaw action is repeated back and forth. Notice that, if one of the neon bulbs is defective, the other bulb will not flash.

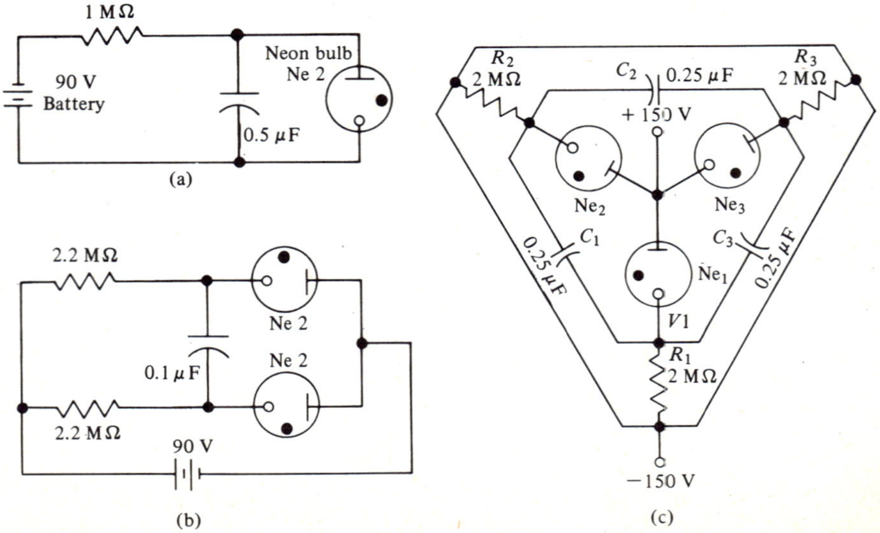

Fig. 10-16
Neon-bulb flasher circuits. (a) Single-bulb flasher. (b) See-saw flasher. (c) Ring flasher.

Finally, in Fig. 10–16(c), the circuit for a three-bulb ring flasher is shown. This is a continuation of the seesaw flasher, with a third neon bulb added into the circuit. In this arrangement, the neon bulbs flash one after another around the ring. If the bulbs are flashing around in the clockwise direction, you can make them flash in the counterclockwise direction by touching any one of the bulbs at a suitable instant. It should be kept in mind that a three-bulb flasher must have neon bulbs that are fairly well matched. In other words, we must select bulbs that "fire" at approximately the same voltage. Otherwise, we will have a seesaw flasher, and one of the bulbs will refuse to flash in the circuit.

Questions: Chapter 10

1. Explain the step-by-step action of the demonstration circuit in Fig. 10–3.
2. Why does the AC current graph in Fig. 10–3 have a square shape?
3. Does any AC current actually flow "through" the capacitor in Fig. 10–3? Explain your answer.
4. How does a capacitor react in a circuit to which a DC voltage is applied?
5. How does a capacitor react in a circuit to which an AC voltage is applied?
6. If a small capacitor is replaced in an AC circuit with a larger capacitor how is the circuit action affected?
7. What is the name given to the opposition that a capacitor offers to an AC current, and what are the units of this opposition?
8. How are paper capacitors constructed?
9. What is the advantage of an electrolytic capacitor?
10. Why should large capacitors be shorted before beginning to work on electronic equipment?
11. Describe a variable capacitor such as is used to select stations in an AM radio.
12. What is a capacitance decade box?
13. Define a voltage variable capacitor.
14. Explain the operation of a neon flasher.

True–False:

1. Capacitors are sometimes called condensers.
2. The opposition that a capacitor offers to an AC current flow is called resistance.
3. Capacitance is measured in farads.
4. Reactance is measured in ohms.
5. In an AC circuit, a capacitor will block the flow of current.
6. A capacitor stores energy in the form of a charge.
7. The symbol for capacitive reactance is X_C.
8. Some capacitors are polarized.

9. Capacitors that are used in laboratory experiments should be put away with a charge.
10. A polarized capacitor can be safely connected in an AC circuit.
11. High-voltage capacitors should be discharged before touching any part of an electrical circuit.
12. A capacitor can be operated at any voltage level.
13. A variable capacitor may have rotor plates and stator plates.
14. A variable capacitor has the greatest capacitance when the plates are wide open.
15. Adjustments of trimmer capacitors are usually made with a screwdriver.

Multiple choice: choose the best answer

1. Capacitors are sometimes called
   a. resistors
   b. tubes
   c. condensers
   d. reactors
2. The opposition that a capacitor offers to an AC current flow is called
   a. capacitance
   b. conductance
   c. resistance
   d. reactance
3. The unit of capacitance is the
   a. farad
   b. ampere
   c. volt
   d. ohm
4. The unit of reactance is the
   a. farad
   b. ampere
   c. volt
   d. ohm
5. Some of the safety precautions that should be observed when working with capacitors in high-voltage circuits are
   a. observe the maximum voltage rating
   b. observe polarity in electrolytic capacitors
   c. short high-voltage capacitors
   d. all of the above
6. Small variable capacitors that are usually adjusted with a screwdriver are called ____________ capacitors.
   a. stator
   b. trimmer
   c. diode
   d. padder

11

RESONANT-CIRCUIT PRINCIPLES

11.1 Series-Resonant Circuits

Resonant circuits are often used in electronic equipment. The word *resonant* comes from "resound." It means a build-up of a tone or a signal by means of suitable devices. For example, if you speak in the open air, your voice does not sound unusually loud. However, if you speak into a barrel, your voice sounds loud and boomy. The build-up of a sound by the barrel also tends to die out more slowly than when you speak in the open air. We say that the barrel is a sound *resonator.* The same general action is found in series circuits that have both inductance and capacitance. Let us observe a demonstration that shows the action of *series resonance.*

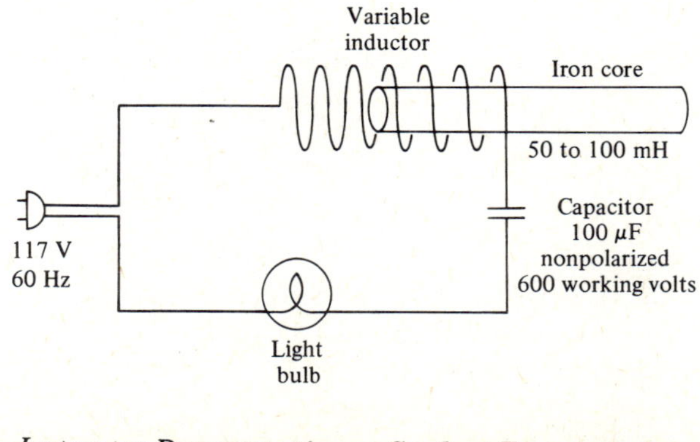

Fig. 11-1 Demonstration of series resonance.

Instructor Demonstration or Student Experiment:

Connect the circuit shown in Fig. 11-1. The variable inductor and the fixed capacitor are connected in series with a light bulb, and the circuit is plugged into a 117-volt 60-Hertz outlet. As the iron core is moved into the coil, the bulb glows more brightly. However, as a certain point is passed, the bulb then becomes dimmer as the core is moved further into the coil. This demonstration proves that the greatest amount of current flows when a certain amount of inductance is connected in series with a certain amount of capacitance.

We say that this circuit is series-resonant at a frequency of 60 Hertz. If we operate the circuit at some other frequency, a different amount of inductance will be required to make the circuit resonant. The reason for this fact is that the circuit is resonant only when the amount of inductive reactance is equal to the amount of capacitive reactance. The inductive reactance of a coil becomes greater as the frequency becomes higher. On the other hand, the capacitive reactance of a capacitor becomes less as the frequency becomes higher.

The graph in Fig. 11-2 shows how resonance occurs in a series *LC* circuit, as in a radio tuner. An inductance of 300 microhenrys (μH) is connected in series

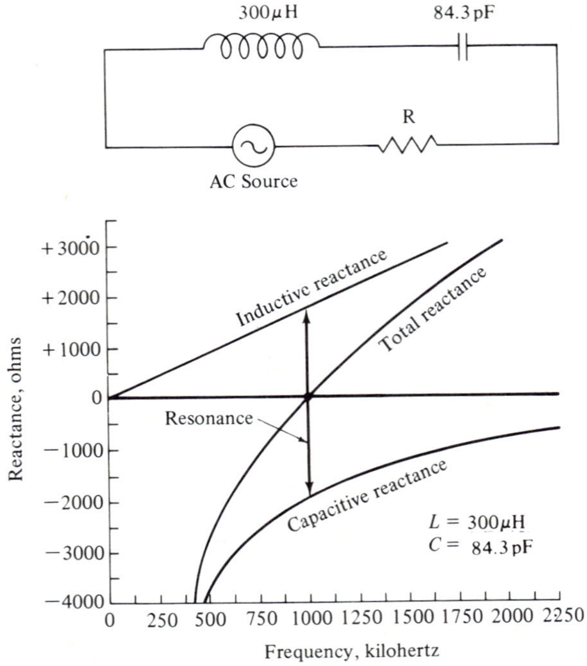

Fig. 11-2
Resonance occurs at a frequency which makes the inductive reactance equal to the capacitive reactance.

with a capacitance of 84.3 picofarads, and the circuit is connected to an AC source. As shown in the graph, the inductive reactance increases with frequency, and the capacitive reactance decreases with frequency. At a frequency of 1,000 kilohertz (kHz), the inductive reactance is +1884 ohms, and the capacitive reactance is −1884 ohms. The voltage across the capacitor is 180° out of phase with the voltage across the inductor, so that the voltage across both components adds up to zero. All the voltage is dropped across the resistor. Therefore, the inductive reactance cancels out the capacitive reactance at this frequency when the circuit becomes series-resonant. At 1,000 kilohertz, the total reactance of the circuit is zero, and the total opposition to current flow is equal to the value of R.

Instructor Demonstration or Student Experiment:

Connect the circuit shown in Fig. 11-3. This is a basic radio receiver that is

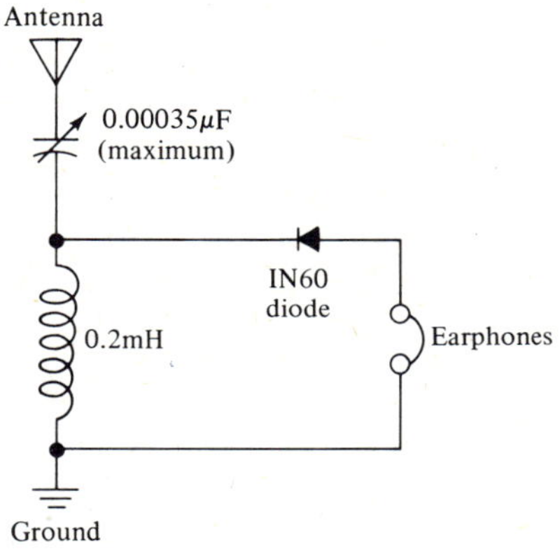

Fig. 11-3
Experiment in series resonance using a basic radio receiver.

wired up by connecting a 0.2-millihenry (mH) coil in series with a 0.00035-microfarad variable capacitor. The series *LC* circuit is operated from an antenna and a ground connection. Audible reception is obtained by means of a detector diode and a pair of earphones connected in series across the coil terminals. The antenna may consist of up to seventy-five feet of wire suspended between two buildings or from a building to a pole or tree. The ground connection is easily made to a water pipe. Notice that, when the variable capacitor is operated, the circuit becomes resonant at various radio station frequencies, and that when one station is "tuned in," the other stations are "tuned out."

Notice that the inductive reactance was varied in Fig. 11-1 in order to obtain resonance, whereas the capacitive reactance was varied in Fig. 11-3 to obtain resonance. These experiments show that there are two basic methods of making inductive and capacitive reactances equal. The method that is chosen is merely one of convenience or of economy. The formula below demonstrates that either L or C can be varied to vary the resonant frequency (f_0)

$$f_0 = \frac{1}{2\pi\sqrt{LC}}$$

where $\pi = 3.14$
L = inductance in henrys
C = capacitance in farads

We will now examine Fig. 11-3 to see why the series-resonant circuit is able to tune in a given radio station signal and to tune out the other radio station signals. In Fig. 11-4, we see a graph for the series-resonant tuning circuit. This graph shows the amount of current through the earphones when the circuit is tuned to a radio signal which has a frequency of 1,000 kilohertz. The graph also shows the amount of current through the earphones for other radio signal frequencies. For example, a radio signal frequency of 500 kilohertz will cause very

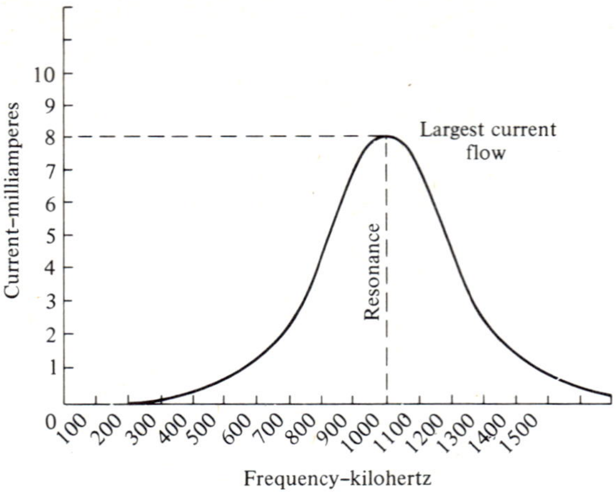

Fig. 11-4
Action of series-resonant tuning circuit.

little current through the earphones. As another example, a radio signal frequency of 1,500 kilohertz will cause very little current through the earphones. This is called the *tuning action* or selectivity of the *LC* circuit in the radio receiver.

We observe that if the variable capacitor were tuned to make the series circuit resonant at a frequency of 500 kilohertz, a radio signal with a frequency of 500 kilohertz would produce the largest current flow through the earphones. However, a radio signal with a frequency of 1,000 kilohertz would produce very little current through the earphones. As another example, if the variable capacitor was tuned to make the series circuit resonant at a frequency of 1,500 kilohertz, a radio signal with a frequency of 1,500 kilohertz would then produce the largest current through the earphones. A radio signal, however, with a frequency of 1,000 kilohertz would produce very little current through the earphones.

The question is as follows: Why does the largest amount of current flow through the earphones at the resonant frequency of the series *LC* circuit? This happens because the total reactance of the *LC* circuit is near zero at the resonant frequency. This fact is seen in Fig. 11-2. Since the total reactance of the *LC* circuit is zero at its resonant frequency, we know that a given signal voltage from the antenna will produce the greatest signal current through the coil at this resonant frequency. When the signal current through the coil is greatest, we also know from Ohm's law that the signal voltage across the coil will have its greatest value, and in turn, the largest possible current flows through the diode and earphones.

11.2 Parallel-Resonant Circuits

A parallel-resonant circuit has an action which is opposite to that of a series-resonant circuit. That is, when a parallel circuit has an inductive reactance which is equal to its capacitive reactance, the least amount of current flows in the circuit. Let us observe a demonstration which shows the action of *parallel resonance.*

Instructor Demonstration or Student Experiment:

Connect the circuit shown in Fig. 11-5; a variable inductor and a fixed capacitor are connected in parallel with each other. This parallel arrangement is

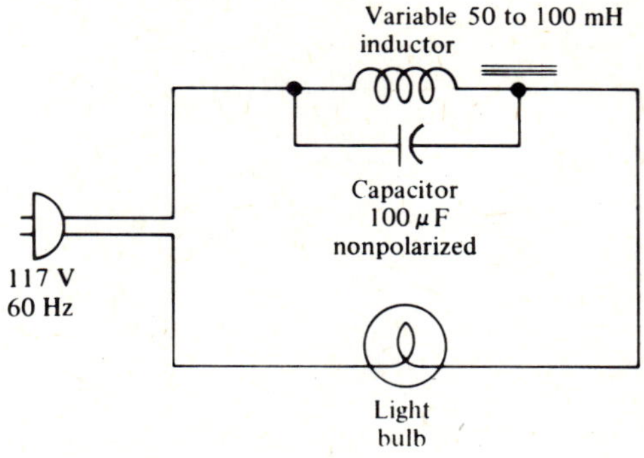

Fig. 11-5
Demonstration of parallel resonance.

connected into the circuit with the light bulb and is plugged into a 117-volt 60-Hertz outlet. As the iron core is moved into the coil, the bulb becomes dimmer. However, as a certain point is passed, the bulb then glows brighter. This demonstration shows that when a certain amount of inductance is connected in parallel with a certain amount of capacitance, the least amount of current then flows in the circuit.

Notice that this circuit is parallel-resonant at a frequency of 60 Hertz. If we operate the circuit at some other frequency, a different amount of inductance or capacitance will be required to make the circuit resonant. This is the same principle that we noticed previously in the action of series-resonant circuits. We find that the same product of inductance and capacitance is used to obtain either series resonance or parallel resonance at a given frequency. Let us observe an experiment which shows the action of parallel resonance in a simple radio receiver.

Instructor Demonstration or Student Experiment:

Connect the circuit as shown in Fig. 11-6. This is a simple radio receiver

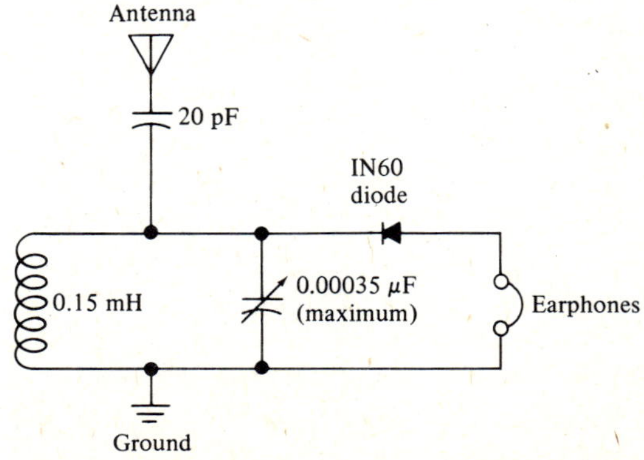

Fig. 11-6
Experiment in parallel resonance for a simple radio receiver.

wired up by connecting a 0.15-millihenry coil in parallel with a 0.00035-microfarad variable capacitor. The parallel *LC* circuit is operated from an antenna and a 20-picofarad coupling capacitor with a ground connection at the other coil terminal. Audible reception is obtained by means of a detector diode and a pair of earphones connected in series across the coil terminals. Notice that when the variable capacitor is operated, the circuit becomes resonant at various radio station frequencies, and that when one station is "tuned in," the other stations are "tuned out."

The principle of operation of the tuner in Fig. 11-6 is as follows. When the parallel-resonant circuit is tuned to a radio-station frequency, the circuit has a high amount of total reactance (see Fig. 11-7). Therefore, the signal currents

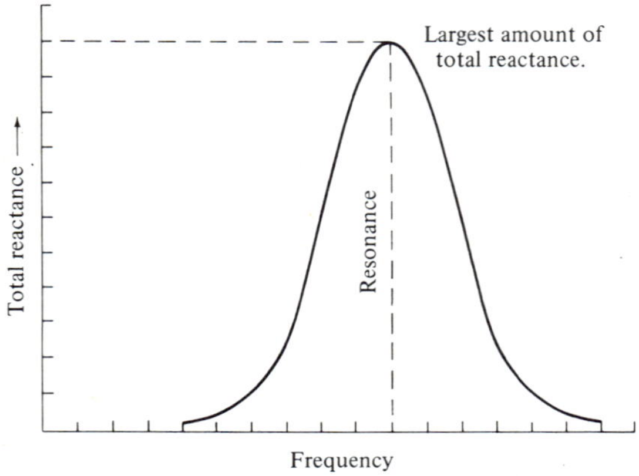

Fig. 11-7
Action of parallel-resonant tuning circuit.

from the antenna follow the path of least resistance through the diode and earphones and to ground. On the other hand, the radio-station frequencies to which the parallel-resonant circuit is not tuned are conducted directly to ground. That is, a nonresonant signal frequency from the antenna has a path of least resistance to ground through the coil and the variable capacitor. This is a path of low resistance for the nonresonant signal frequency because the parallel circuit has a small amount of total reactance at off-resonance frequencies.

11.3 Combination Series and Parallel-Resonant Circuits

Series and parallel-resonant circuits are combined in various types of electronic equipment. One of the simplest examples is seen in the *wavetrap* tuner for a radio receiver shown in Fig. 11-8. In this arrangement, a series *LC* circuit is used to tune the receiver to a given station. However, the parallel *LC* circuit is used to eliminate an interfering station. For example, we may wish to receive a radio station signal at a frequency of 1,000 kilohertz and to reject an interfering radio station signal at a frequency of 750 kilohertz. Therefore, the series *LC* circuit would be tuned to resonate at 1,000 kilohertz, and the parallel *LC* circuit would be tuned to resonate at 750 kilohertz. In turn, a high reactance is placed in the path of the 750-kilohertz signal from the antenna to the tuner. However, much less

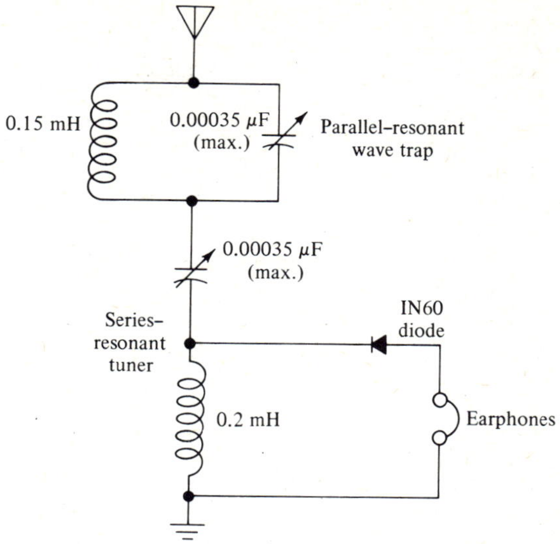

Fig. 11-8
Series-resonant tuner with a parallel-resonant wavetrap.

reactance is placed in the path of the 1,000-kilohertz signal which passes from the antenna to the tuner.

If both the series *LC* circuit and the parallel *LC* circuit in Fig. 11-8 are tuned to 1,000 kilohertz, little or none of the radio station signal will produce current flow through the earphones. This happens because the parallel *LC* circuit stops the 1,000-kilohertz signal from passing from the antenna to the tuner, even though the series *LC* circuit is tuned to accept a 1000-kilohertz signal.

11.4 The Radio-Frequency Transformer

Another widely used radio tuning circuit that illustrates the principle of resonance is shown in Fig. 11-9. We should consider the action of this tuner very carefully.

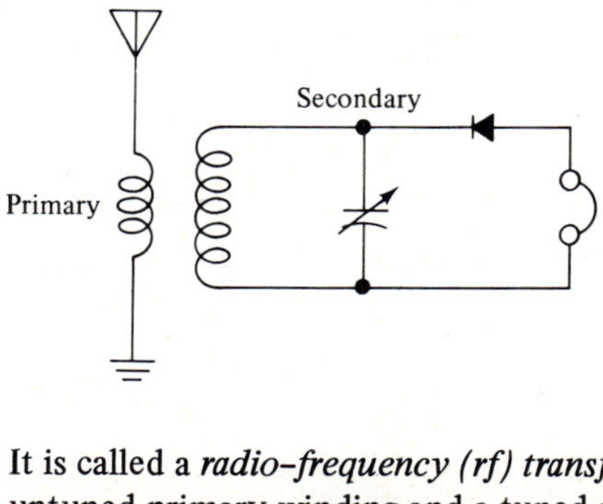

Fig. 11-9
A radio-frequency transformer circuit.

It is called a *radio-frequency (rf) transformer*. This type of rf transformer has an untuned primary winding and a tuned secondary winding. Figure 11-10 shows the appearance of this device, often called an *rf input transformer* or an *antenna coupling transformer.* Notice in Fig. 11-9 that the radio signal currents from the antenna flow through the primary winding, and to ground. Several radio signal

Fig. 11-10
Appearance of rf input transformers for a radio receiver.
Courtesy: J.W. Miller Co.

currents with different frequencies may be flowing in the primary circuit. Each of these signal currents produces a magnetic field which moves through the secondary winding, thereby inducing signal currents in the secondary winding.

Although several radio signals with different frequencies are induced in the secondary winding (Fig. 11-9), the secondary accepts only the signal frequency to which it is tuned. On the other hand, the secondary rejects all signal frequencies to which it is not tuned. Let us follow the resonant-circuit action in the secondary. We notice that the magnetic field from the primary cuts each turn of wire in the secondary winding. In other words, a voltage is induced in each turn of the secondary. These voltages add up in series to produce the total output voltage of the secondary. Therefore, the secondary operates as a *series-resonant* circuit.

In the circuit shown in Fig. 11-9, the primary winding is said to be *coupled* to the secondary winding. This is an example of *inductive coupling.* If the primary winding is close to the secondary winding, the rf transformer is said to be *tightly coupled.* But if the primary winding is separated at a distance from the secondary winding, the rf transformer is said to be *loosely coupled.* Loose coupling has the advantage that the secondary circuit becomes more selective. This means that the secondary circuit tunes sharply, and is better able to tune out interfering radio-station signals. However, loose coupling has the disadvantage that the audible output is weaker than if tight coupling is used.

Better selectivity can be obtained with tight coupling of the rf transformer if a wave trap is used, as is shown in Fig. 11-11. In Fig. 11-11(a), a parallel *LC* circuit is used as a wave trap. It operates in the same way as the wave trap shown in Fig. 11-8. The series LC type of wave trap is shown in Fig. 11-11(b). It consists of an inductor and a variable capacitor connected in series across the primary winding of the rf transformer. Notice that this arrangement tunes out an interfering radio station signal by providing a path of least resistance from antenna to ground.

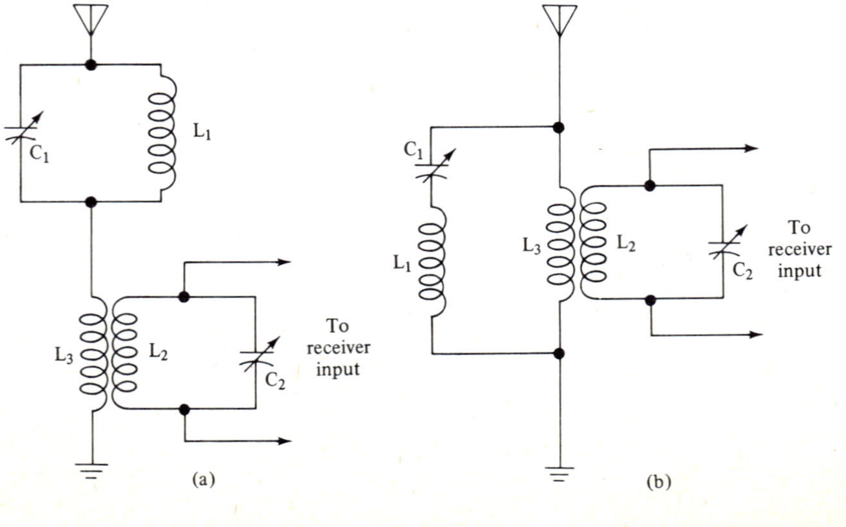

Fig. 11-11
Wavetraps used with an rf transformer.
(a) parallel trap. (b) series trap.

Questions: Chapter 11

1. What is the difference between a series-connected resonant circuit and a parallel-connected resonant circuit?
2. Inspect the formula for resonance shown below; how many values of L and C would resonate at one megahertz?

$$f_0 = \frac{1}{2\pi\sqrt{LC}}$$

3. Explain the term "tuning action" as it applies to a radio receiver.
4. Why are radio stations not at the resonant frequency of the circuit shown in Fig. 11-3 "tuned out"?
5. Is the current maximum or minimum when a radio station is "tuned in" with the circuit shown in Fig. 11-3?
6. Which of the components can you vary in a series LC circuit to obtain resonance?
7. What is the relationship between the values of the inductive and the capacitive reactances in a resonant circuit?
8. Explain the basic difference between the values of the inductive and the capacitive reactances in a parallel-resonant circuit.
9. What is the relationship of the current in the capacitor and the current in the inductor in a parallel LC circuit at resonance?
10. Define inductive coupling.
11. Compare the advantages and the disadvantages of loose and tight coupling.
12. What is a wave trap?

True-False:

1. At resonance, the current is maximum in a series LC circuit.
2. At resonance, the total current is maximum in a parallel LC circuit.
3. For any practical value of capacitance, there is a value of inductance that will cause resonance.
4. A radio is usually tuned to a radio station by varying the inductance of an inductor.
5. In a transformer, a signal is coupled from one set of windings to another set of windings by capacitive coupling.
6. There are thousands of values of inductance and capacitance that will produce resonance at ten kilohertz.
7. There is only one product of L and C values that will produce resonance at one megahertz.
8. A parallel-resonant circuit appears as a low AC resistance, and a series-resonant circuit appears as a high AC resistance.
9. A wave trap is used to block a certain frequency.
10. A series LC circuit could be used in parallel with a load to pass an unwanted signal to ground.

Multiple choice: choose the best answer.

1. In a series *LC* circuit at resonance
   a. inductance is equal to capacitance
   b. inductive reactance is equal to capacitive reactance
   c. inductance is minimum
   d. capacitance is minimum
2. In a series *LC* circuit at resonance, current is
   a. maximum
   b. zero
   c. infinity
   d. minimum
3. To resonate an *LC* circuit, we may
   a. vary the capacitance
   b. vary the inductance
   c. vary either the capacitance or the inductance
   d. all the above are correct
4. In a transformer, the signal is coupled from one winding to another by
   a. inductive coupling
   b. the coefficient of coupling
   c. capacitive coupling
   d. current flow
5. An *LC* circuit that is used to block a certain frequency is called a
   a. frequency trap
   b. trapper
   c. wave trap
   d. serial
6. Increasing the inductance in a series *LC* circuit will
   a. decrease the resonant frequency
   b. increase the capacitance
   c. increase the resonant frequency
   d. decrease the capacitance
7. Selecting a certain radio station with a variable capacitor is called
   a. changing
   b. trapping
   c. stationing
   d. tuning

12

AC AND DC MOTORS

12.1
General Types of Motors

Factories, shops, business houses, and residences make very wide use of AC and DC motors. We find a DC starter motor in every automobile. Refrigerators and washing machines are operated by AC motors, and electric clocks are operated by a special type of AC motor that runs at an exact speed. The motor principle is used in many devices that might not be thought of as motors; for example, the watt hour meters (Fig. 12–1) that are installed in almost all buildings are a form of AC motor. All voltmeters and ammeters operate on the motor principle. We

Fig. 12–1
A watt hour meter uses an AC motor for its operation.
Courtesy: Sangamo Electric Company

find that the loudspeakers used in radio, television, and high-fidelity receivers also operate on the motor principle. It is important, therefore, for every technician to have a good understanding of motor action.

12.2 DC Motor Action

A motor has the function of changing electricity into motion. In most cases, an *armature* or rotor is turned on a shaft. The principle of motor action is shown in Fig. 12-2. The armature consists of an electromagnet mounted on a shaft.

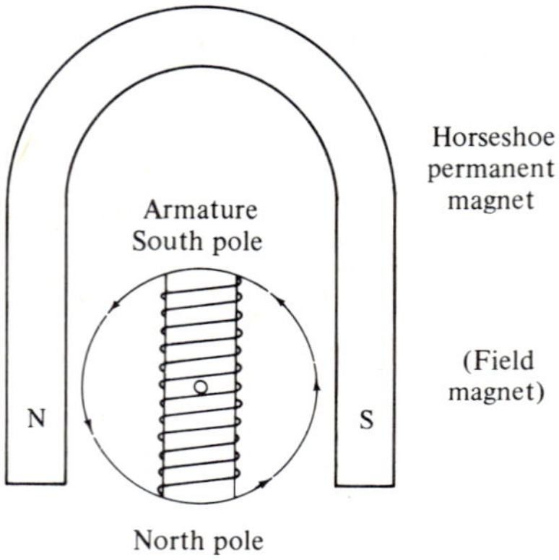

Fig. 12-2
The armature rotates between the poles of the field magnet.

This armature can turn around between the poles of a permanent magnet called the *field magnet.* When the armature is in the position shown in the diagram, without a current through the winding, it does not turn. This is because it is attracted as much by the north pole as by the south pole of the field magnet. However, if we pass a current through the winding on the armature, the south pole of the armature will be attracted by the north pole of the field magnet, and the north pole of the armature will be attracted by the south pole of the field magnet. Therefore, the armature will turn one-quarter of a revolution.

Notice that with current through the winding as shown in Fig. 12-2, the armature will turn in a counterclockwise direction because of the direction of the induced magnetic field in the armature. The magnetic attraction on the armature will produce no more turning action when the sets of opposite magnetic poles are nearest each other. However, because the armature has weight, and has been placed in motion, it continues to turn past this point. Notice carefully in Fig. 12-3 that we have reversed the direction of current flow in the armature winding, and that the magnetic poles of the armature are also reversed. The north pole of the armature is repelled by the north pole of the field magnet, and the south pole of the armature is repelled by the south pole of the field magnet. The result is that the armature continues to turn another one-quarter of a revolution to again arrive at the condition shown in Fig. 12-2.

By reversing the current through the armature at the correct time, we see that the armature continues to turn around as a result of attraction and repulsion

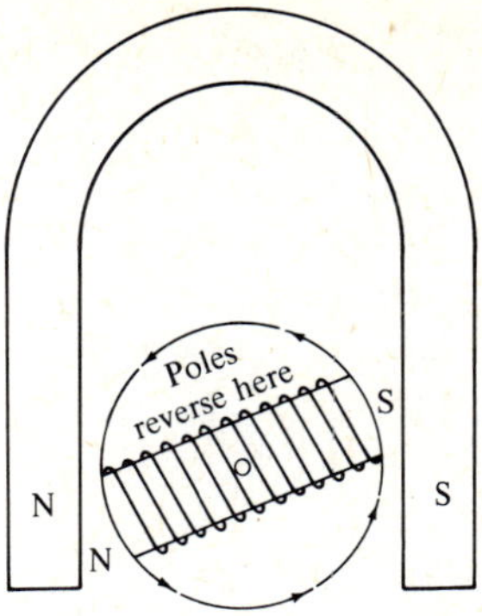

Fig. 12-3
The current changes direction and reverses the polarity of the armature.

of the magnetic poles. These current reversals are made in a DC motor by means of an automatic switch called a commutator. The arrangement of a commutator is shown in Fig. 12-4. The commutator consists of two semicircles of copper, and

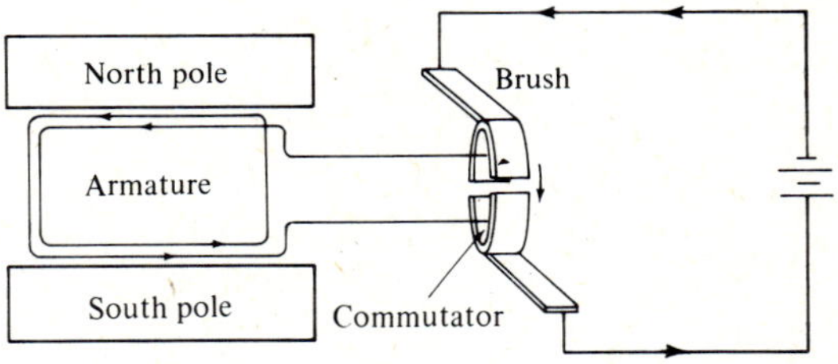

Fig. 12-4
A commutator operates as an automatic switch.

is mounted on the shaft that turns with the armature. Connections to the commutator are made by the armature winding, and by brushes that slide against the commutator as it rotates. Therefore, the current flow through the armature winding is always reversed at the correct time to keep the forces of attraction and repulsion moving the armature in the same direction. Brushes are usually made of

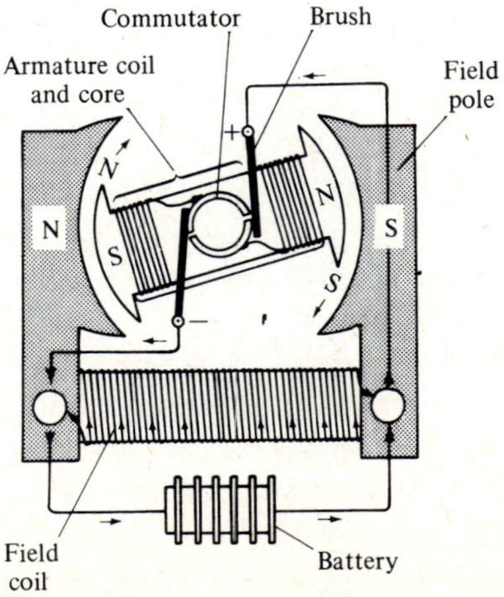

Fig. 12-5
Basic motor arrangement with an electromagnetic field.

carbon, as in automobile starter motors, because a longer life is obtained than if the brushes were made of metal. However, we sometimes find metal brushes used in toy motors.

Automobile starter motors, and most large types of DC motors, use an electromagnetic field which is produced as shown in Fig. 12-5. Notice that the DC current in the field winding flows in the same direction continuously. That is, the field winding is not switched by the commutator. Thus, the electromagnetic field takes the place of the permanent-magnet field which is used in small motors.

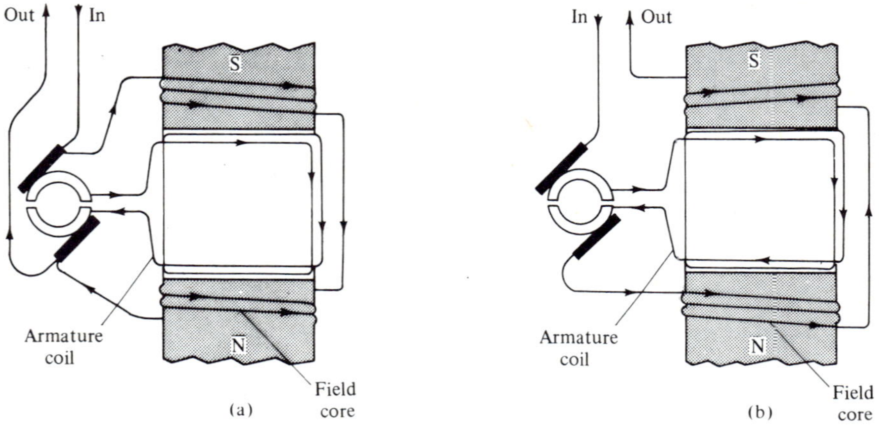

Fig. 12-6
Motor circuits: (a) Shunt. (b) Series.

The motor circuit shown in Fig. 12-6(a) is called a *shunt-field* circuit because the field winding shunts the armature winding. In other words, the field winding and the armature winding are connected in parallel.

Another widely used DC motor circuit is shown in Fig. 12-6(b). This is called a *series-field* circuit because the field winding is connected in series with the armature winding. Automobile starter motors use the series-field circuit. A series-field circuit gives more turning force when the weather is cold. This type of winding works well when the motor turns over very slowly. But a series-field circuit is not used when a motor turns rapidly, as in a blower fan. It is not very satisfactory in such applications because its armature will speed up under light loads. The result is that the bearings and commutator are likely to wear out rapidly, and the winding may even fly off the armature.

The most satisfactory motor circuit for light-load applications at moderate and fairly high speeds is the shunt-field circuit. This circuit does not give a strong turning force when it starts up, but its turning force increases as the armature nears normal operating speed. Also, the armature keeps a fairly even speed, even if the load is removed entirely. This steady speed feature is useful under average loads in order to avoid rapid wear of the bearings and commutator. It also avoids possible damage to the armature in case the motor is called upon to run under no-load conditions occasionally.

Most motors use more than two armature poles. For example, most toy motors use three armature poles, as shown in Fig. 12-7. Two field poles are used, as

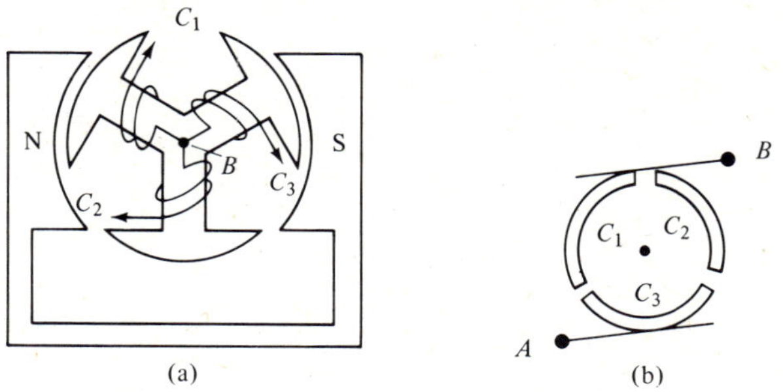

Fig. 12-7
Three-pole armature, as used in toy motors. (a) Armature and field arrangement. (b) Three-segment commutator.

with an armature that has two poles. The advantage of three armature poles is (a) smoother operation and less jerkiness at low speeds, and (b) ability of the armature to start from any position (elimination of "dead center"). As shown in Fig. 12-7(a), each armature pole has a winding. The three windings are all connected together at one end. The other ends, marked C_1, C_2, and C_3, are connected to the three commutator segments shown in Fig. 12-7(b). Notice that one of the brushes B in Fig. 12-7(b) connects to that part at which all three windings are connected together at B in Fig. 12-7(a). The battery that runs the motor is connected from brushes A to B in Fig. 12-7(b).

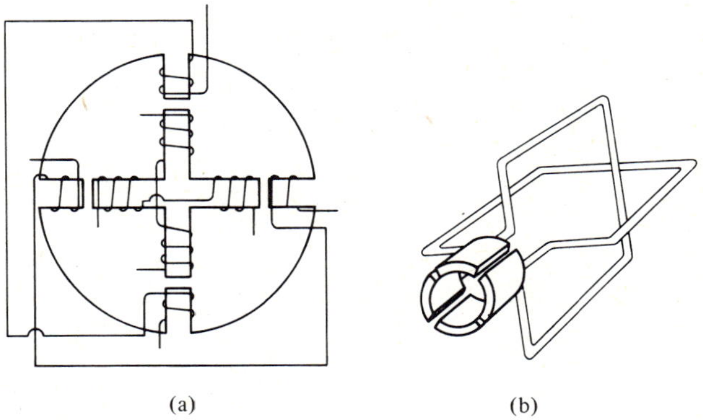

Fig. 12-8
A simple four-pole motor plan. (a) Armature and field arrangement. (b) Commutator has four segments.

Larger DC motors sometimes use four armature poles and four field poles, as shown in Fig. 12-8(a). A four-segment commutator is used, as seen in Fig. 12-8(b). Four poles provide smoother operation than three poles, particularly when starting and during low-speed operation. An electromagnetic field is shown in Fig. 12-8. All the larger types of motors use electromagnetic fields. A starter motor in an automobile has an armature with numerous segments, such as shown in Fig. 12-9. Two field poles are used. Because of the many poles on the armature, the armature has the appearance of a slotted drum. As shown in the diagram, a drum armature has a continuous winding instead of individual windings on each pole.

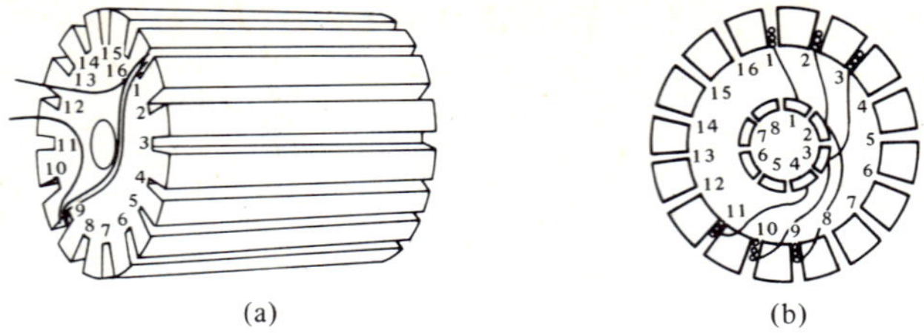

Fig. 12-9
Drum-type armature. (a) Slot arrangement, and plan of winding. (b) Commutator connections.

This drum winding and the numerous commutator segments reduce the sparking at the brushes and commutator so that long life is obtained even when the motor is operated under heavy loads.

12.3 AC Motor Action

Electric clocks are one of the extensive fields of application for AC motors. Since a clock must keep accurate time, the motor must run at an exact speed. This is an outstanding advantage of the *synchronous* motor. Figure 12-10 shows the basic principle of the synchronous motor. The rotor is often in the form of a

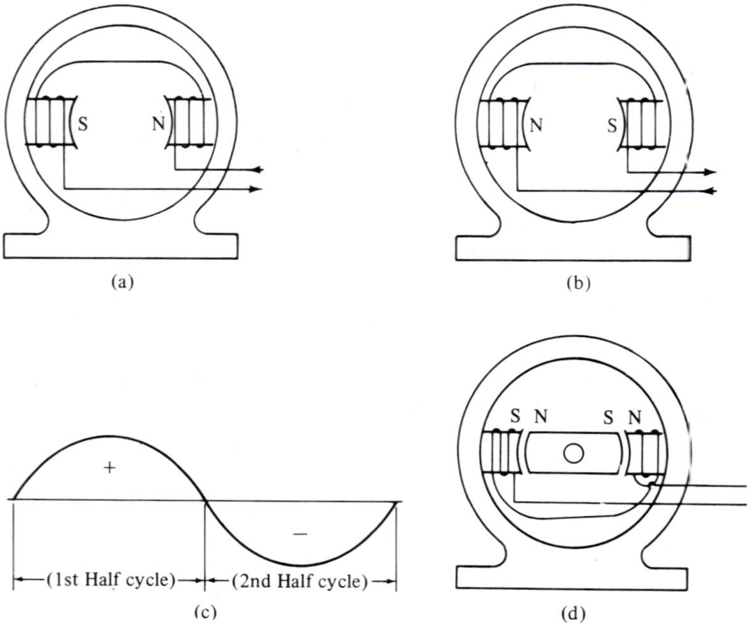

Fig. 12-10
The basic synchronous motor. (a) First field polarity. (b) Second field polarity. (c) AC current reverses its polarity over each cycle. (d) Rotor is attracted by stator in this position.

permanent magnet which revolves on a shaft. In the simplest type of synchronous motor, a pair of electromagnets are used as field poles. The north and south poles of the armature are attracted to the south and north poles of the field. However, since alternating current flows through the field coils, the field poles change their polarity on each cycle of the alternating current, and the armature is revolved by the changing field.

In Fig. 12-10(a), we see that the two field or *stator* windings are connected in series. The positive half cycle of AC current (Fig. 12-10 [c]) flows through the stator winding. During the time that the right-hand winding is producing a north pole, the left-hand winding is producing a south pole. Next, the AC current in Fig. 12-10(c) passes through zero, and then flows in the opposite direction through the stator windings. Therefore, the right-hand winding now produces a south pole, and the left-hand winding produces a north pole as shown in Fig. 12-10(b). From this electrical action, it is seen that each field pole becomes a north and a south pole once during each cycle.

The drawing in Fig. 12-10(d) shows the permanent-magnet armature or rotor in place between the field poles. We see that, since AC current flows through the stator windings, the rotor poles will be alternately attracted and repelled. With the position shown in the diagram, the rotor poles are attracted by the stator poles, and tend to remain in this position. However, during the next half cycle of alternating current, the stator poles reverse their polarity, which causes the rotor poles to be repelled away form the stator poles. Notice that, if the rotor is given a spin, this magnetic repulsion helps the rotation for one-quarter turn, and then magnetic attraction helps the rotation for the next one-quarter turn. This action continues, and the rotor is "locked in synchronism," or, it keeps exactly in step with the frequency of the AC current.

Since the frequency of ordinary house current is 60 cps, or 60 Hertz, we recognize that the synchronous motor in Fig. 12-10(d) must rotate at the rate of 60 rps, or 3,600 rpm. When a four-pole synchronous motor is used, as shown in Fig. 12-11, the rotor moves only one-half as far over one cycle. Therefore,

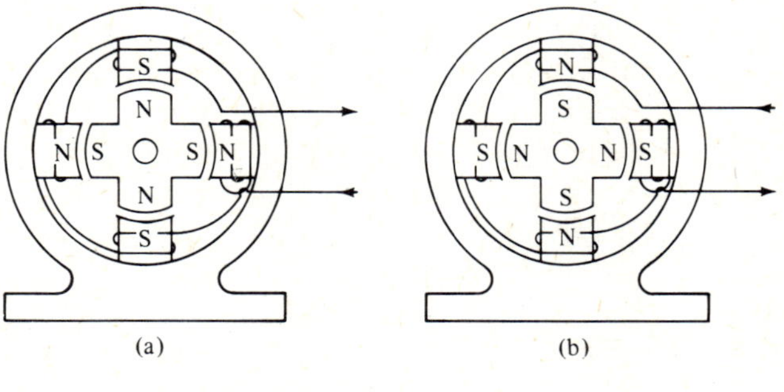

Fig. 12-11
Relationship between the rotor and stator of a four-pole synchronous motor during one cycle of AC. (a) First half-cycle. (b) Second half-cycle.

this type of synchronous motor rotates at the rate of 1,800 rpm. When a six-pole arrangement is used, the motor rotates at the rate of 1,200 rpm. If twenty-four poles are used, a synchronous motor rotates at the rate of 300 rpm. We find that no synchronous motor is self-starting. That is, some means must be provided to start the rotor turning when AC voltage is applied to the motor. The manner in which this is done will be explained in the next several pages.

Most AC motors used in refrigerators, washers, and so on are *induction* motors. An induction motor does not run at an exact speed, but it is simple and comparatively economical to manufacture. The basic induction motor uses a pair of electromagnetic poles, as in a synchronous motor. However, the rotor of an induction motor is an assembly of copper bars and copper rings that look like a cage. Therefore, the rotor of an induction motor is called a *squirrel cage.* Figure 12-12 shows the appearance of a squirrel-cage rotor. Notice that a squirrel

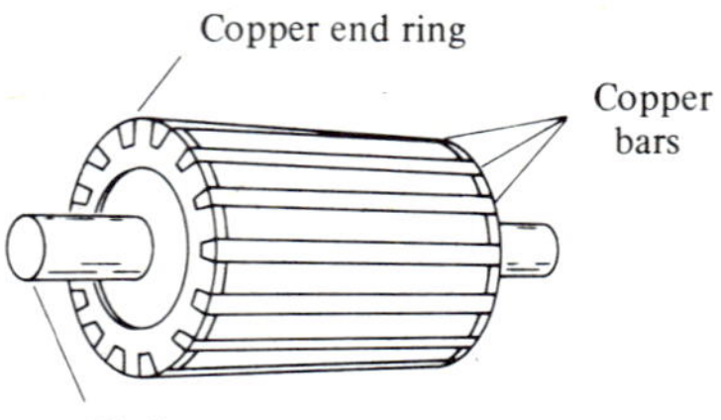

Fig. 12-12
Construction of a squirrel-cage rotor.

cage is basically a short-circuited winding, having no connection with any outside circuit.

An induction motor operates on the basis of transformer action. In other words, AC currents are induced in the squirrel cage by the magnetic field from the stator poles. These AC currents that are induced in the rotor assembly also produce magnetic poles. The induced magnetic fields in the rotor are attracted and repelled by the stator poles, thereby keeping the rotor turning around. An induction motor can be compared with a synchronous motor in that the rotor is basically a pair of magnetic poles. The difference, however, is that the induced rotor poles are not fixed exactly in place as with a permanent magnet. This means that the rotor of an induction motor always has a slight amount of *slip,* and the rotor turns almost, but not quite, at synchronous speed.

The induction motor, like the synchronous motor, is not self-starting. Some arrangement must be provided to start the armature turning around in the proper direction when the AC current is applied. There are various means for starting both induction motors and synchronous motors. One of the methods shown in Fig. 12-13 is called the *shaded-pole* method. This method uses a slot in each stator

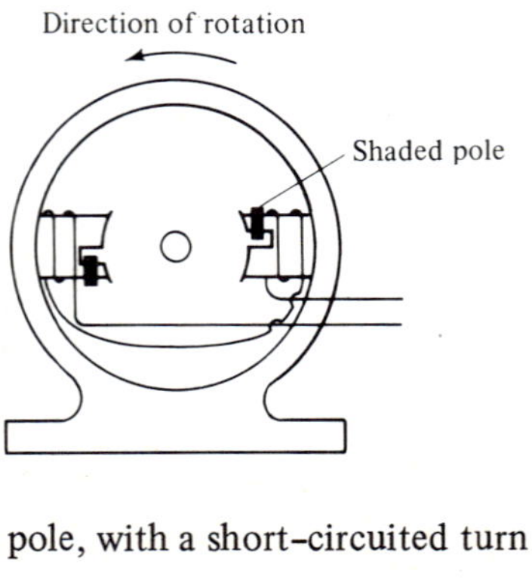

Fig. 12-13
A shaded-pole stator.

pole, with a short-circuited turn of copper strip around one-third of the pole.

Since AC current flows in the windings, an AC magnetic field is produced through the short-circuited turns. Accordingly, AC current is induced in the short-circuited turns. Notice carefully that the induced current produces a magnetic field which is delayed, compared with the magnetic field in the other two-thirds of the stator pole. That is, we say that the magnetic field in the shaded-pole *lags* the magnetic field in the other part of the pole. Now, let us see how this lagging magnetic field starts the rotor turning.

Suppose that we say that the AC current in Fig. 12-13 is producing a north pole in the winding of the right-hand field assembly. That is, the main part of this field has a north polarity. It induces AC current in the squirrel cage, and the rotor has a south polarity next to the main part of the stator pole. But a moment later, a north pole builds up in the shaded part of the stator pole. Therefore, this delayed north pole attracts the south polarity in the rotor, and the rotor starts to turn counterclockwise. This action continues until the rotor is turning fast enough so that the main part of the stator pole "takes hold" of the induced field in the squirrel cage. Then the rotor develops full power and runs like a synchronous motor, but with some small amount of slip. For example, when an induction motor is lightly loaded it may run at 1,760 rpm, instead of 1,800 rpm, as would be the case with a synchronous motor.

Although a shaded-pole starting arrangement is simple and is economical to manufacture, it does not produce much starting power; therefore, the induction motor must be started under light load or no load. When an induction motor must be started under a fairly heavy load, as in a refrigerator or air conditioner, a *capacitor-start* arrangement is generally used. This type of induction motor has the same squirrel-cage rotor as described above; however, capacitor starting requires four field poles, as shown in Fig. 12-14. One pair of field poles is used

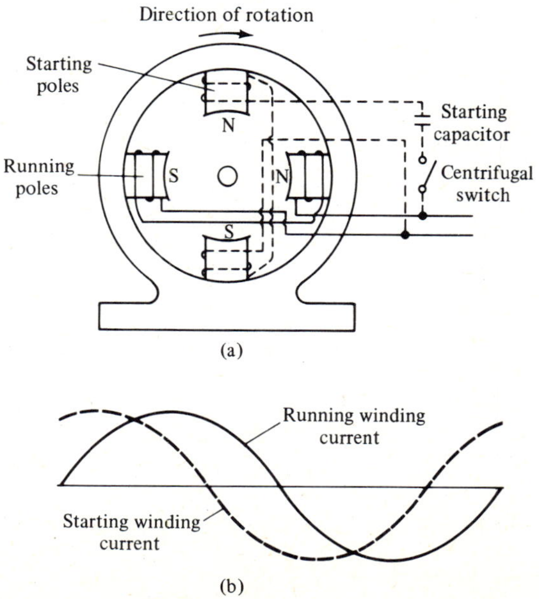

Fig. 12-14
Arrangement of the starting and running windings of a single-phase capacitor-start motor. (a) Winding configurations. (b) Current waveforms.

only for starting. The other pair of field poles is used for running the motor after it comes up to speed. Let us consider how a capacitor-start motor is brought up to running speed.

A capacitor-start motor can be compared with a shaded-pole arrangement in that the magnetic field from one pair of poles is delayed with respect to the magnetic field from the other pair of poles. As shown in Fig. 12-14(b), the running-winding current is delayed with respect to the starting-winding current. This is just another way of saying that the starting-winding current leads the running-winding current. This leading current is produced by a capacitor which is connected in series with the starting winding. We find that the current in an inductive circuit (such as the running winding) always lags the current in a comparatively capacitive circuit (such as the series capacitor and starting winding).

The rotor in Fig. 12-14(a) turns clockwise when the motor is starting. For example, to show how the direction of rotation is determined, we take the time when the top starting pole has north polarity and a current is induced in the squirrel cage. At the next moment, the lagging current in the right-hand running pole builds up a north polarity, and it attracts the squirrel cage in a clockwise direction. After the induction motor comes up to normal speed, a centrifugal switch on the shaft automatically opens the starting circuit. This is done to avoid heating up the starting capacitor and damaging it. The starting capacitor is usually a large nonpolarized electrolytic capacitor. It is designed only for starting purposes, and must never be left in the circuit for a long period of time.

Questions: Chapter 12

1. What are some of the common uses of motors that operate on AC voltage?
2. Name some of the common uses of motors that operate on DC voltage.
3. What is the basic principle of magnetism upon which motor action is based?
4. Explain the purpose of the commutator in a DC motor.
5. What is the purpose of the brushes in a DC motor?
6. State the material that is most often used for motor brushes.
7. What is the advantage of a series motor?
8. Discuss the disadvantage of a series motor.
9. What is the advantage of a shunt motor?
10. Explain the disadvantage of a shunt motor.
11. Why is a three-pole armature often used in small motors?
12. State the advantage of a synchronous motor.
13. What part of a motor is the stator?
14. Why does an induction motor have a certain amount of slip?
15. Explain the purpose of the shaded-pole in an induction motor.
16. What is the purpose of the centrifugal switch in a capacitor-start motor?

True–False:

1. A motor has the function of changing electricity into motion.
2. A watt-hour meter uses a DC motor for its operation.
3. A motor operates on the principle of attraction and repulsion on magnetic fields.
4. A motor will usually run in either direction that it is started.
5. Automobile starter motors use a shunt-type winding.
6. A commutator changes the direction of current in the stator field coil of a motor.
7. A shunt-field motor usually has a fairly constant speed.
8. A synchronous motor is a constant-speed motor.
9. A series motor may go so fast as to destroy itself without a load.
10. A series motor is used where a large starting power is required.
11. A shunt motor has a large starting power.
12. Clocks use synchronous motors.
13. The induction motor is a type of synchronous motor.
14. The squirrel-cage motor is a type of induction motor.
15. The shaded-pole method is used to develop a constant speed in an induction-type motor.

13

ELECTRICAL INSTRUMENTS AND MEASUREMENTS

13.1 The d'Arsonval Meter Movement

Many electrical instruments, such as the multimeter shown in Fig. 13-1 use a d'Arsonval meter movement. This device operates on the motor principle, and

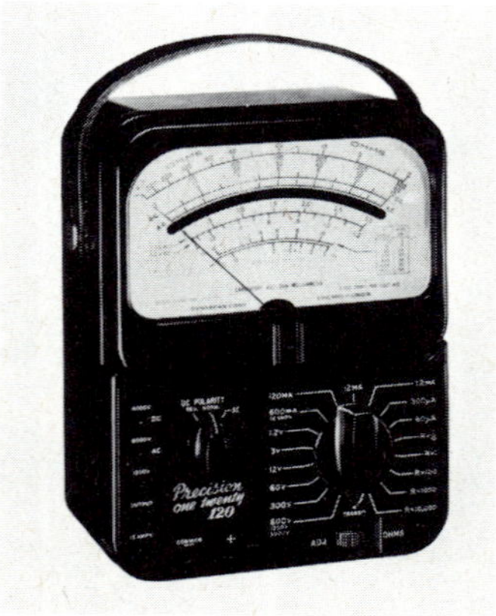

Fig. 13-1
A simple multimeter.
Courtesy: Precision Instrument Company

produces a pointer deflection over a scale. The parts of a d'Arsonval movement are seen in Fig. 13-2. It consists of a permanent magnet with a pair of soft-iron

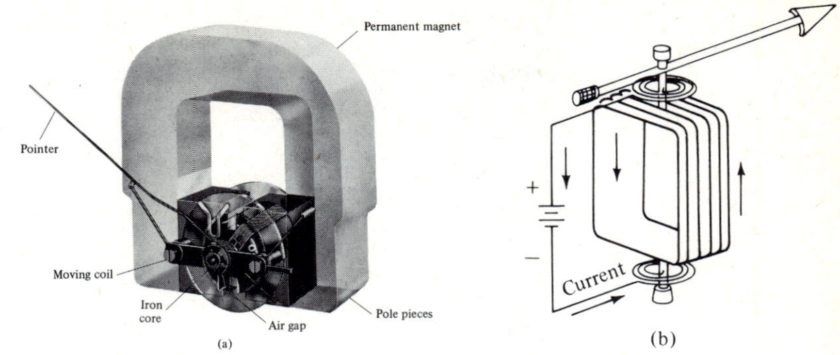

Fig. 13-2
Meter movement: (a) Arrangement of a d'Arsonval movement.
Courtesy: Weston Instrument Co.
(b) Construction of the moving coil.

pole pieces. A round iron core is mounted between the pole pieces. An air gap is left between the pole pieces and the core so that a moving coil can turn. This moving coil rides in pivot bearings, and carries a pointer. Two thin spiral springs keep the coil in its starting position, and also conduct electric current to the winding of the moving coil. This coil consists of many turns of very fine wire. When current flows through the winding, the coil produces a magnetic field which is attracted by the poles of the permanent magnet causing the coil to turn on its bearings against the force of the spiral springs. A larger current through the coil causes it to rotate further against the spring force.

The horseshoe-magnet arrangement shown in Fig. 13-2 was used in all d'Arsonval movements until recent years. Today, more compact permanent-magnet arrangements are used fairly extensively. An example is the concentric-magnet movement shown in Fig. 13-3 which has two bar-type magnets in circular-sector form inside a soft-steel ring. The advantage of this arrangement is that the

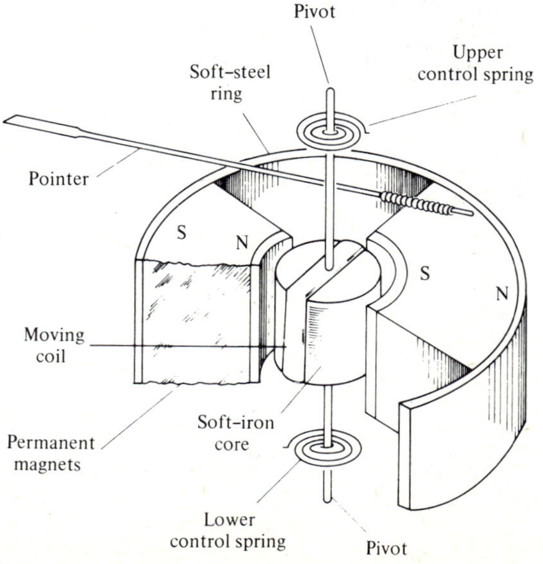

Fig. 13-3
Diagram of a concentric-magnet movement.
Courtesy: General Electric Company.

movement occupies less space, so that a smaller case can be used for the electrical instrument.

Notice that a soft-iron core is used in Fig. 13-3 in the same way as shown in Fig. 13-2. On the other hand, the core-magnet arrangement seen in Fig. 13-4 utilizes a permanent magnet inside the moving coil instead of an iron core. Basically, the core-magnet construction is the concentric-magnet arrangement "turned inside out." These newer designs have made possible the manufacture of compact "pocket multimeters," such as shown in Fig. 13-5. Regardless of the type, most d'Arsonval movements use jeweled bearings, as shown in Fig. 13-6. There-

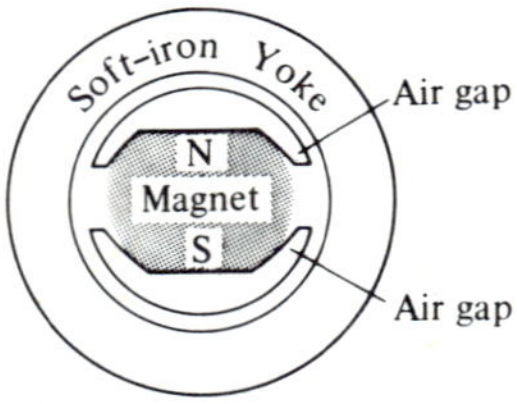

Fig. 13-4
Plan of the core-magnet arrangement.

Fig. 13-5
A compact multimeter.
Courtesy: Triplett Electrical Instrument Company

Fig. 13-6
A jeweled bearing arrangement.

fore, the movement can be compared with a watch that has jeweled bearings. Like a high-quality watch, a meter should never be dropped or jarred unnecessarily.

13.2 The d'Arsonval DC Voltmeter

A voltmeter indicates electrical pressure in volt units. The number of volts being measured is indicated on a scale, such as shown in Fig. 13-7. In this example, the

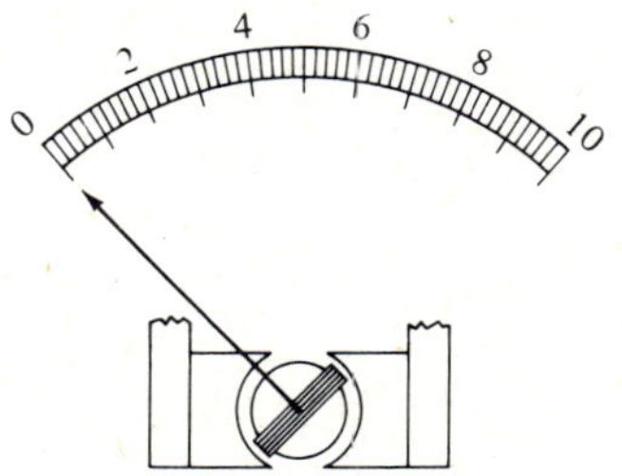

Fig. 13-7
Scale intervals for a moving-coil voltmeter.

range of the scale is from 0 to 10 volts. The same scale may be used in a multimeter to indicate over the range from 0 to 1 volt. That is, each number on the scale is divided by 10 when the multimeter is operated on its 0-to-1-volt range. Likewise, the same scale may be used to indicate over the range from 0 to 100 volts. In this case, each number on the scale is multiplied by 10. If this same scale is used to indicate over the range from 0 to 1,000 volts, each number on the scale is multiplied by 100. This multiplication is done "in the head" by the technician when reading the scale indication on a given range.

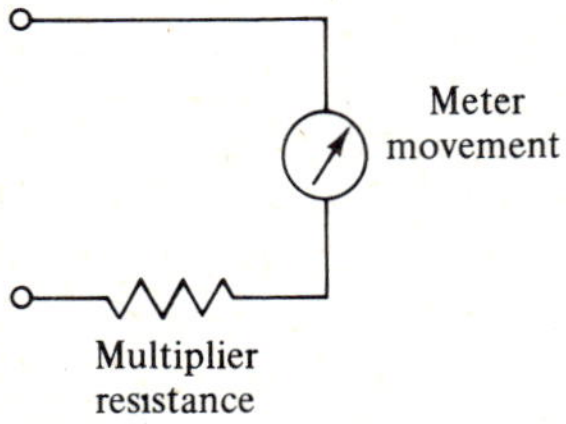

Fig. 13-8
The full-scale voltage indication depends on the size of the multiplier resistance.

The basic voltmeter circuit consists of a d'Arsonval movement and a *multiplier* resistor (Fig. 13-8). The number of ohms used in the multiplier resistance depends upon the voltage range that is in use on the voltmeter. A voltage range is described in terms of *full-scale* indication. For example, in Fig. 13-7, we say that the full-scale indication is 10 volts. However, we could also state that the full-scale indication is 100 volts in case we are using this scale on the 100-volt range of the voltmeter. Now let us consider the circuit in Fig. 13-9 that is used in a

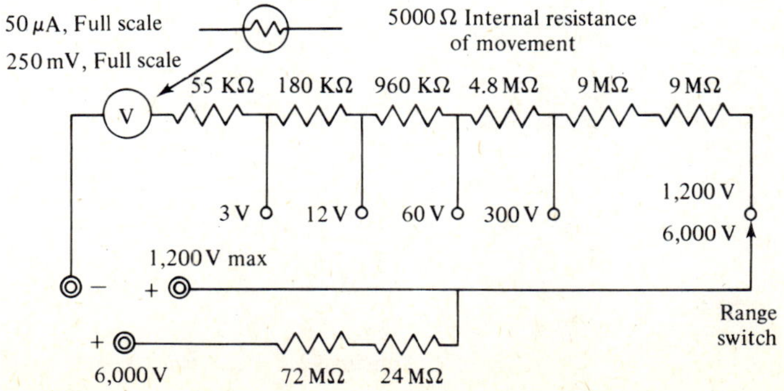

Fig. 13-9
DC voltmeter circuit with six ranges.
Courtesy: Triplett Electrical Instrument Co.

voltmeter having six ranges. A five-position *range switch* is used with an extra terminal for the 6,000-volt range.

The d'Arsonval movement used in Fig. 13-9 has a full-scale current of 50 microamperes (μA). In other words, the pointer will indicate full scale when 50 μA flow through the moving coil. At full scale, there are 250 millivolts (mV), or 1/4 volt, across the coil winding. By the use of Ohm's law we can determine that the winding has a resistance of 5,000 ohms. If the range switch is set to its 3-volt position, a 55,000-ohm resistor is connected in series with the movement, limiting the full-scale current to 50 μA. Therefore, the total resistance of the meter and series-multiplier resistor of the meter is 60,000 ohms when operating on the 3-volt range. Notice how this fact is shown by Ohm's law:

$$I = \frac{E}{R}$$

or,

$$I = \frac{3}{60{,}000} = 50\ \mu\text{A}$$

In other words, when a resistance of 60,000 ohms is connected across a 3-volt source, 50 μA of current will flow through the resistance. Therefore, the full-scale indication of the meter with this multiplier resistor is 3 volts.

We see that the total resistance of the meter and multiplier resistors on each of the higher voltage ranges is as follows:

12-volt range, 240,000 ohms.
60-volt range, 1,200,000 ohms (1.2 megohms).
300-volt range, 5,000,000 ohms (5 megohms).
1,200-volt range, 14,000,000 ohms (14 megohms).
6,000-volt range, 110,000,000 ohms (110 megohms).

13.3 Ammeter and Milliammeter Arrangement

An ammeter indicates current flow in ampere units, and a milliammeter indicates current flow in 0.001-ampere units. The basic current-meter arrangement consists of a d'Arsonval movement and a *shunt resistor,* as shown in Fig. 13-10. In

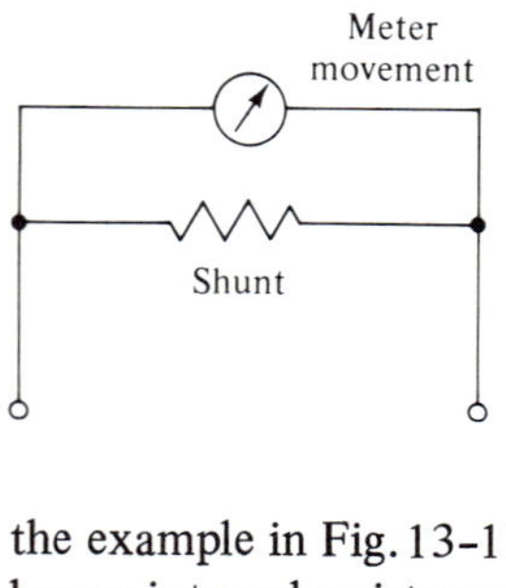

Fig. 13-10
The full-scale current indication depends on the size of the shunt resistance.

the example in Fig. 13-11(a), we are using a 1-milliampere (mA) movement which has an internal resistance of 100 ohms. The movement by itself provides a current range from 0 to 1 mA. Now if we connect a 100-ohm shunt resistor across the movement, as shown in Fig. 13-11(b), we will get a current range from 0 to 2 mA. This fact is easily seen if we notice that one-half the incoming current must

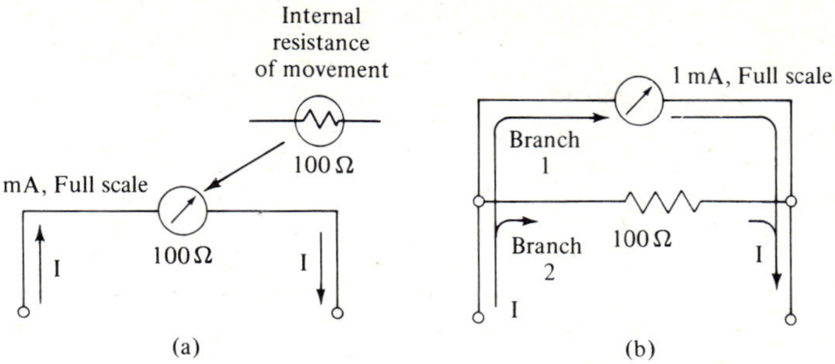

Fig. 13-11
Example of milliammeter ranges. (a) Movement alone has a range of 1 mA. (b) Movement and shunt provides a range of 2 mA.

flow through the shunt resistance and the other one-half must flow through the movement. Therefore, it will take 2-mA input current to get full-scale indication on the meter.

Suppose that we use a very small size of shunt resistance in Fig. 13-11(b), so that only 1 mA flows through the movement when 1 ampere flows into the meter circuit. In this case, the pointer will show full scale for 1 ampere of current. That is, the ammeter will then operate on a 1-ampere range. Next, let us suppose that we use a still smaller size of shunt resistance in this ammeter circuit, so that 1 mA flows through the movement when 10 amperes flow into the meter circuit. In turn, the pointer will indicate full-scale for 10 amperes of current; that is, the ammeter will then operate on a 10-ampere range.

The most common mistake that is likely to be made in the use of an ammeter is an attempt to measure current by connecting the ammeter across a circuit component. Remember that a *current meter* must always be connected in *series* with the circuit, as shown in Fig. 13-12(a). If a mistake is made and the current meter

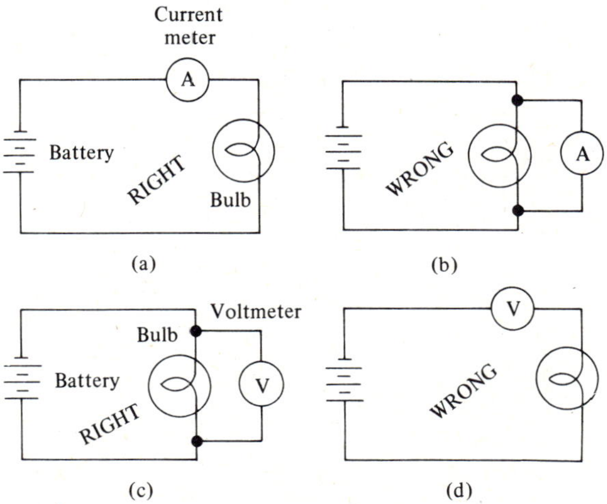

Fig. 13-12
Right and wrong applications of ammeters and voltmeters. (a) Right way to apply an ammeter. (b) WRONG way to apply an ammeter. (c) Right way to apply a voltmeter. (d) WRONG way to apply a voltmeter.

is connected across the light bulb, as shown in 13-12(b), the meter will be damaged or completely burned out. On the other hand, a *voltmeter* must be connected *across* the circuit component, as shown in Fig. 13-12(c). If a mistake is made and the voltmeter is connected in series with the light bulb, as shown in 13-12(d), the indication will be too low, and the bulb will not glow.

Another common mistake that is often made in the use of an ammeter or a voltmeter is to start the measurement on too low a range. For example, we might surmise that there is approximately an ampere flowing in a circuit. In turn, if we set an ammeter to its 1-ampere range and then connect it into the circuit, the meter will be burned out if the current is actually 10 amperes. Therefore, the safe rule to remember is to always set the ammeter to its highest current range before connecting it into the circuit. The range setting can then be lowered as needed to get a convenient reading.

The same safety rule applies to the application of a voltmeter. Most technicians make the mistake of starting a measurement on too low a range. For example, we might assume that there are about 10 volts across a certain circuit component. If we set a voltmeter to its 10-volt range and connect it across the component, the meter will be burned out if the voltage is actually 100 volts. The same danger is present when working with a milliammeter, if we forget that a milliampere is equal to 0.001 ampere. For example, if we set a multimeter to its 1-mA range and connect it into a circuit that has 1 ampere flowing, the meter will be instantly burned out.

13.4 Ohmmeter Arrangement

An ohmmeter indicates resistance values or sizes in ohm units. On its higher ranges, an ohmmeter indicates resistance values in kilohm or megohm units. The basic ohmmeter arrangement consists of a d'Arsonval movement, a multiplier resistor, and a battery connected in series, as shown in Fig. 13-13. Notice that the

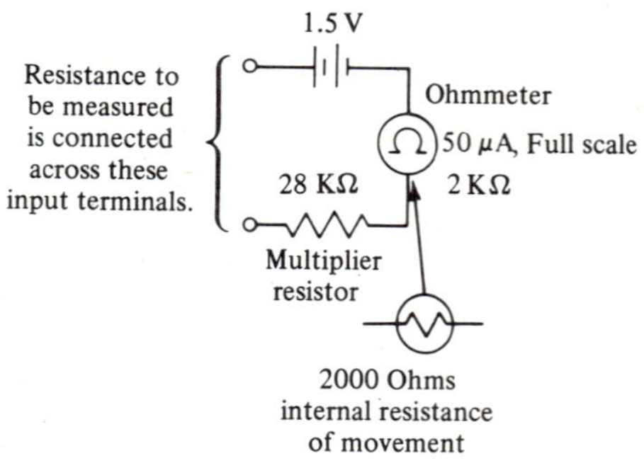

Fig. 13-13
Basic ohmmeter arrangement.

difference between an ohmmeter and a voltmeter is that the ohmmeter contains a battery. Therefore, an ohmmeter always applies voltage across the resistance that is to be measured. This causes current to flow through the meter circuit, and the number of ohms of resistance being measured is indicated on a scale. Figure 13-14 shows an ohmmeter scale above the voltage and current scales of a multimeter.

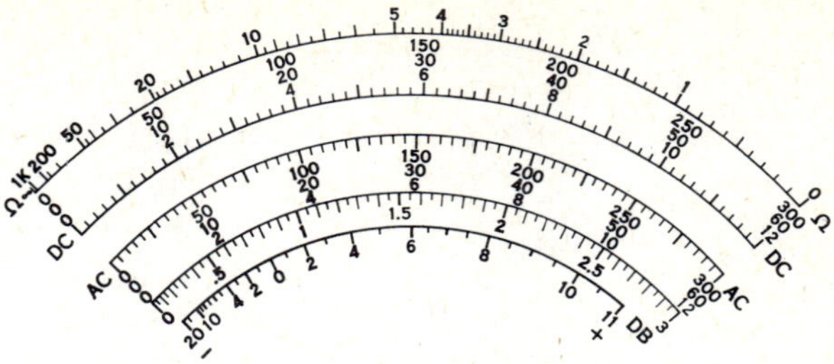

Fig. 13–14
Comparison of an ohms scale with dc voltage and current scales.

We note in Fig. 13-14 that 0 ohm is marked at the right-hand end of the ohms scale. In other words, *0 ohm* is indicated when *full-scale current* is flowing through the movement. We note also in Fig. 13-13 that full-scale current will flow through the movement when the input terminals of the ohmmeter are short-circuited. That is, the multiplier resistance has a value such that the pointer deflects full-scale when the input terminals of the ohmmeter are short-circuited. In this example, the multiplier resistance has a value of 28,000 ohms. When the input terminals of the ohmmeter are open-circuited, the pointer rests at the left-hand end of the ohms scale in Fig. 13-14. This means that the pointer indicates an infinite number of ohms (∞), which means an open circuit.

We observe in Fig. 13-15(a) that, if a 30,000-ohm resistor is connected across the input terminals of the ohmmeter, only one-half as much current flows through the meter as when the input terminals are short-circuited. In other words, the

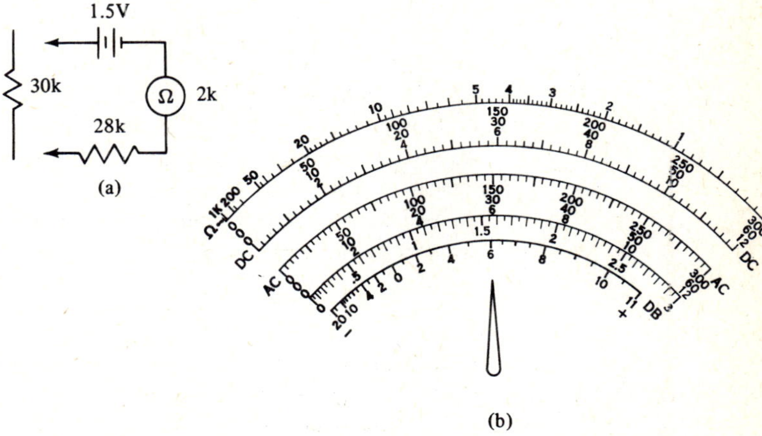

Fig. 13–15
The ohmmeter. (a) Ohmmeter connection to measure a 30 kilohm resistor. (b) Scale indication for a 30 kilohm reading.

internal resistance of the ohmmeter is 30,000 ohms, and we are connecting another 30,000 ohms in series with this resistance. Therefore only one-half as much current can flow, and the pointer deflects to half-scale. If we were drawing a scale for this basic ohmmeter arrangement, we would mark "30,000" at the half-scale point, or we might mark "30" at the half-scale point and state that we are reading the scale in kilohm units.

Returning to Fig. 13-14, we observe that the ohms scale becomes quite cramped at the left-hand end. This crowding of the scale divisions makes it difficult to read the indication accurately. For this reason, it is advisable to use an ohmmeter range that lets us read resistance values near the half-scale location, instead of the full-scale location. Multimeters (Fig. 13-16) provide several ohmmeter ranges for this reason. One way of providing high-resistance ranges is to use more bat-

Fig. 13-16
A basic multimeter that can be used to measure volts, amperes and ohms.
Courtesy: Simpson Electric Co.

tery voltage and a larger value of multiplier resistance. To provide low-resistance ranges, both a multiplier resistance and a shunt resistance may be used. We do not need to consider these details of ohmmeter arrangements at this time.

The most common mistake that is likely to be made in the application of an ohmmeter is an attempt to measure resistance in a "live" circuit. That is, resistance measurements must be made in "dead" circuits only. Otherwise, the ohmmeter is very likely to be damaged or burned out completely. Remember that capacitors will often hold a charge for some time. Therefore, a capacitive circuit might be "alive" although the source of power has been turned off. A safe rule is always to discharge the capacitors in a circuit before attempting to make resistance measurements.

13.5 AC Voltmeter Arrangement

An AC voltmeter measures AC electrical pressure in volt units. Figure 13-17 shows the basic arrangement. Notice that an AC voltmeter is very similar to a DC voltmeter; however, a semiconductor rectifier is placed in series with the meter circuit

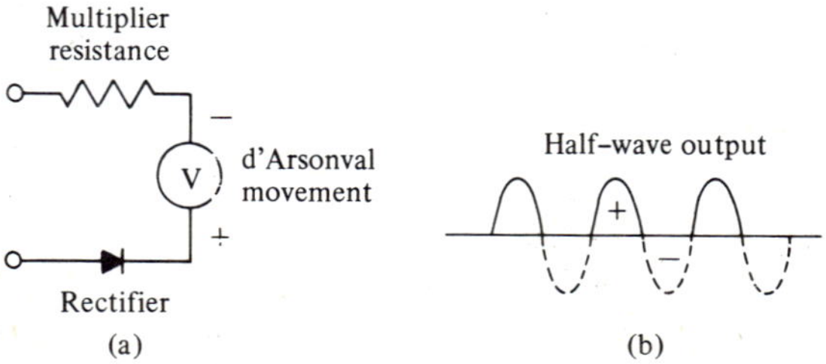

Fig. 13-17
Basic AC voltmeter arrangement. (a) Connections. (b) Rectifier action.

in the case of an AC voltmeter. As shown in Fig. 13-17(b), an AC voltage has both positive and negative half-cycles. Therefore an AC voltage cannot deflect a d'Arsonval meter because each positive half cycle pulls the pointer in an opposite direction from that of the following negative half cycle. However, when a rectifier is placed in series with the circuit, only the positive half cycles can flow through the movement. This rectification allows an AC voltage to deflect the pointer. The scale is marked in AC volts.

As with the DC voltmeter, several ranges are provided for an AC voltmeter. Higher ranges are obtained by the use of higher values of multiplier resistance. If we use an AC voltmeter to measure DC voltage, the meter will indicate incorrectly. The same general application rules are followed for both AC and DC voltmeters. Thus, an AC voltmeter is always connected across a component in a circuit. It is important to set the AC voltmeter to its highest range before it is connected into the circuit, and it is a good rule to always set either an AC or a DC meter to its highest range after disconnecting it from a circuit. With this procedure, we are less likely to damage the meter by overload the next time it is used.

Questions: Chapter 13

1. What is the name of the basic meter movement that is used in most multimeters?
2. Explain the operation of the meter movement shown in Fig. 13-2.
3. Why are jeweled bearings used in d'Arsonval meter movements?
4. What is the difference between a voltmeter and an ammeter?
5. Explain the purpose of the multiplier resistor in a voltmeter.
6. What is the purpose of the range switch on a voltmeter?
7. Define the meaning of the term "full-scale" deflection as it applies to to a voltmeter.
8. What is the purpose of the shunt resistor in an ammeter?
9. How is a current meter connected in a circuit?
10. What are some of the safety precautions that must be followed when using voltmeters and ammeters?
11. How is an ohmmeter used to measure resistance?
12. What is the basic difference between a voltmeter and an ohmmeter?
13. How does an AC voltmeter differ from a DC voltmeter?
14. What is a good rule to use when storing a voltmeter after it has been in use?

True–False:

1. The d'Arsonval meter movement is the basic component in voltmeters, ammeters, and ohmmeters.
2. The coil of a d'Arsonval meter movement is made from a few turns of coarse wire.
3. The pointer is mounted on the moving coil in a d'Arsonval meter movement.

4. Meter movements are very rugged and will not be damaged when dropped.
5. A voltmeter uses shunt (parallel) resistors to develop several ranges.
6. A voltage range is described in terms of full-scale indication.
7. Regardless of the voltage range of the meter in Fig. 13-9, the maximum current through the meter is 50 μA.
8. The shunt resistor takes the extra current in an ammeter circuit.
9. A current meter is connected in parallel to the component to measure current.
10. An ohmmeter must never be connected to a circuit to which current or voltage is applied.

Multiple choice: choose the best answer.

1. To measure current through a circuit element, an ammeter should be connected in
   a. parallel with the element
   b. series with the voltage source
   c. parallel with the voltage source
   d. series with the element
2. To measure voltage drop across a circuit element, a voltmeter should be connected in
   a. parallel with the voltage source
   b. parallel with the element
   c. series with the voltage source
   d. series with the element
3. The type of meter that uses an internal battery is called a/an
   a. watt-meter
   b. ammeter
   c. voltmeter
   d. ohmmeter
4. A voltmeter is used to measure
   a. current
   b. voltage
   c. resistance
   d. wattage
5. Before measuring resistance in a circuit, first
   a. set the range switch to the highest scale
   b. set the range switch to the lowest scale
   c. observe polarity
   d. turn off circuit power
6. A multimeter can be used to measure
   a. current
   b. resistance
   c. voltage
   d. all the above

7. An ohmmeter is used to measure
   a. resistance
   b. voltage
   c. wattage
   d. current
8. To increase the range of a voltmeter, we must use a
   a. shunt resistor
   b. multiplier resistor
   c. battery
   d. parallel resistor
9. To increase the range of an ammeter, we must use a
   a. shunt
   b. multiplier resistor
   c. battery
   d. series resistor

14 SEMICONDUCTOR DEVICES

14.1 Semiconductor Principles

As we noted earlier, a semiconductor is a substance which is neither a good conductor nor a good insulator. We can compare a semiconductor with a resistor, although we will find that it is a special kind of resistor. The most important use of semiconductor substances is in *control devices.* Control devices are components that are used to control the direction of current flow or to increase the strength of a current or voltage. For example, in Chapter 13 we discussed the action of a semiconductor diode in changing an alternating current into a current that flows in one direction only.

The two most often used semiconductor substances are *germanium* and *silicon.* Figure 14-1 shows the appearance of a germanium ingot which has been "grown" from specially prepared germanium. This ingot will be cut into tiny

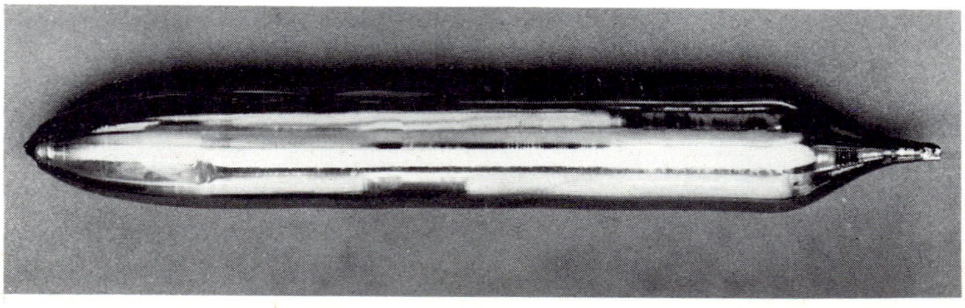

Fig. 14-1
A germanium ingot.
Courtesy: Westinghouse Corp.

wafers for manufacturing various semiconductor devices. Pure germanium (or silicon) is not an ordinary resistive substance. For example, if we apply 1.5 volts across a wafer of pure germanium, we might measure a small current flow of 1 μA, as shown in Fig. 14-2(a). Then, if we increase the battery voltage to 6 volts,

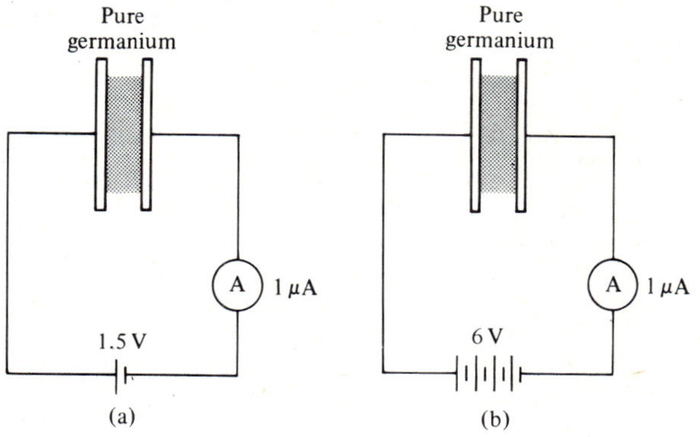

Fig. 14-2
Current conduction in pure germanium. (a) 1.5 volts causes 1 microampere to flow. (b) 6 volts also causes 1 microampere to flow.

we will still measure a current flow of 1 μA, as shown in Fig. 14-2(b). In other words, pure germanium does not obey Ohm's law in the usual sense. Therefore, we call the current that flows in Fig. 14-2 *saturation current.*

Although there are a few uses for saturation-current devices, most of the semiconductor devices found in electronic equipment do not operate on the basis of saturation current. Most of our semiconductor devices, therefore, do not use pure germanium or silicon; instead, they make use of "impure" germanium or silicon. This type of semiconductor is said to be *doped.* At this time, it is helpful to return to a review of atomic structure in order to see how a pure semiconductor substance is changed by doping, and how it conducts electric current.

The materials germanium and silicon are crystals. This means that their atoms are lined up exactly in rows and columns like a battalion of soldiers. A simplified picture of a germanium atom with its four outer electrons is shown in Fig. 14-3.

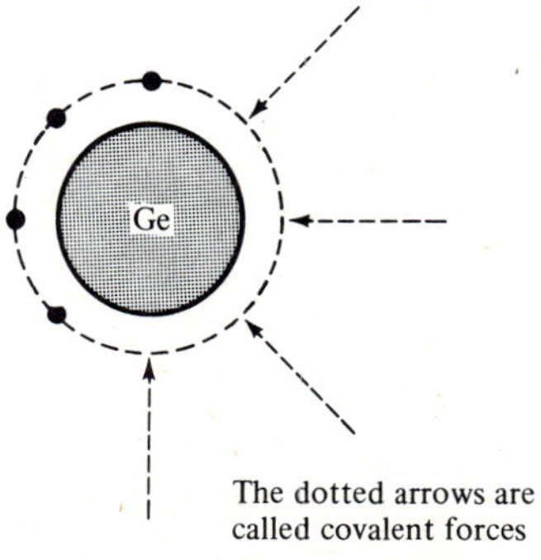

Fig. 14-3
To form a crystal, a germanium atom will share four electrons from nearby germanium atoms.

When a germanium material is melted and allowed to cool, it crystallizes into a solid in which each atom shares (attracts) four electrons from four nearby germanium atoms. That is, the outer shell of each atom is filled with eight electrons, although four of the electrons are not its exclusive property. This electron sharing is said to be caused by *covalent forces*. These covalent forces cause a lineup of germanium atoms in the crystal as shown in Fig. 14-4.

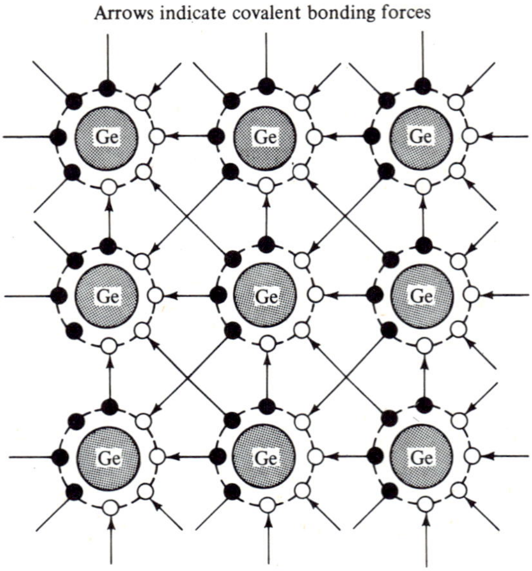

Fig. 14-4
Structure of a germanium crystal indicating the covalent forces between atoms.

Now let us see how a saturation current flows in a pure germanium crystal. The drawing in Fig. 14-5 represents a slab of pure germanium with a pair of metal electrodes on each side. A battery is connected to apply a voltage across the crystal. If every electron in the crystal happens to be in its correct location, as in Fig. 14-4, there can be no current flow through the crystal. However, an occasional electron can be forced or freed from its correct location by energy in the form of heat, light, or other kinds of electromagnetic waves. Even at room temperature, there are a few so-called *free electrons* in a crystal of pure germanium. Whenever an electron anywhere in the crystal happens to be forced out of its correct location, it can jump to the next atom. When a battery voltage is applied so that all the free electrons move in the same direction, saturation current flows through the crystal.

There is more to saturation-current flow than simple electron movement from one atom to the next. That is, we find that an understanding of semiconductor devices requires that *holes* and *hole flow* must be included, as noted in Fig. 14-5.

A careful study of Fig. 14-5 shows that the positive electrode attracts an electron from the left-hand germanium atom. Now, this atom is left with a missing electron, and this missing electron is called a hole. We see that if an electron jumps into this hole from the next atom to the right, this jump will leave a hole

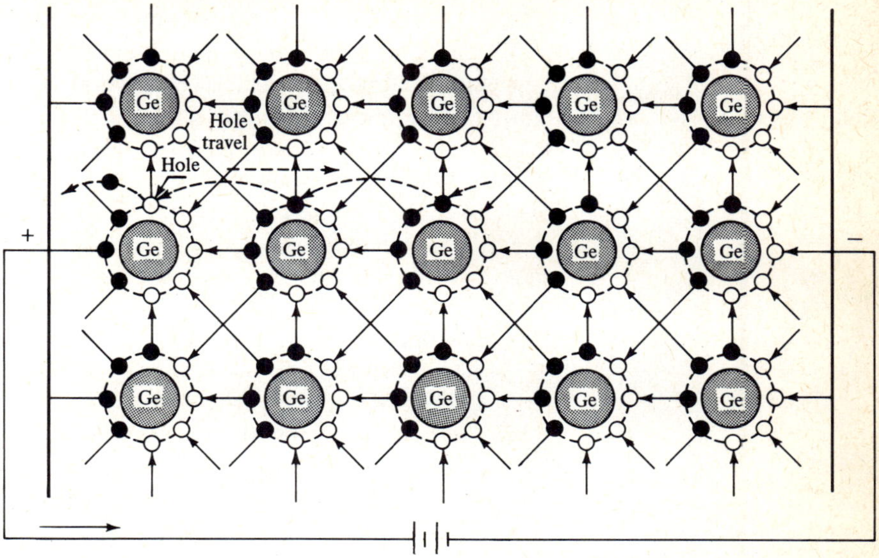

Fig. 14–5
Saturation-current flow in a pure germanium crystal.

behind. Therefore, we have hole travel in one direction in a germanium crystal, as well as electron travel in the other direction. The result of electron travel and hole travel is *drift current,* or saturation current. In the case of a pure crystal, we speak of saturation current, because there are comparatively few free electrons produced by the surrounding heat and light. So only a small number of electrons and holes are available to produce current flow, and the available current cannot be increased by increasing the battery voltage. That is, "saturation current" means that the crystal is conducting as much as possible. The only way to increase the amount of saturation current is to heat the crystal to a higher temperature. Let us see how this works.

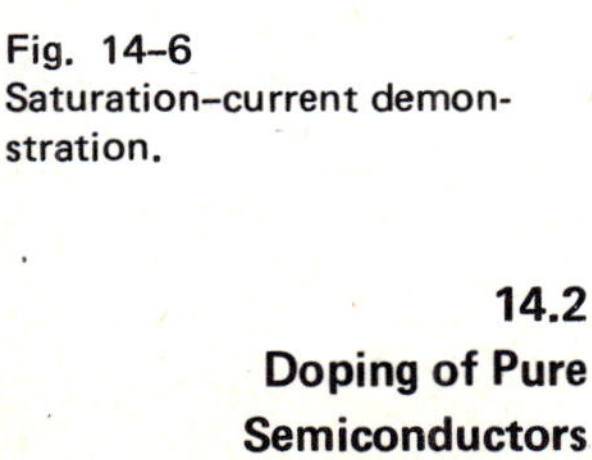

Fig. 14–6
Saturation-current demonstration.

Instructor Demonstration or Student Experiment:

Connect the germanium diode in series with the microammeter and 3-volt battery as shown in Fig. 14–6. It is important that the diode be connected into the circuit with reverse polarity. Under this condition, only a small reverse current flows, which is chiefly saturation current. Notice that when the diode is warmed between your fingers or heated by a lamp, the saturation current increases. This is the same effect as that which would occur in pure germanium.

14.2 Doping of Pure Semiconductors

Most semiconductor devices use doped germanium, or doped silicon. The semiconductor may be doped with a trace of phosphorus, a trace of boron, or various other "impurities." In this case, we use the word "impurity" to mean any substance other than the semiconductor material. That is, semiconductor devices would not operate properly if they did not contain "impurities." Two widely

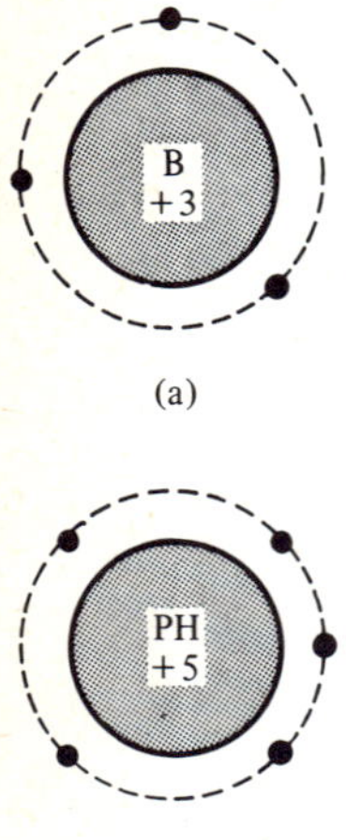

Fig. 14-7 Semiconductor doping substances. (a) The boron atom has three outer electrons. (b) The phosphorous atom has five outer electrons.

used doping substances are boron and phosphorus. To understand the doping action, let us see how a doping substance differs from semiconductor material.

We note from the plan of a boron atom in Fig. 14-7 that boron has three outer electrons. A boron atom has one outer electron less than a germanium atom. On the other hand, phosphorus has five outer electrons. A phosphorus atom has one more outer electron than a germanium atom. If a trace of boron is added to a melt of pure germanium, the crystal that is formed will have a fairly low resistance compared to a pure germanium crystal. Also, if a trace of phosphorus is added to a melt of pure germanium, the resulting crystal will have a fairly low resistance compared to a pure germanium crystal. Be aware that germanium doped with boron is quite different from germanium doped with phosphorus. Let us see why this is so.

Again, observe in Fig. 14-4 that each germanium atom fits perfectly into the crystal arrangement. Each atom fits perfectly because each of the atoms has four outer electrons. However, if a phosphorus atom is in the place of a germanium atom, the fit is not perfect because the phosphorus atom has an extra outer electron which does not fit into the crystal arrangement (see Fig. 14-8). Therefore,

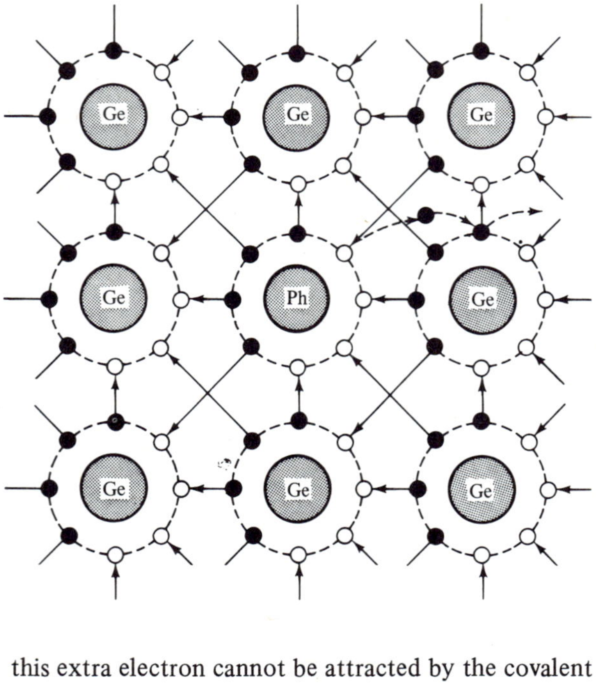

Fig. 14-8 Germanium doped with phosphorus, showing a free electron F.

this extra electron cannot be attracted by the covalent forces. Instead, this extra electron floats through the crystal as a *free electron.* This free electron can move easily from one end of the crystal to the other end if a voltage is applied across the crystal. This type of semiconductor is called *N-type germanium.* N-type means that the crystal has free negative charges, or free electrons as current carriers.

When a voltage is connected across a piece of N-type germanium, electrons flow in much the same way as marbles pushed into a pipe. Figure 14-9 shows how this comparison works. As in the case of pure germanium, the amount of

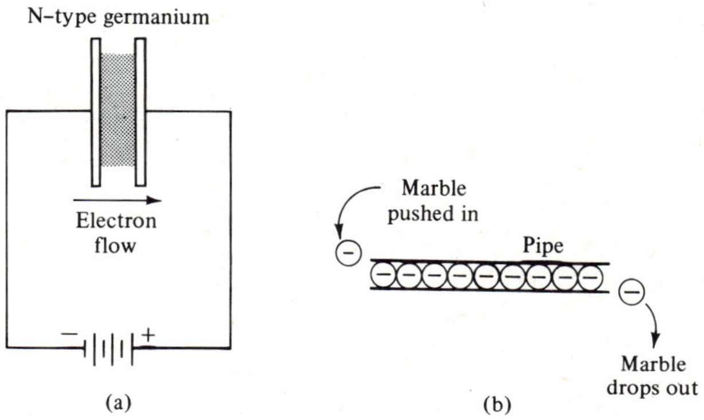

Fig. 14-9
Conduction in N-type germanium. (a) Basic circuit. (b) Marble-and-pipe comparison.

current increases if we increase the temperature of the semiconductor material because more electrons are forced loose from their atoms.

Now let us replace a germanium atom in a pure germanium crystal with an atom of boron, as shown in Fig. 14-10. Notice that the fit is not perfect, because

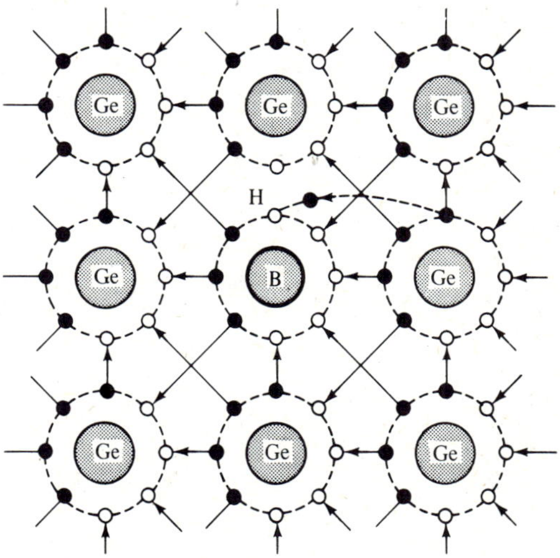

Fig. 14-10
Germanium doped with boron, showing a hole, H.

the boron atom has a missing outer electron. Therefore this missing electron, or hole, is like a floating positive charge in the crystal. This hole can move easily from one end of the crystal to the other end if a voltage is applied across the crystal. This type of semiconductor is called *P-type germanium.* P-type means that the crystal has free positive charges, or holes, as the current carriers. When a voltage is connected across a piece of P-type germanium, holes flow in much the same way as marbles pushed into a pipe, as shown in Fig. 14-11.

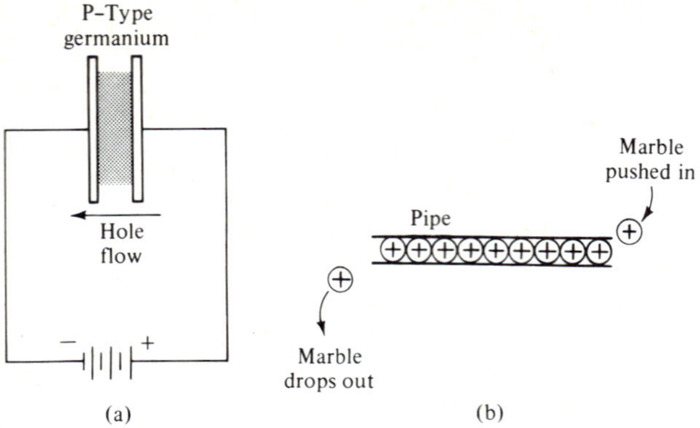

Fig. 14-11
Conduction in P-type germanium. (a) Basic circuit. (b) Marble-and-pipe comparison.

14.3 Principle of a Junction

A *junction diode,* or *rectifier,* consists of a layer of N-type semiconductor and a layer of P-type semiconductor, as shown in Fig. 14-12. That is, the N-type crys-

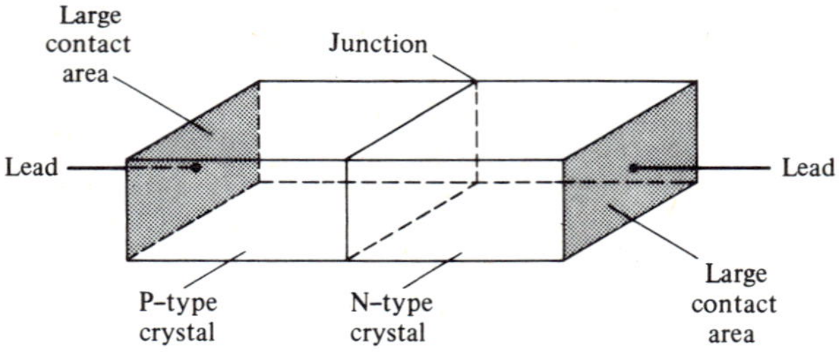

Fig. 14-12
Plan of a semiconductor diode.

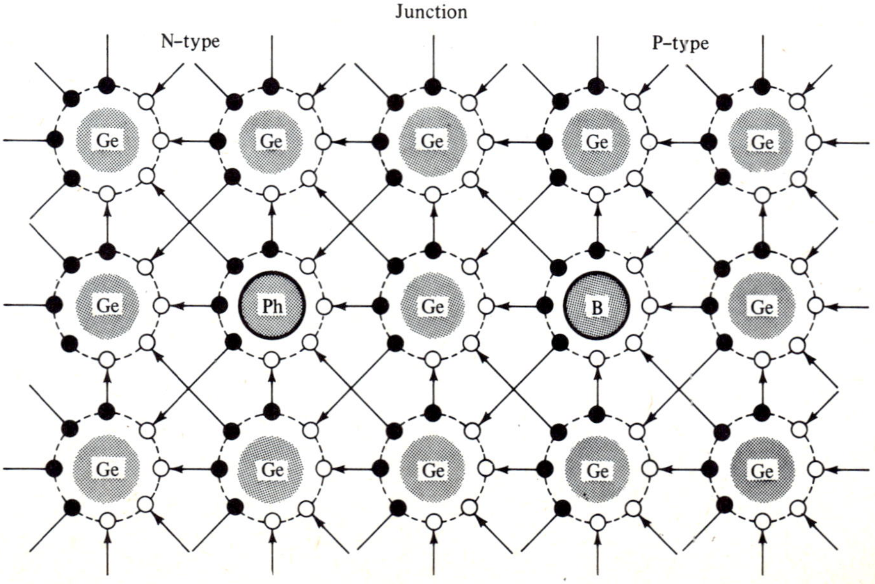

Fig. 14-13
Atomic arrangement in a junction diode.

tal and the P-type crystal meet at a junction. In manufacturing a semiconductor diode, the junction is continuous, and there is no space or break at the junction. The junction is simply a place at which the N-type material changes into P-type material.

The atomic arrangement in a junction diode is shown in Fig. 14-13. Notice that there is a phosphorus atom on the left-hand side of the junction, and there is a boron atom on the right-hand side of the junction. We also see that the extra electron from the phosphorus atom has moved into the outer shell of the boron atom. This is the condition of the crystal when there is no voltage applied across it.

Now, let us follow the *forward conduction* of electricity through a junction diode. In Fig. 14-14, a battery is connected to a diode; the negative polarity is applied to the N-type material, and the positive polarity is applied to the P-type material. This causes current flow through the diode. Let us consider the path

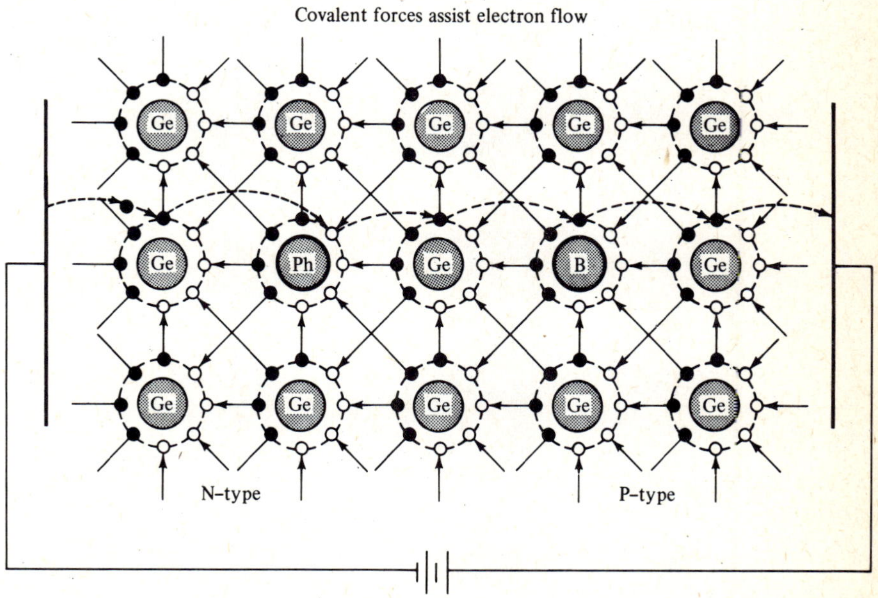

Fig. 14-14
Forward conduction through a junction diode.

of several current carriers through the diode. In the forward-conduction action, the negative pole of the battery supplies an electron to the N-type material. The electron is passed along to the phosphorus atom, which normally has five outer electrons. At the same time, the battery takes an electron from the P-type material. This electron is provided by the boron atom, which normally has three outer electrons. Finally, the boron atom takes a fourth electron provided by the phosphorus atom because of the covalent forces in the crystal arrangement. This process is repeated as long as the battery voltage is applied across the diode.

A very small amount of voltage across a junction diode causes a large forward current. This is another way of saying that the resistance of a diode biased in the forward direction is near zero.

Now let us follow the *reverse conduction* of electricity through a junction diode. Suppose that the battery polarity is reversed in Fig. 14-14. It is then, very difficult to make electrons flow backward across the junction. This is true because the boron atom would have to give up its fourth electron to the phosphorus atom. We know that the covalent forces in the crystal arrangement push an electron from the phosphorus atom to the boron atom. Therefore, we have to work against these covalent forces when we try to move an electron from the phosphorus atom to the boron atom. This is just another way of saying that there is very little current flow in reverse conduction of a diode.

We find that the reverse-current flow in a diode is about the same as the saturation current flow in a pure semiconductor. That is, the reverse-current flow is due only to electrons that happen to be knocked out of place by heat or light. Since comparatively few electrons are knocked out of place under ordinary conditions, it follows that reverse-current flow is very small. This is another way of saying that the resistance of a diode biased in the reverse direction is very great. The diode acts as a one-way switch, passing current in one direction and blocking current in the other direction.

Instructor Demonstration or Student Experiment

Connect a DC voltmeter to the output terminals of an ohmmeter to determine the polarity of the voltage in the ohmmeter (Fig. 14-15). Be sure that

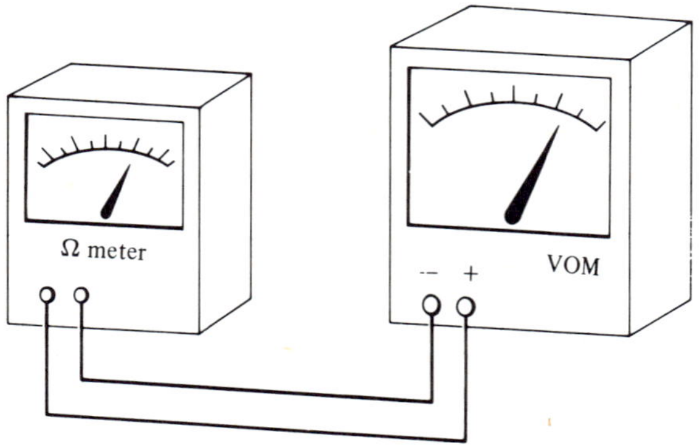

Fig. 14-15
Checking the polarity of the voltage out of an ohmmeter.

you label the leads of the ohmmeter. Now connect the anode of a junction diode with the positive lead of the ohmmeter and the cathode of the diode to the negative lead of the ohmmeter (see Fig. 14-16 [a]). Record the forward resistance of the diode.

Now reverse the leads of the ohmmeter so that the positive lead is connected to the cathode of the diode and the negative lead is connected to the anode of the diode. Record the reverse or back resistance of the diode. With these two values of resistance, determine the front-to-back ratio of the diode.

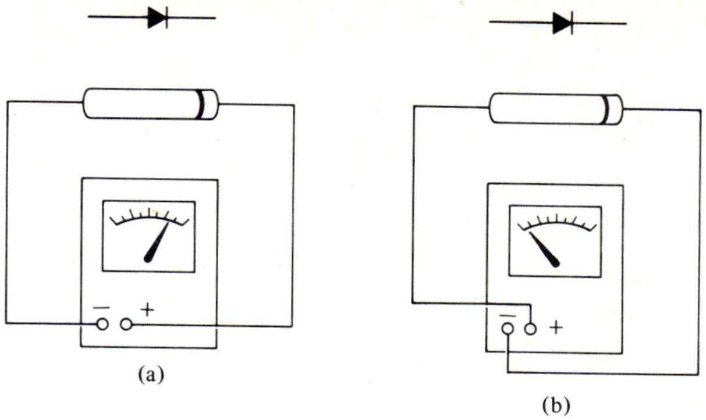

Fig. 14-16
Testing the resistance of a diode with a VOM. (a) Forward resistance. (b) Reverse resistance.

14.4 Principles of Transistor Action

The basic transistor consists of a semiconductor crystal with two *diode junctions,* as shown in Fig. 14-17. Notice that a transistor can be regarded as a slab of semi-

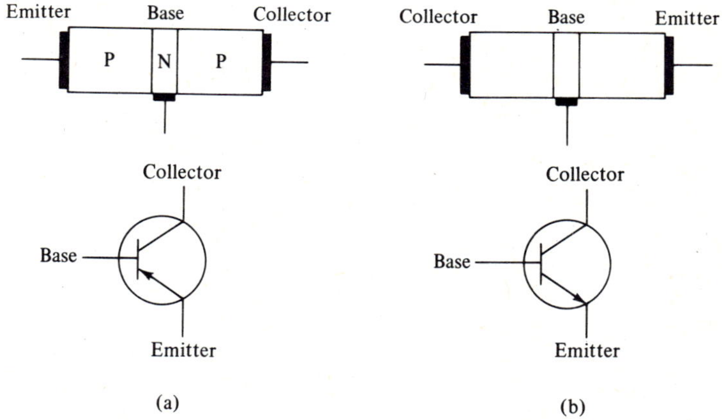

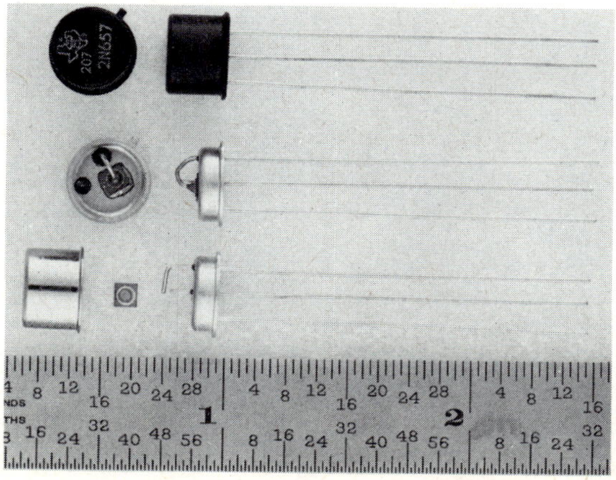

(c)

Fig. 14-17
Basic transistor arrangements and symbols. (a) PNP transistor. (b) NPN transistor. (c) Photograph of a transistor. Courtesy: General Electric Company

conductor material sandwiched between two blocks of opposite-type material. Thus, we have *PNP* and *NPN* arrangements. We will find that a PNP transistor has the same action as an NPN transistor, and that the only difference is in the polarity of the battery voltages in the circuit. Notice that a transistor has a junction between its emitter and base, and another junction between its base and collector. These are called the *emitter junction* and the *collector junction.*

In most transistor amplifier circuits, the emitter junction is *forward-biased,* and the collector junction is *reverse-biased.* The method of forward and reverse biasing is shown in Fig. 14-18. When a junction is forward-biased, the polarity

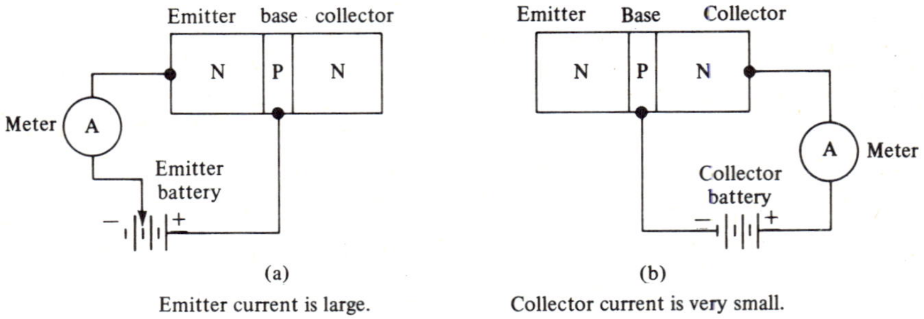

Fig. 14-18
The transistor appears as two diodes. (a) Current flow from emitter to base. (b) Current flow from base to collector.

of the battery voltage is such that it causes current flow through the junction. When a junction is reverse-biased, the polarity of the battery voltage is such that it causes practically no current flow through the junction. However, we find a very small flow of saturation current through a reverse-biased junction. We note that an increase of battery voltage across a forward-biased junction results in the forward current increasing. On the other hand, if the battery voltage is increased across a reverse-biased junction, the very small saturation-current flow does not increase.

Combining the two diode circuits in Fig. 14-18 to give the transistor circuit shown in Fig. 14-19, we discover the surprising fact that there is very little current in the base lead, although the current in the emitter is large. This is because most

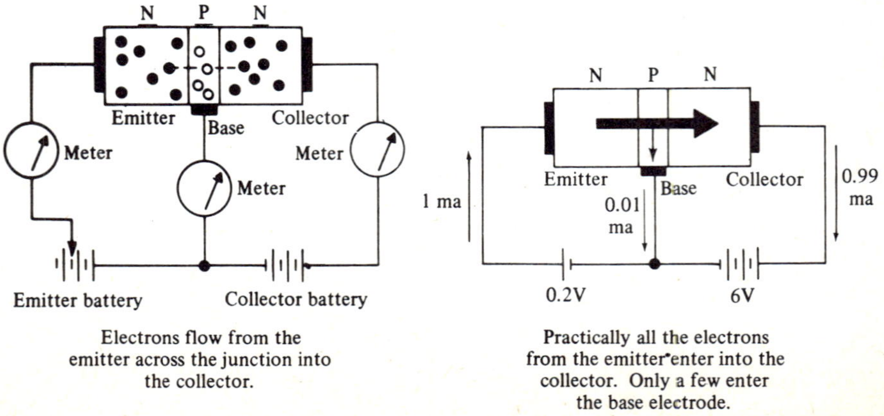

Fig. 14-19
(a) Electron-current flow in an NPN transistor; (b) Basic common-base circuit arrangement.

of the large current drawn by the emitter passes through the base into the collector. We might also note that if the emitter voltage is increased, the emitter current increases and the collector current increases. However, there is very little change in the small base current. Although the current in the base lead is very small, it is important in the circuit action. That is, if we disconnect the base lead, the large current into the emitter and out of the collector drops to zero.

We notice in Fig. 14-19(b) that the emitter battery has a low voltage and the collector battery has a high voltage. We also see that a low voltage change in the emitter circuit can cause large voltage in the collector circuit. This is the basic principle of the *amplifier,* or step-up action in a transistor. At this point let us list the important facts in the amplifier action.

1. In the circuit in Fig. 14-19, the collector battery has a much higher voltage than the emitter battery.
2. If the emitter circuit is opened, no current will flow in the collector circuit. The collector circuit is automatically opened.
3. If the emitter circuit is closed, a large current flows in the collector circuit. In other words, the collector circuit is automatically closed.
4. When there is low voltage change in the emitter circuit, a high voltage change is automatically developed in the collector circuit.

These reactions are the result of amplifier action, and this type of amplifier is called a voltage amplifier.

Figure 14-20 shows the battery polarities that are used in a PNP transistor circuit. Carriers in the form of holes flow into the emitter at the positive terminal,

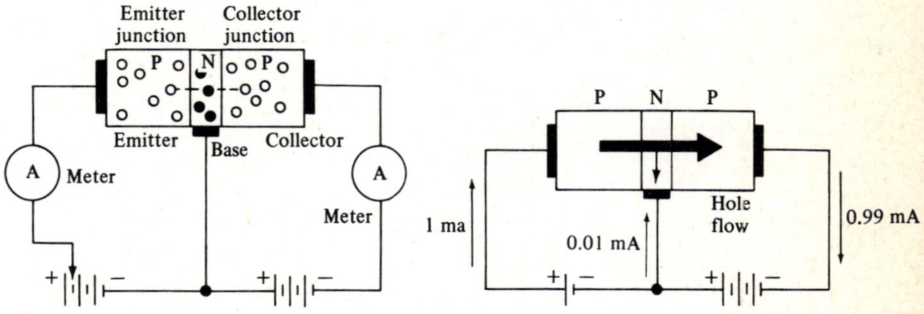

(a) Holes flow from the emitter across the junctions into the collector.

(b) Practically all the electrons from the emitter enter into the collector. Only a few enter the base electrode.

Fig. 14-20
Hole-current flow in a PNP transistor; common-base connection.

and a forward current of holes flows through the emitter junction. Very few of these holes go to the base electrode. Instead, the holes pass directly into the collector. Electrons flow into the collector and base from the external circuit to combine with the holes. Electrons flow out of the emitter to the external circuit. The circuit action is basically the same as for an NPN transistor. That is, a large voltage change in the collector circuit can be caused by a small voltage change in the emitter circuit. We also note that an increase in the emitter current causes an increase in the collector current.

14.5 Basic Amplifier Arrangements

The arrangements shown in Figs. 14-19 and 14-20 are called *common-base* circuits. Another basic amplifier arrangement, called the *common-emitter* circuit, is shown in Fig. 14-21. Notice that the difference between the common-base (CB)

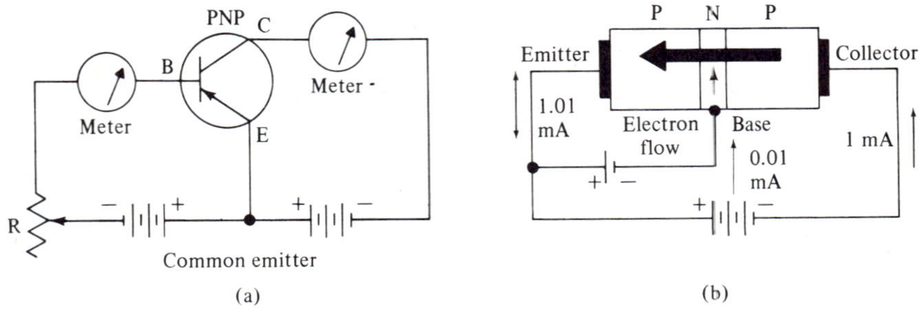

Fig. 14-21 Basic common-emitter arrangement. (a) The two batteries are connected together at the emitter electrode. (b) A small current in the emitter circuit produces a large current in the collector circuit.

circuit and the common-emitter (CE) circuit is in the electrode at which the batteries are connected together. That is, in Fig. 14-19, the batteries are connected together at the base electrode. For this reason, we call this circuit the common-base circuit. On the other hand, in Fig. 14-21, the batteries are connected together at the emitter electrode. Therefore, we call this circuit the common-emitter circuit. In the common-emitter circuit a small voltage change in the base circuit causes a large voltage change in the collector circuit. We also notice that a small current change in the base circuit causes a large current change in the collector circuit. So the common-emitter circuit is a *current amplifier* as well as a *voltage amplifier.* The CE arrangement is the most widely used transistor circuit in modern electronic equipment because it amplifies both current and voltage.

The third basic transistor arrangement, called the *common-collector* (CC) circuit, is shown in Fig. 14-22. Notice that the batteries are connected together

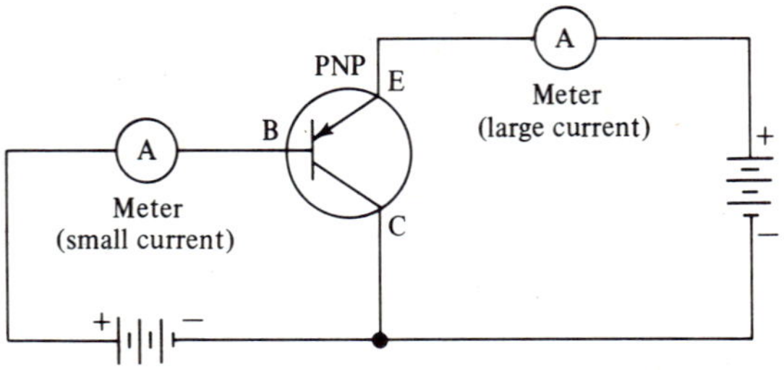

Fig. 14-22 Common-collector (CC) arrangement.

at the collector electrode. In this circuit, current flows in the emitter at all times. However, a small current change in the base circuit produces a large current change in the emitter circuit. Therefore, the common-collector circuit is a current amplifier. Notice that the CC circuit is not a voltage amplifier. In other words, if we change the voltage at the base, there is a smaller voltage change in the emitter circuit. Hence, the CC circuit can be used only as a current amplifier.

Instructor Demonstration or Student Experiment:

Connect the circuit shown in Fig. 14-23 with R_2 set to zero resistance. Slowly increase the value of R_2 until the ammeter I_C reads 5 mA. Record the values of I_C, I_B, and I_E. Now change I_B by 10 μA each side of its recorded value and record the values of I_C and I_E in each case. The current gain of the circuit is equal to the change of the collector current divided by the change of the base current (20 μA).

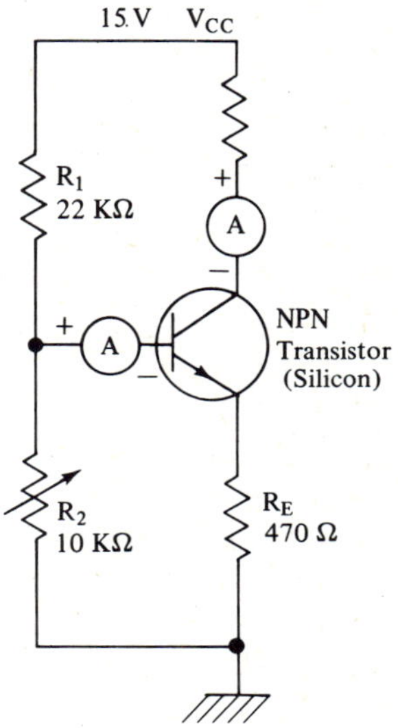

Fig. 14-23
Demonstration amplifier circuit using a common-emitter amplifier.

Questions.

1. What was the value of the current gain of the circuit?
2. How did the value of the collector current compare to the value of the emitter current during each step?
3. How did the value of the base current compare to the value of the collector current in each step?

Questions: Chapter 14

1. What is the most important application of semiconductor materials?
2. Name the two most often used semiconductor materials.
3. Define the term "saturation current."
4. Why are there few free electrons in a piece of pure semiconductor material?

5. State your definition of hole flow as it applies to semiconductor materials.
6. Define the term "impurity" as it applies to semiconductor materials.
7. How does a doping substance differ from a semiconductor material?
8. Explain how N-type germanium is formed.
9. How is P-type germanium formed?
10. What is the junction in a junction diode?
11. Draw a diagram of a junction diode.
12. Draw a diagram of a resistor, battery, and a diode with the diode connected in the forward direction.
13. Draw a diagram of a resistor, battery, and diode connected in the reverse direction.
14. Draw a diagram of an NPN transistor and label each lead.
15. Define the terms "forward bias" and "reverse bias."
16. Draw an amplifier with a transistor connected in the common-base arrangement.
17. Draw an amplifier with the transistor connected in the common-emitter arrangement.
18. Draw an amplifier with the transistor connected in the common-collector arrangement.

True-False:

1. The resistance of a piece of pure semiconductor material is very low.
2. Doping a pure semiconductor with an impurity that has five electrons in its valence shell forms a P-type semiconductor.
3. An N-type semiconductor is formed by doping a pure semiconductor with an impurity that has five valence electrons.
4. A junction diode is formed from N-type and P-type semiconductor materials.
5. A diode has a low resistance to current flow when biased in the forward direction.
6. Very little current flows in a reverse-biased diode.
7. A transistor may be formed with NPN material sandwiched together.
8. A junction diode is made by placing a piece of N-type semiconductor material against a piece of P-type semiconductor material.
9. A diode can be used as a rectifier to change AC current into DC current.
10. In the common-emitter transistor amplifier, the emitter is usually connected to ground.
11. In a CE amplifier, a small change in base current causes a large change in collector current.
12. The CE amplifier is a current as well as a voltage amplifier.

Multiple choice: choose the best answer.

1. The two most often used semiconductor materials are
   a. silicon and carbon
   b. holes and electrons
   c. silicon and germanium
   d. N and P
2. To form an N-type semiconductor, we add material with a valence of ____ .
   a. 3
   b. 4
   c. 5
   d. 6
3. The materials added to a pure semiconductor are called
   a. dope
   b. melts
   c. additives
   d. impurities
4. At a very low temperature, a pure semiconductor material would be a
   a. high resistance
   b. good conductor
   c. diode
   d. impurity
5. In P-type semiconductor material, current is primarily by
   a. electron flow
   b. hole flow
   c. heat
   d. light
6. A common-emitter amplifier is
   a. a current amplifier
   b. a voltage amplifier
   c. both a current and a voltage amplifier
   d. neither a current nor a voltage amplifier
7. In the three basic transistor amplifier connections, we will find that the input is ________ biased and the output is ________ biased.
   a. reverse, forward
   b. forward, reverse
   c. reverse, reverse
   d. forward, forward

15

BASIC PRINTED CIRCUITS AND INTEGRATED CIRCUITS

15.1 How Printed Circuits Are Made

In former years, all circuits were connected with copper wire from one component to another. This is called *point-to-point* wiring. However, during the 1940s a new wiring system called a *printed circuit* was used in special equipment. The use of printed circuits grew rapidly, and this method of connection is now preferred by manufacturers of radio, television, and high-fidelity equipment. A large percentage of military electronics equipment uses printed-circuit boards to mount and wire components together. A printed-circuit board consists of metal conduction strips on a thin insulating board. The appearance of printed wiring is shown in Fig. 15-1.

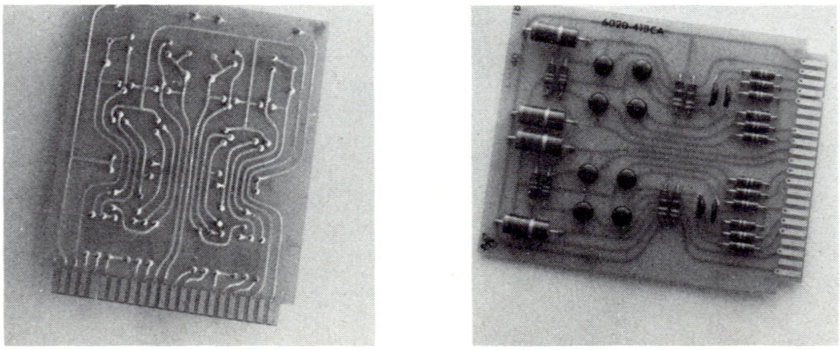

Fig. 15-1
Appearance of printed wiring.

After a printed-circuit board, or PC board, has been formed, the components are added, and the connections are all made at the same time by dipping the underside of the PC board into molten solder or by hand soldering. Since the components are often mounted on the board by automatic machinery, the cost of manufacture is much less than with ordinary copper-wire connections. Also, the amount of space necessary for a given number of components is considerably less because less space is required for connecting wires and terminals.

After the PC board is etched, holes are drilled for mounting components. After the component leads have been inserted through the holes, all the connections can then be soldered in a single dipping process. Figure 15-2 shows the appearance of a printed circuit with some parts mounted. Notice that each component has its leads terminated on the underside of the board. These leads are spaced to match the holes that have been placed in the PC board for them. Therefore, each component can be dropped into place, and its terminals are then ready to be soldered to the conducting strips on the underside of the board. In the laboratory, or in small production quantities, the components are soldered to the board using a low-wattage pencil-type soldering iron. Details of printed-circuit soldering are covered in the Appendix, page 256.

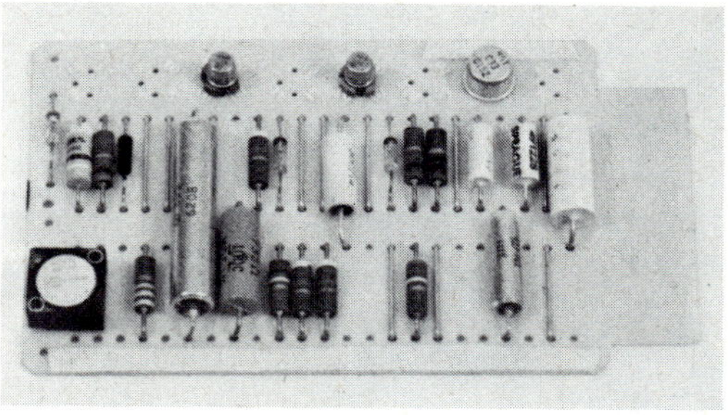

Fig. 15-2
Printed circuit board with some parts mounted.

15.2 Packaged Circuits and Integrated Circuits

Packaged circuits and integrated circuits are small-sized assemblies of electronic parts contained in an insulated case with several terminal leads. *Packaged circuits* usually contain resistors and capacitors connected together in various ways. A packaged circuit may also contain one or more semiconductor diodes. However,

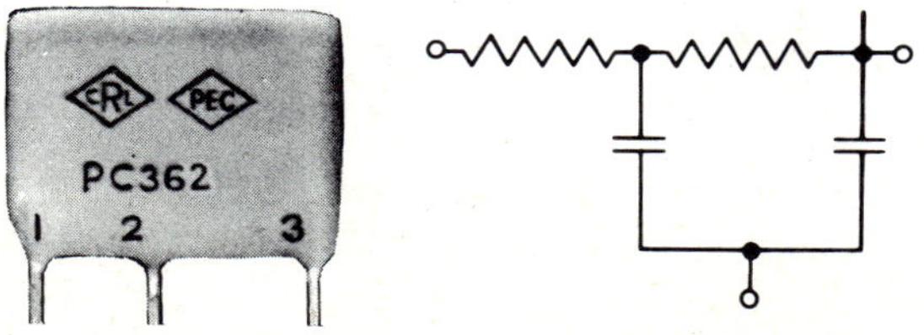

(a) (b)

Fig. 15-3
A packaged electronic circuit. (a) Appearance. (b) Circuit.
Courtesy: Centralab

a packaged circuit does not contain transistors. An assembly of electronic parts which contains one or more transistors is called an *integrated circuit.* Both packaged and integrated circuits are made in slab, block, and round shapes. Figure 15-3 shows the appearance of a packaged circuit. Figure 15-4 shows typical integrated circuits.

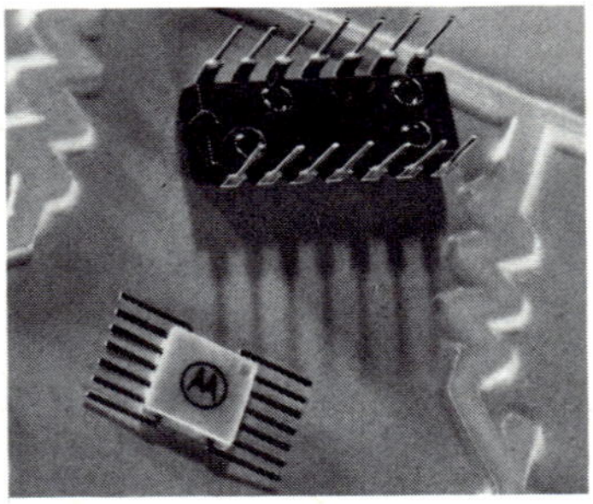

Fig. 15-4
Integrated-circuit packages.
Courtesy: Motorola

Manufacture of an integrated circuit starts with a single wafer of N-type or P-type silicon, as shown in Fig. 15-5. Next, a doping substance is *diffused* into

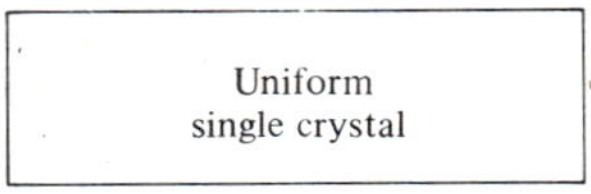

Fig. 15-5
Silicon wafer used as starting material for an integrated circuit.

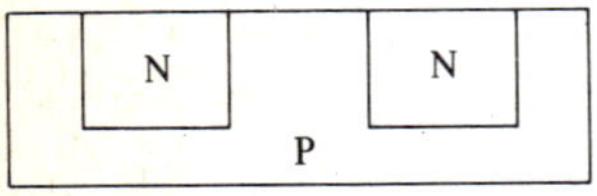

Fig. 15-6
Diffusion of N-type areas into P-type material.

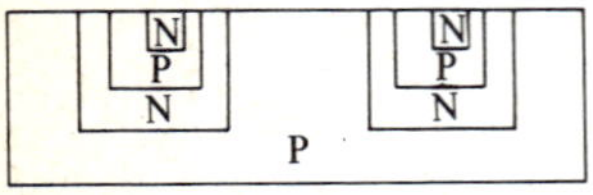

Fig. 15-7
Diffusion of additional P-type and N-type regions to form transistors.

the silicon to produce diode junctions at certain places. For example, Fig. 15-6 shows two N-type areas which have been diffused into a P-type wafer or *substrate.* Each of these N-type areas forms a diode junction with the P-type substrate. Notice that if a voltage is applied between the N-type areas, no current will flow. In other words, one of the diode junctions will be reverse-biased. This is called electrical isolation of the N-type areas.

Transistors are formed in an integrated circuit by the diffusion of additional P-type and N-type regions, as shown in Fig. 15-7. Two transistors have been formed by producing a pair of NPN junctions. Each transistor can be operated separately, because the P-type substrate provides electrical isolation between the two transistors. Connections are made to the integrated transistors as shown in Fig. 15-8. The exposed surface of the integrated circuit is first coated with an

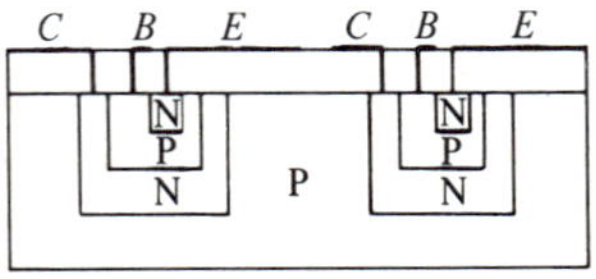

Fig. 15-8
Addition of metallized contacts to transistor elements.

insulating material, and the surface of this insulated coating is then opened at certain points. These openings are metallized to make connection to the emitter, base, and collector elements of the transistors.

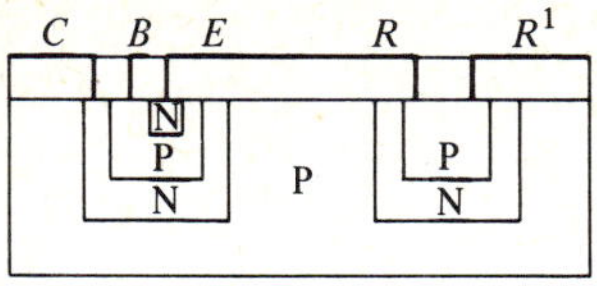

Fig. 15-9
Connection of contacts to P-type region to form integrated resistor.

Integrated resistors are formed as shown in Fig. 15-9. In this example, there is an integrated transistor on the left-hand side and an integrated resistor on the right-hand side. The integrated resistor is similar to a transistor, except that no emitter is diffused into the P-type region. Instead, two metallized contacts are made to the P-type region. In turn, R and R^1 represent the terminals of the integrated resistor. The amount of resistance between the terminals depends upon the area and the depth of the P-type region.

Integrated capacitors are formed as shown in Fig. 15-10. In this example, there is an integrated transistor on the left-hand side, and an integrated capacitor on the right-hand side. The capacitor is formed by an N-type region and a metallized area C^1. An insulating oxide is used between the N-type region and a metallized area to serve as a dielectric. Thus, C and C^1 represent the terminals of an integrated capacitor.

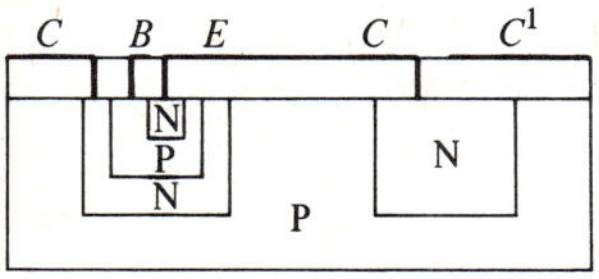

Fig. 15-10
Use of oxide as a dielectric to form integrated capacitor.

Let us suppose that an integrated circuit is to be formed which consists of an NPN transistor, a resistor, and a capacitor. The resistor and capacitor are to be connected in series, and the emitter of the transistor is to be connected to the resistor. In turn, the integrated circuit arrangement will be as shown in Fig. 15-11.

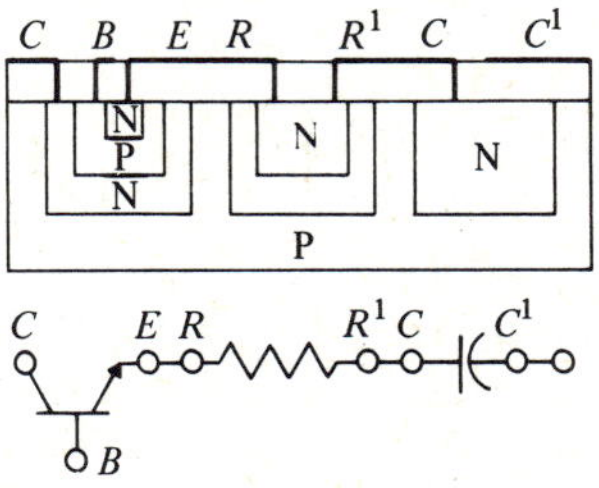

Fig. 15-11
Completed silicon chip containing transistor, resistor, and capacitor.

The integrated transistor is at the left-hand side, the integrated resistor is in the middle, and the integrated capacitor is at the right-hand side. The series connections are made by the metallized contacts between E and R, and between R^1 and C. The metallized contacts C, B, and C^1 represent the points of connection for the external leads of the integrated circuit.

An important advantage of an integrated circuit is its very small size compared with that of ordinary chassis construction or of printed-circuit construction. This small size saves both space and production cost. Technicians often call an integrated circuit an *IC* or a chip.

Questions: Chapter 15

1. What are the advantages of printed wiring?
2. List three methods used to produce printed wiring.
3. How are holes usually made in a printed-circuit board?
4. On which side of a printed-circuit board should you mount the components?
5. Why must solder never be held with the teeth?
6. What is an integrated circuit?

7. How is an integrated circuit formed?
8. What are the advantages of an integrated circuit?

True–False:

1. Printed wiring is popular because it makes circuit assembly faster and more economical than other methods.
2. A printed wiring board is made up of a metal board in which metal conductors are bonded.
3. The mounting holes in a printed-circuit board are drilled to the size of the leads on the components that are to be mounted.
4. A heat sink should be used whenever possible when soldering semiconductors.
5. Power transistors are usually mounted on heat sinks.
6. An integrated circuit may consist of capacitors, resistors, transistors, and diodes.
7. Packaged circuits contain transistors.

Multiple-choice: choose the best answer.

1. The reason for the rapid expansion of the use of printed wiring is that it
   a. looks better than point-to-point wiring
   b. is easier to repair than point-to-point wiring
   c. is easier to assemble units made with printed wiring
   d. is easier for technicians to make repairs on printed circuits.
2. Printed circuits in comparison to ordinary chassis assemblies are
   a. more likely to be damaged by heat
   b. more economical to build
   c. more compact
   d. all of the above

16

ELECTRONIC POWER SUPPLIES

16.1 How a Power Supply Works

Most electronic power supplies change alternating current into direct current. For example, a power supply in a television receiver changes the 117-volt 60-Hertz house current into direct current. Thus, a power supply is used in the same way that batteries are used in electronics equipment. A power supply takes the place of a battery. All electronic power supplies use *rectifiers* to change alternating current into direct current. A rectifier is basically a diode. Rectifier diodes (Fig. 16-1) are made from selenium, germanium, silicon, and other semiconductor substances.

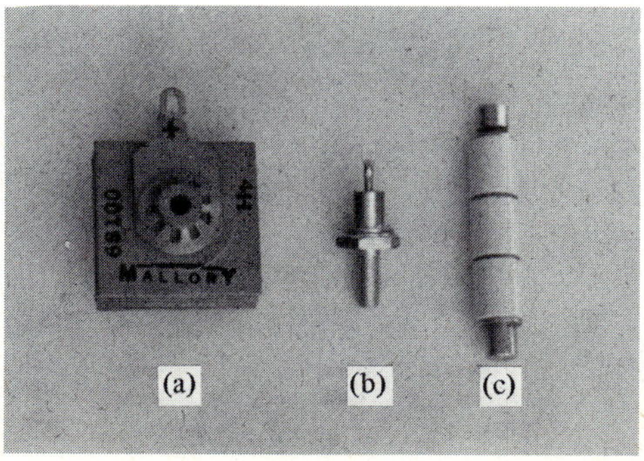

Fig. 16-1
Typical rectifier diodes. (a) Selenium. (b) Silicon. (c) Germanium.
Courtesy: International Rectifier Corp.

Figure 16-2 shows the basic operation of a rectifier. The AC input voltage

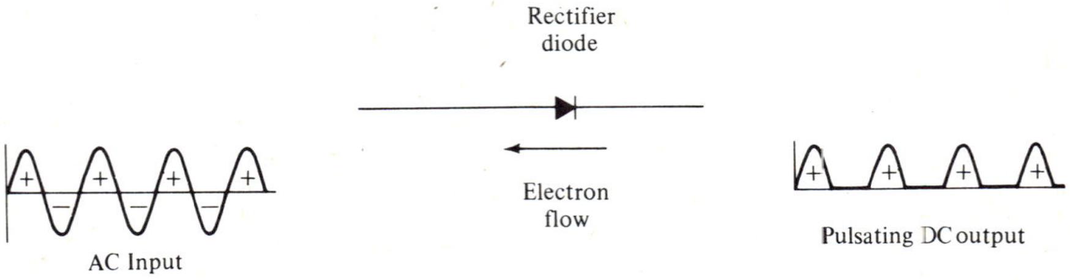

Fig. 16-2
Operation of a rectifier.

goes through positive and negative half cycles. However, current (electrons) can flow through the rectifier diode in one direction only. Therefore, the rectifier output consists of half cycles that have positive polarity. This is called a *pulsating-DC* output because the output voltage rises and falls. Before the pulsating-DC output can be used by a radio receiver, for example, the output voltage must be smoothed or *filtered* so that it becomes practically pure DC. Let us consider this filter action.

As shown in Fig. 16-3, a basic filter consists of a capacitor connected in series with the rectifier diode. Current is drawn from the filter capacitor by a load re-

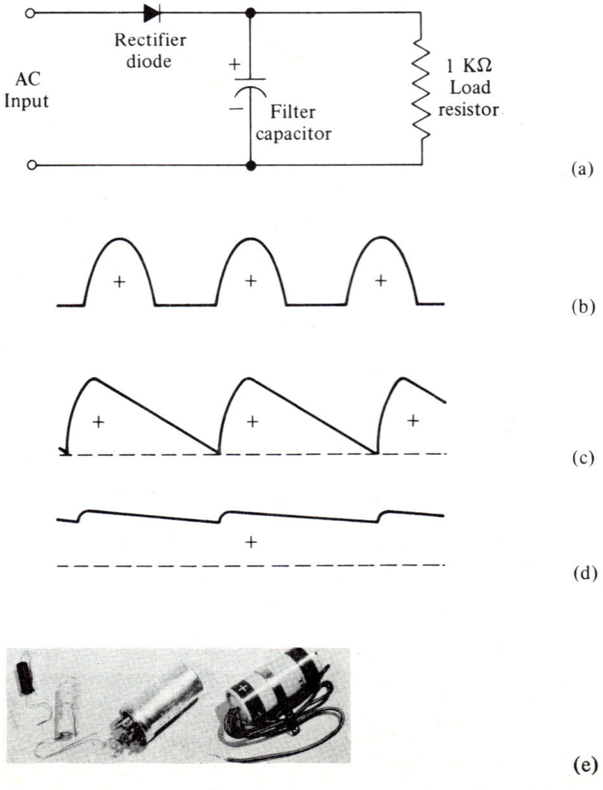

Fig. 16-3
Basic filter action. (a) Filter circuit. (b) Output voltage, very small filter capacitance. (c) Output voltage, medium filter capacitance. (d) Output large filter capacitor. (e) Typical electrolytic filter capacitors.
Courtesy: Sprague Electric Company

sistor. We find that the smoothness of the voltage across the load resistor depends on how much filter capacitance is used. When the filter capacitance is very small, the voltage across the load resistor will have the rise and fall shown in Fig. 16-3(b). That is, the voltage output will consist of positive half cycles. This poor filtering action is the result of the very small charge that is taken by a small capacitance. In turn, it is almost as if there were no filter capacitance in the circuit.

Next, Fig. 16-3(c) shows the output voltage that is produced when the filter capacitance is large enough to have a noticeable effect. In this example, there is enough capacitance to take a fair amount of charge each time that the rectifier conducts. This charge is large enough to slow up the voltage fall between the half cycles of conduction by the rectifier. However, there is not enough filter capacitance in this example to provide good filtering action. That is, this filter does not produce smooth DC. This rise and fall is called *ripple voltage.*

Notice in Fig. 16-3(d) that a large filter capacitance produces an output voltage that is more nearly like pure DC. That is, there is enough capacitance in this example to take a considerable amount of charge. Because this heavy charge acts as a reservoir or storage "tank" for electricity, only some of its charge is drained off between half cycles of conduction by the rectifier. In other words, the voltage fall between the times of conduction is slowed up to such an extent that the output voltage looks almost like pure DC. If a very large filter capacitance were used, we can see that the output voltage would be practically pure DC; the ripple voltage would be practically zero.

16.2 Full-Wave Rectification

Less filter capacitance is required if the capacitor is charged up twice as often. As shown in Fig. 16-4, *full-wave rectification* produces half cycles which follow each other without any waiting time between them. Notice that rectifier 1 provides a half-wave output, and that rectifier 2 also provides a half-wave output. These outputs are combined in the load resistor, and the output from rectifier 2 fills in the waiting time of the output from rectifier 1. In other words, rectifier 2 conducts during the time that rectifier 1 is not conducting.

Figure 16-5 shows the filter action that is produced by a capacitor connected across the load resistor. The basic filtering action is the same as in the case of a half-wave filter. However, the filter capacitor in a full-wave circuit is being charged twice as often. The charge, therefore, has less time to drain off before the capacitor is charged up again. The practical result is that less filter capacitance is necessary in a full-wave power supply than in a half-wave power supply.

Next, let us consider the transformer in Fig. 16-5. This is called a *power transformer.* It may have a voltage step-up action in some power supplies, or it may have a voltage step-down action in other power supplies. Since the secondary of the power transformer is center-tapped, the charging voltage from each rectifier diode is the same. That is, the voltage from (1) to the center tap is the same as the voltage from (2) to the center tap in Fig. 16-5. We see that the entire secondary voltage, from (1) to (2) must be twice as great as the DC output voltage from the power supply.

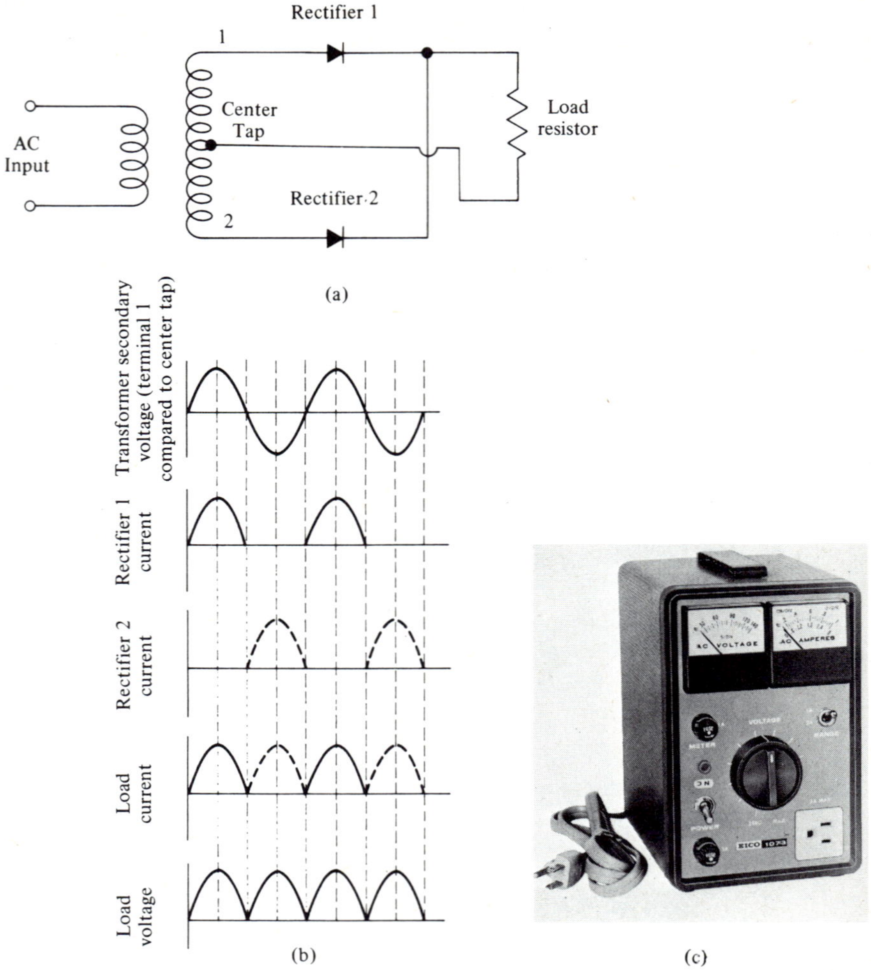

Fig. 16–4
Full-wave rectification. (a) Basic circuit. (b) Circuit voltages and currents. (c) Appearance of bench power supply.
Courtesy: EICO

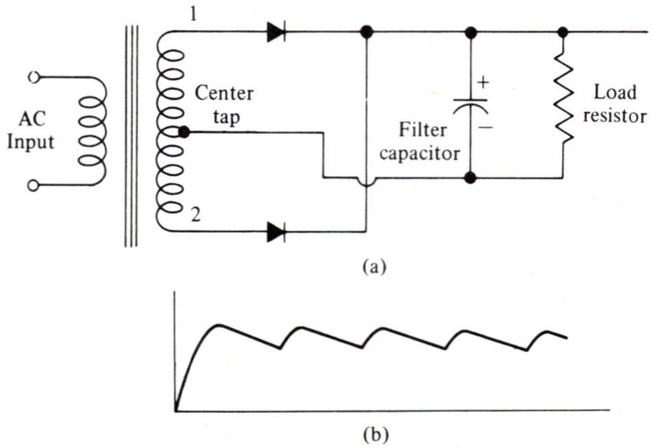

Fig. 16–5
Full-wave filter action. (a) Basic filter circuit. (b) Output voltage.

Instructor Demonstration or Student Experiment:

Connect the circuit in Fig. 16-3(a) to an audio generator that has an output of at least 12 volts at 60 Hertz. First, with an oscilloscope, observe the output across the load resistor with the filter capacitor removed. Add the capacitor to the circuit and observe the output with the oscilloscope. Repeat each step as above, but with a DC voltmeter connected in place of the oscilloscope. Finally, reverse the diode and repeat the above.

Questions:

1. What is the purpose of the filter capacitor?
2. What effect does adding the capacitor have on the DC voltage level of the output?
3. Why does the polarity of the output voltage change when the diode is reversed?

16.3 Full-Wave Bridge Rectification

In case the power transformer does not have a center-tapped secondary, full-wave rectification can be obtained if we use four rectifier diodes. This arrangement is shown in Fig. 16-6 and is called a *bridge rectifier.* Let us consider the half cycle

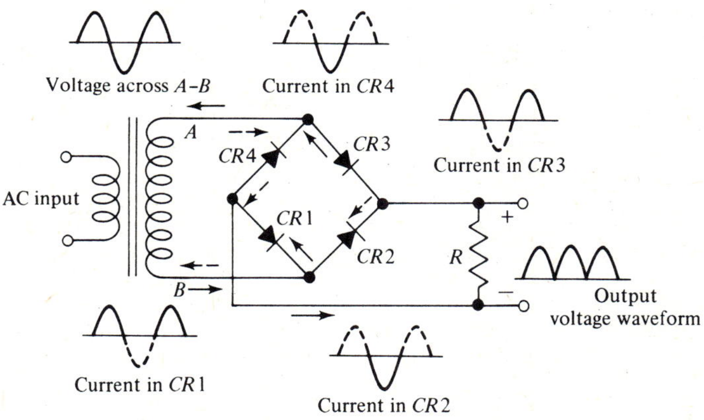

Fig. 16-6
Full-wave bridge rectifier circuit.

when terminal *A* is positive and terminal *B* is negative. In this case, rectifiers *CR*1 and *CR*3 conduct. That is, electrons flow upward through load resistance *R*. On the next half cycle, terminal *A* is negative and terminal *B* is positive. In turn, rectifiers *CR*2 and *CR*4 conduct. Again, electrons flow upward through load resistance *R*. The bridge-rectifier action is basically the same as that of a full-wave circuit with a center-tapped transformer. If a filter capacitor is connected across the load resistor *R* in Fig. 16-6, the pulsating DC output is smoothed in the same manner as shown in Fig. 16-5(b).

16.4 Resistance-Capacitance Filters

Resistance–capacitance, or *RC*, *filters* have better smoothing action than a single capacitor. The most common type of *RC* filter is called a *pi filter*. It consists of two filter capacitors and a resistor, connected as shown in Fig. 16-7. The filtering action becomes better if the resistance of *R* is increased. However, since there is also a loss of DC voltage across *R*, there is a practical limit to the amount of resistance that can be used. The amount of filter capacitance can be made as large as necessary. However, very large filter capacitors are comparatively expensive. Therefore, when an unusually well-filtered output voltage is required, we use another type of filter, called an *LC* filter.

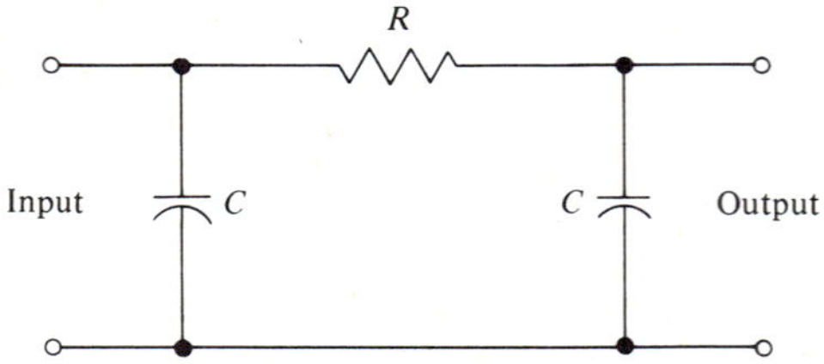

Fig. 16-7
RC pi-type filter.

16.5 Inductance-Capacitance Filters

Inductance–capacitance, or *LC*, filters are the most practical type to use when a power supply must provide considerable current with an unusually well-filtered output. The most common type of *LC* filter is a pi filter, shown in Fig. 16-8.

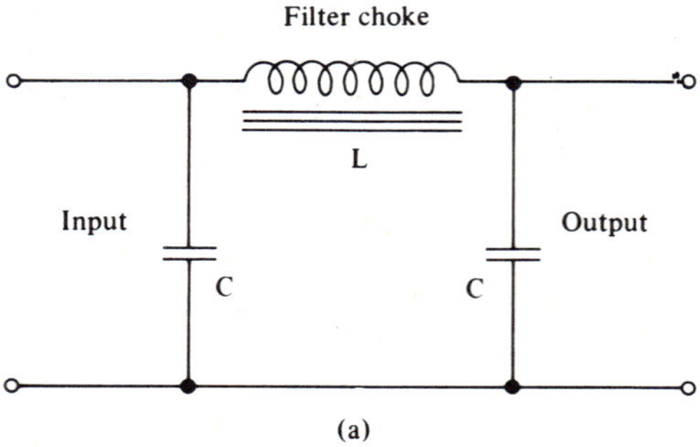

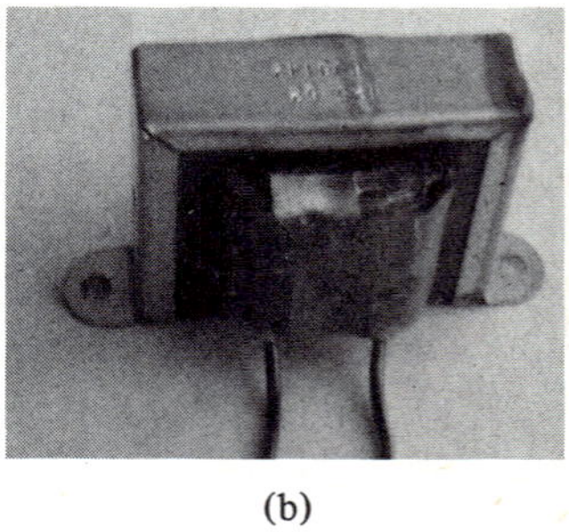

Fig. 16-8
LC pi-type filter. (a) Circuit diagram. (b) Typical filter choke.

Its filtering action becomes better if the inductance of *L* is increased. There is an advantage to using a filter choke instead of a filter resistor, because there is comparatively little loss of DC voltage across a choke. On the other hand, a filter choke costs considerably more than a power resistor.

Instructor Demonstration or Student Experiment:

Connect the circuit shown in Fig. 16-9, with the capacitors removed. Observe the load voltage with an oscilloscope. (Set to AC input). Observe the output with first C_1 connected, then only C_2 connected. Finally, connect both capacitors and measure the ripple voltage with the oscilloscope at the output.

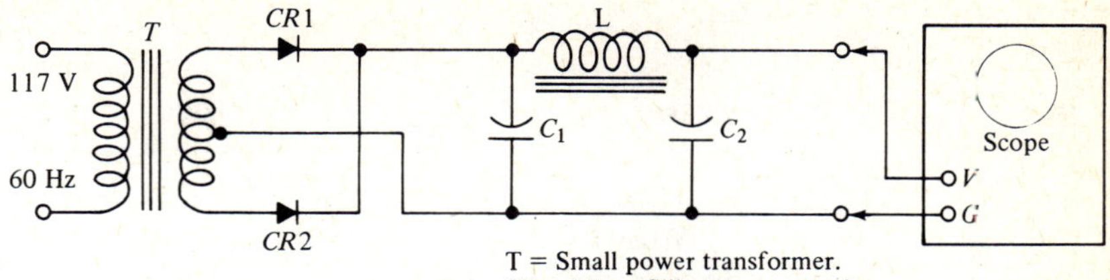

T = Small power transformer.
CR_1, CR_2 = Silicon or germanium rectifiers.
L = 1 Hy, approx.
C_1, C_2 = 40 μF, approx., rated for transformer output voltage.

Fig. 16-9
Demonstration full-wave power supply with pi-type LC filter.

1. How does the full-wave rectifier differ from the bridge rectifier?
2. Which of the capacitors had the most effect in reducing ripple voltage?

16.6 Power Supplies With Positive and Negative Output Voltages

Different types of electronics equipment require positive, negative, or sometimes both positive and negative supply voltages. Therefore, power-supply circuits are arranged to provide the necessary polarity or polarities, as well as the necessary voltage or voltages. For example, Fig. 16-10(a) shows how a half-wave power supply is connected to produce a positive output voltage. Notice that a ground

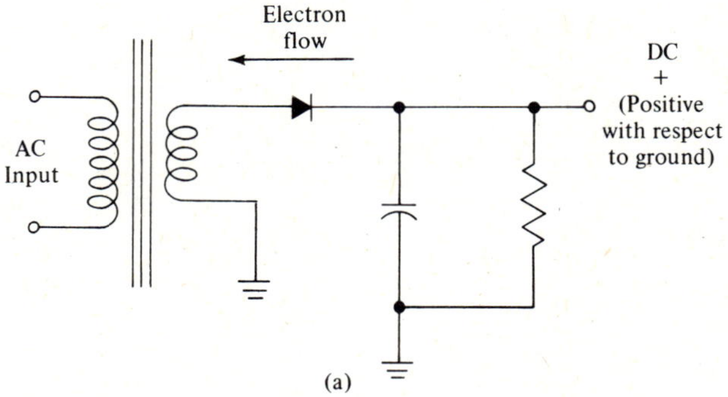

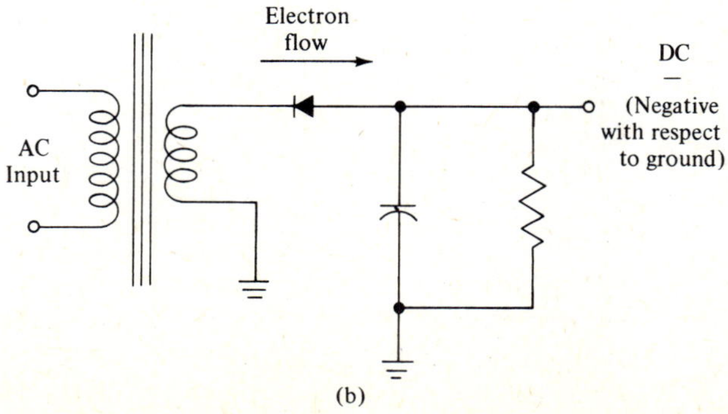

Fig. 16-10
Power supply arrangements for either polarity of output. (a) Positive output with respect to ground. (b) Negative output with respect to ground.

circuit is used in this example. The ground may be either a metal chassis to a common wire called a ground bus, or common ground. Next, Fig. 16-10(b) shows how the half-wave power supply is connected to produce a negative output voltage. When the rectifier polarity is reversed, the polarity of the output voltage is also reversed. Output polarity is stated with respect to ground.

When both positive and negative supply voltages are required, the arrangement shown in Fig. 16-11 is often used. Notice that this circuit is basically a

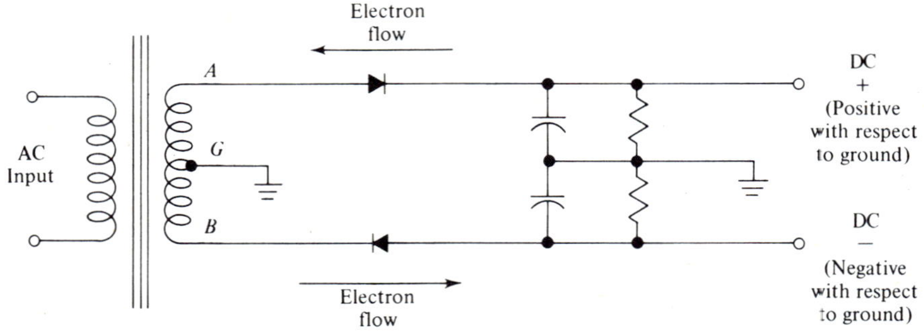

Fig. 16-11
Power supply with both positive and negative output polarities.

combination of the two circuits shown in Fig. 16-10. In many cases, different voltages are required at the positive and negative outputs. Therefore, in Fig. 16-11, the secondary of the power transformer will not be center tapped. That is, we might have more turns from A to B, than from B to G. In this case, the positive output voltage will be greater than the negative output voltage.

16.7 Voltage-Doubler Arrangement

An input AC voltage can be doubled at the DC output of a power supply without using a step-up transformer. If a voltage-doubler circuit is used, as shown in Fig. 16-12, the DC output voltage will be twice as great as the AC input voltage. Let

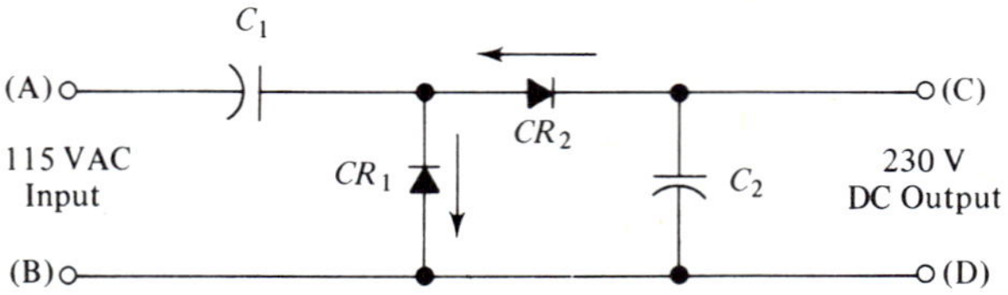

Fig. 16-12
A voltage-doubler circuit.

us see how this circuit works. If the AC voltage applies a negative polarity to (A) and a positive polarity to (B), rectifier $CR1$ conducts, and filter capacitor C_1 charges up to 115 volts with the polarity shown in the diagram. At the next moment, the AC voltage applies a positive polarity to (A) and a negative polarity to (B). Therefore, rectifier $CR2$ conducts, and C_2 charges up to 230 volts. This double charge is produced because C_1 is already charged, and is in series with the AC line and C_2.

The voltage–doubler arrangement is basically a half–wave power supply. However, it consists of two half–wave circuits. The second circuit is connected so that the second filter capacitor charges in series with the first filter capacitor. The practical result is the same as when two batteries are connected in series. In other words, the two voltages are added together. Voltage–doubler power supplies are often used in television receivers, and in other kinds of electronics equipment. The advantage of a voltage–doubler arrangement is that a step–up transformer is not used. Power transformers are comparatively expensive components. Figure 16–13 shows a complete circuit for a voltage–doubler power supply used in a portable TV receiver.

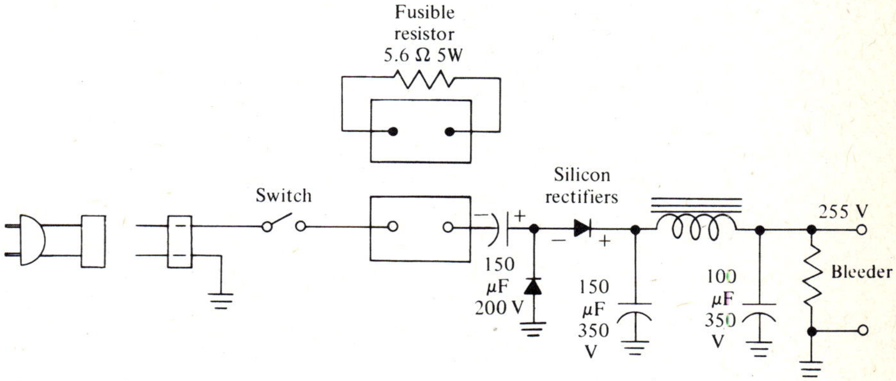

Fig. 16–13
Schematic of typical low-voltage supply used in portable TV receivers.

In Fig. 16–13, an off–on switch is connected in series with the power line. A fusible resistor is also connected in series with the switch. This resistor is a protective device. In case of an accidental short circuit in the receiver, the resistor will burn out before the silicon rectifiers are damaged. The resistor is a comparatively low–cost component. In this example, an *LC* pi filter is used with the voltage–doubler circuit. There are 255 volts DC across the 100–μF filter capacitor. Notice that the 3K resistor is not a load resistor; it is a dropping resistor. The bleeder resistor connected across the output terminals helps to stabilize the output voltage under conditions of load variation (change in current drain).

16.8 Voltage–Tripler Arrangement

Voltage triplers are used in some television receivers. To *triple* means to multiply three times. Figure 16–14 shows a typical circuit. Notice that *CR*1 and *CR*2 are in a voltage–doubler circuit such as explained previously. Therefore, filter capacitor C_2 is charged in series with C_1. In turn, C_2 is charged to double voltage. We see that, when C_1 charges up, C_3 will also charge up because *CR*3 conducts. However, since C_3 is connected in series with C_2, the output voltage will be three times the line voltage, or $3E$.

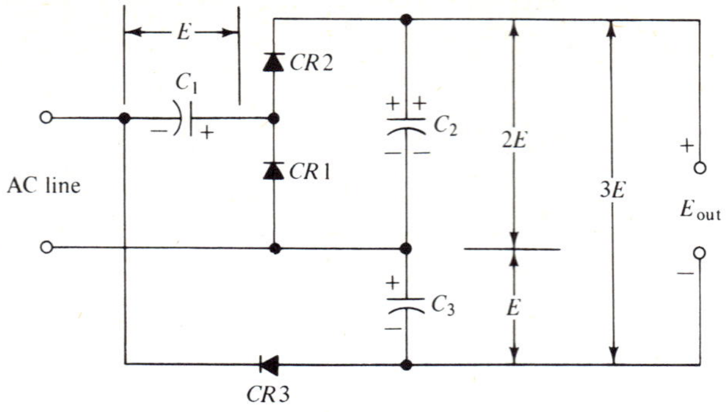

Fig. 16-14
Voltage tripler circuit.

The important facts to observe in this type of power supply are:

1. The first filter capacitor C_1 charges up to the line voltage E.
2. The second filter capacitor C_2 charges up to $2E$.
3. The third filter capacitor C_3 charges up to E.
4. Since C_2 and C_3 are connected in series, the total output voltage is $3E$.

16.9 Voltage-Quadrupler Arrangement

We will find voltage quadruplers used in some TV receivers, and other kinds of electronics equipment. To *quadruple* means to multiply four times. A typical circuit is shown in Fig. 16-15. Notice that this is basically an arrangement which

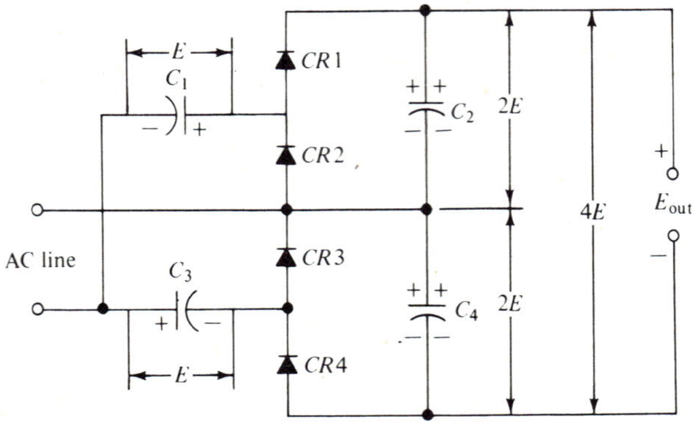

Fig. 16-15
Voltage quadrupler circuit.

uses two voltage doublers connected in series. That is, the first voltage doubler consists of $CR1$, $CR2$, C_1, and C_2. The second voltage doubler consists of $CR3$, $CR4$, C_3, and C_4. Filter capacitor C_2 is charged up to double voltage, and C_4 is also charged up to double voltage. Because C_2 and C_4 are connected in series, the total output voltage is four times the line voltage.

Questions: Chapter 16

1. What is the purpose of most electronic power supplies?
2. State the purpose of a rectifier.
3. What are three types of rectifiers?
4. Describe the output wave form of a rectifier.
5. Discuss the action of a filter.
6. Why does the output from a power supply contain a ripple voltage?
7. Explain the meaning of the term "ripple voltage" as it applies to a filter.
8. Draw a circuit of a full-wave rectifier.
9. Explain the difference between a half-wave and a full-wave rectifier.
10. What does step-up and step-down mean in reference to a transformer?
11. Draw a circuit of a pi filter.
12. How is the filtering action affected by a change of *R* or *C* in a pi filter?
13. What is the advantage and disadvantage of using an inductor in a pi filter instead of a resistor?
14. What is a voltage doubler?
15. How does a voltage-tripler circuit compare to a voltage-doubler circuit?

True-False:

1. Most power supplies convert DC current into AC current.
2. A power supply takes the place of a battery.
3. A rectifier is basically a diode.
4. Electric current flows easily through a diode from anode to cathode.
5. The output of a rectifier is called pulsating DC.
6. Before a rectified voltage is usable as a DC voltage, it must be filtered.
7. Filtering action removes the DC part of the rectifier output.
8. The small AC voltage that is left in the DC after filtering is called the ripple voltage.
9. The filter capacitor stores energy during the peak of the input and returns it to the circuit during the minimum of the input voltage.
10. The output ripple voltage of a well-filtered power supply is practically zero.
11. A small value of filter capacitance is required with a half-wave rectifier in comparison to a full-wave rectifier.
12. The full-wave bridge does not require a center-tapped transformer.
13. *RC* filters are more effective in reducing ripple voltage than *LC* filters.
14. *LC* filters cost more than *RC* filters.
15. A power supply may have both a positive and a negative output voltage.
16. When a voltage-doubler circuit is used, the output voltage is twice as large as the input voltage.

Multiple choice: choose the best answer.

1. All electronic power supplies use
   a. resistors

b. transistors
c. rectifiers
d. voltage doublers

2. The output of a rectifier circuit is called ____________ voltage.
a. AC
b. DC
c. pulsating DC
d. pulsating AC

3. In order for the output of a rectifier to be useful to power a radio, it must be
a. filtered
b. pulsated
c. amplified
d. doubled

4. The purpose of the capacitor in a filter circuit is to
a. double the voltage
b. rectify the ripple
c. act as a storage for the DC voltage
d. act as a rectifier

5. The output of the full-wave rectifier
a. is not useful as a DC voltage source
b. is easier to filter than the half-wave supply
c. does not require filtering
d. has more ripple voltage than a half-wave supply

6. The primary purpose of a power transformer is to
a. isolate the load from the AC line
b. step-up or step-down the line voltage
c. filter the output of a rectifier
d. rectify the line voltage

7. For best filtering action we should use a
a. single capacitor
b. *LC* pi filter
c. *RC* pi filter
d. capacitor and resistor

8. The advantage of using an inductor over a resistor in a filter is that inductors
a. cost less
b. take less space
c. have less DC resistance
d. weigh less

9. The voltage doubler is basically
a. a half-wave power supply
b. a full-wave power supply
c. a bridge rectifier
d. two half-wave power supplies

17 ELECTRONIC OSCILLATORS

17.1 Principles of Oscillation

An *oscillator* is basically a vibrator, and an *oscillation* is basically a vibration. We are familiar with the vibration of a guitar string when the string is plucked. This is an example of mechanical vibration. Electrical and electronic vibrations are usually called oscillations. A sine wave of voltage, such as shown in Fig. 17-1, is called an oscillation when it is produced by a transistor circuit. We will find that

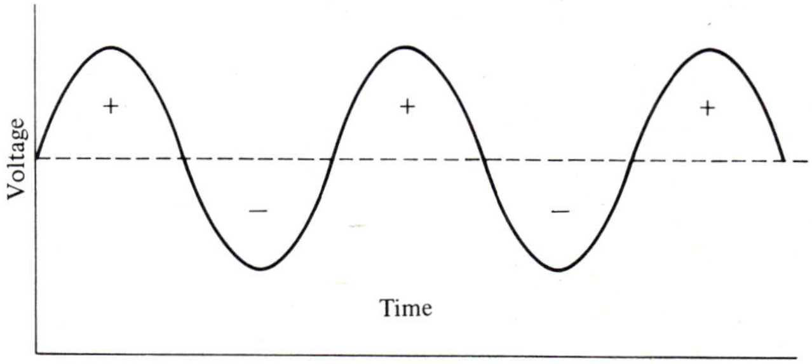

Fig. 17-1
A sine-wave oscillation.

oscillators are closely related to amplifiers. That is, an amplifier steps up an AC voltage—this AC voltage is applied to the amplifier from a separate source. On the other hand, an oscillator steps up an AC voltage which is fed back to its input terminals as shown in Fig. 17-2.

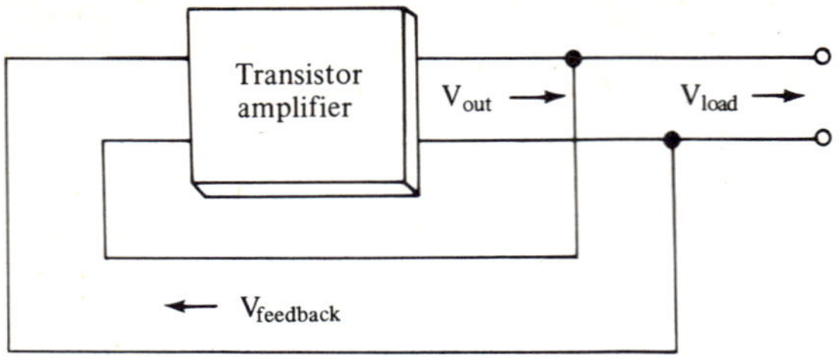

Fig. 17-2
Transistor oscillator showing feedback, block diagram.

Oscillators are in very wide use in electronics equipment. We are familiar with walkie-talkies, such as are shown in Fig. 17-3. The operation of a walkie-talkie is based on transistor oscillator circuits. All radio and television receivers use oscillators. Garage-door openers are basically oscillators. Tape recorders have oscillator circuits. An electronic organ is an assembly of dozens of oscillators. Since electronic oscillators are found in so many types of electronics equipment, it is important for us to learn how they work and to take note of the basic types that are in common use.

Fig. 17-3
A walkie-talkie.
Courtesy: Knight Kit

17.2 Resonant-Circuit Oscillators

The first difference that we find between electronic oscillators is that some produce a sine-wave output, whereas others produce outputs that are not sine waves. We will start our study with sine-wave oscillators. As shown in Fig. 17-4, these sine-wave oscillators make use of resonant circuits. The reason for this is that a resonant circuit responds to a single frequency. In other words, saying that a circuit is operated at one particular frequency, is just another way of saying that the AC voltage in the circuit has a sine-wave form. Therefore, if an amplifier has a resonant circuit for a load, it can operate at only one frequency, and the AC voltage in the amplifier circuit must have a sine-wave form.

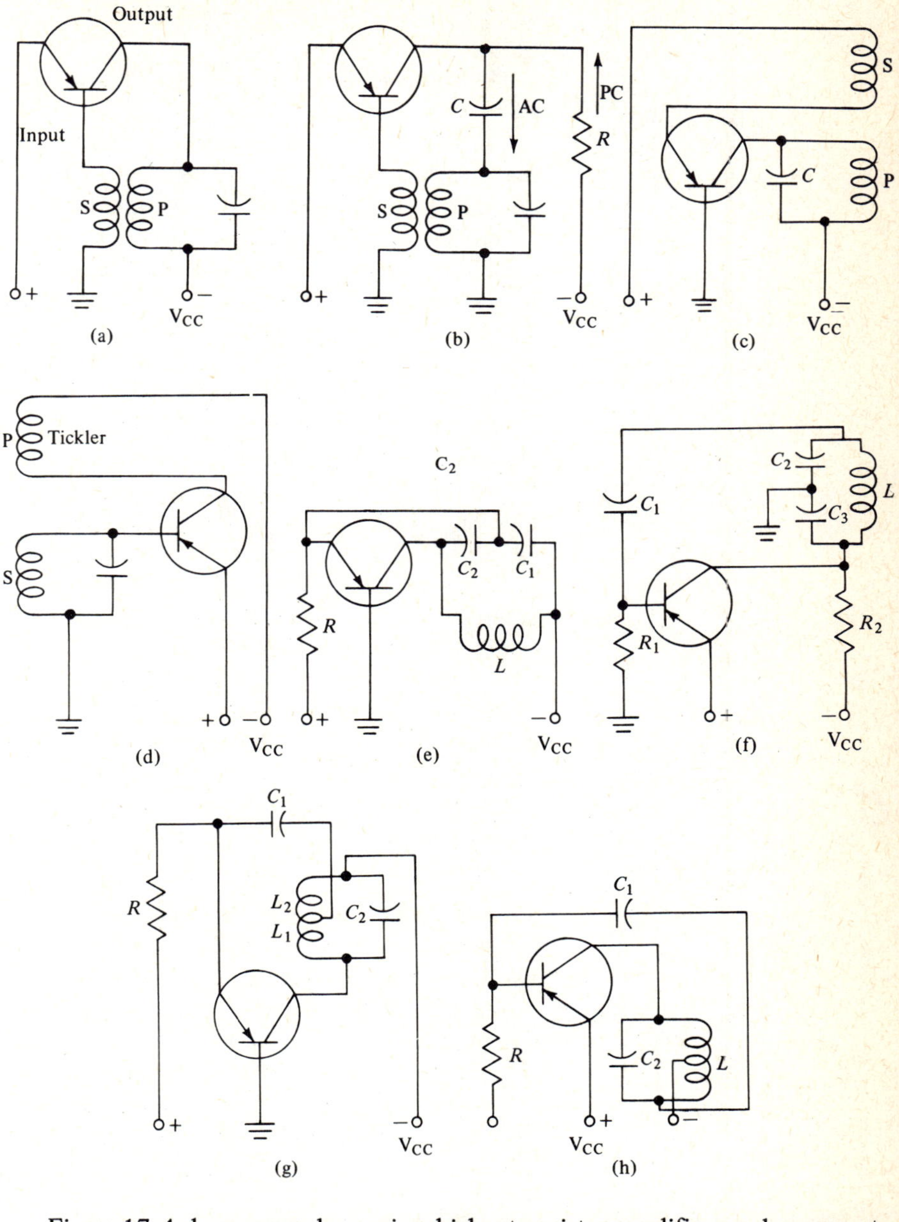

Fig. 17–4
Basic resonant-circuit transistor oscillators.

Figure 17–4 shows several ways in which a transistor amplifier can be connected to a resonant load in order to make the arrangement operate as an oscillator. In Fig. 17–4(a), the amplifier provides its own input by means of a tuned transformer. Notice that the collector is connected to the primary of the transformer, which is tuned by a capacitor. The primary can operate only at the frequency to which it is tuned. In turn, AC voltage of this particular frequency is induced in the secondary winding. Therefore, output of the amplifier (collector circuit) supplies input voltage to the arrangement (base circuit), and the transistor operates as an oscillator.

In Fig. 17–4(b), we have basically the same oscillator arrangement. However, a resistor R is provided here to conduct DC to the collector of the transistor. Notice that a coupling capacitor C is connected in series with the collector and the tuned primary winding. In turn, only AC current flows through the primary winding. This is called a *shunt-fed* oscillator circuit. We find that shunt-fed oscillators are used in high-power radio transmitters for reasons of safety, because high-power AC voltages can produce serious burns. High-voltage DC in the tuned circuits increases the danger to the operator when adjusting the transmitter. However, the shunt-fed arrangement does not entirely eliminate the danger.

The circuit in Fig. 17–4(c) is quite similar to the circuit in (a), except that the emitter is used as the input electrode, instead of the base. As expected, the operation of these two oscillators is very similar. The circuit in (c) is a series-fed arrangement. It should be remembered that, in any oscillator circuit, enough energy must be fed back into the input so that the step-up action of the transistor will cancel out the loss in the tuned transformer. This is just another way of saying that the spacing (coupling) between the primary and secondary windings must not be too great. Unless the primary and secondary are sufficiently near each other, the oscillator will not work.

The oscillator circuit shown in Fig. 17–4(d) is almost the same as the circuit in (a). However, there is the difference that the secondary of the transformer in (d) is tuned, whereas the primary of the transformer in (a) is tuned. The arrangement in (d) is called *tickler feedback.* Whenever the term "tickler" is used, it means that the collector load is untuned, although it is part (the primary) of a tuned transformer. We find that in some applications, such as radio receivers, the tickler-type oscillator is used, but in radio transmitters, the tuned collector type of oscillator is often used.

One important type of sine-wave oscillator, shown in Fig. 17–4(e), does not use a tuned transformer. Instead, this *Colpitts* circuit uses a single coil tuned by two capacitors connected in series. Notice that all the AC voltage produced by the oscillator appears across coil L. Part of this AC voltage appears across capacitor C_1, and part appears across C_2. The part of the AC voltage that appears across C_1 is fed back to the emitter of the transistor. Therefore, the amplifier supplies its own input, and when C_1 has a suitable amount of capacitance, the circuit operates as an oscillator. Notice that R is a bias resistor which applies DC voltage to the emitter. The Colpitts oscillator is used in many transmitters, but is less widely used in receivers.

Another Colpitts oscillator circuit is shown in Fig. 17–4(f). This is a shunt-fed arrangement, in which DC voltage is applied to the collector by means of resistor R_2. By way of comparison, a series-fed Colpitts circuit is shown in (e). Another difference between the two circuits is that the AC voltage is fed back to the base of the transistor in (f), whereas feedback is to the emitter in (e). In either case, we find these circuits used to a much greater extent in transmitters than in receivers.

Another important type of sine-wave oscillator, shown in Fig. 17–4(g), uses a tapped coil and a single tuning capacitor. This is called a *Hartley* oscillator. We will notice that a Hartley oscillator is similar to a tuned-transformer arrangement, except that a tapped coil is like a primary and secondary connected in series. For this reason, a tapped coil which is used in this way is called an *autotransformer.* In this Hartley oscillator circuit, a tuned autotransformer feeds back ac voltage from the collector circuit to the emitter circuit. That is, all the AC voltage produced by the oscillator appears across the tuning capacitor C. However, part of this AC voltage appears across L_1, and part appears across L_2. The part of the AC voltage that appears across L_2 is fed back to the emitter of the transistor.

In Fig. 17–4(h), the same general type of circuit is used as in (g). The only difference is that the feedback in (h) is from collector to base. The tuned autotransformer action is the same. Finally, we should take note of the fact that NPN transistors can be used instead of PNP transistors in any of these oscillator circuits. The only difference in the case of an NPN transistor is the necessity for reversing the polarity of the DC supply voltages.

17.3 Practical Hartley Oscillator Applications

A simple example of an electronic device that uses a Hartley oscillator is the code-practice oscillator shown in Fig. 17-5. It is used by Boy Scouts and beginning

(a)

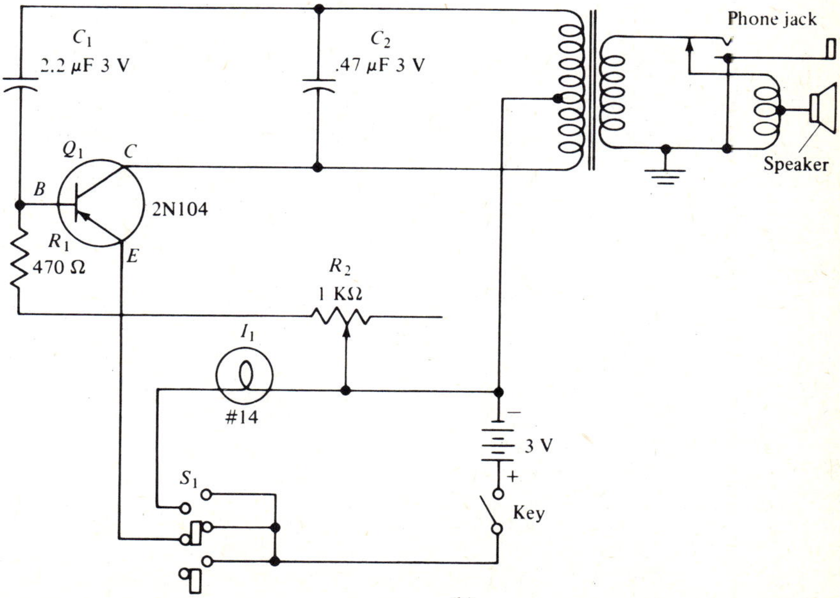

(b)

Fig. 17-5
A code-practice oscillator. (a) Appearance. (b) Circuit.
Courtesy: EICO

amateur radio operators, and can be combined with a walkie-talkie for code transmission by radio. The key is similar to a telegraph key, and operates as a switch. Notice that the circuit of the code-practice oscillator is very similar to the basic Hartley circuit shown in Fig. 17-4(h). However, the tapped coil operates as a primary of the output transformer T_1 in order to couple the sine-wave output to the speaker. In other words, transformer T_1 operates both as a tapped oscillator coil or autotransformer, and as a step-down output transformer to obtain efficient speaker response.

Notice in Fig. 17-5(b) that DC bias voltage is applied to the base of the transistors R_1 and R_2. When the resistance of R_2 is changed, the DC voltage on the base is also changed. In turn, the tone of the sound from the speaker can be changed from a low pitch to a high pitch. We will find at a later point in our study how this change in pitch is produced by a change in base voltage. At this time, we will simply note that R_1 has a comparatively high resistance, which prevents the oscillator from producing a true sine wave output.

Switch S_1 in Fig. 17-5(b) can be set to connect the light bulb I_1 into the circuit. In the first position of the switch, shown in the diagram, the light bulb is not connected into the circuit. When the key is closed, battery voltage is applied to the emitter of the transistor, and the oscillator produces a tone output. Next, if the switch is moved up to its second position, the light-bulb circuit is completed each time that the key is closed, but no voltage is applied to the emitter. Therefore, the light flashes but the speaker does not produce a tone output. Finally, if the switch is moved up to its third position, closure of the key applies voltage to the emitter, and also completes the light-bulb circuit. In turn, the light flashes and the speaker produces a tone output whenever the key is closed.

17.4 Resistance-Capacitance Oscillators

There is an important group of sine-wave oscillators that does not use coils. Since the oscillator circuit consists of a suitable arrangement of resistors and capacitors, this kind of oscillator is called the *RC type.* One of the most common circuit arrangements is the *phase-shift oscillator* shown in Fig. 17-6. To understand how

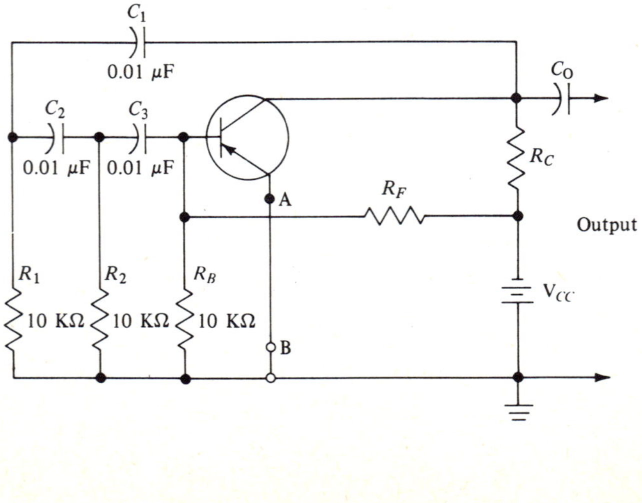

Fig. 17-6 Transistor RC phase-shift oscillator.

this oscillator works, we must note that "phase" means "timing." In other words, AC voltage must not only be fed back from the collector to the base of the transistor, but it must also be fed back with correct timing. If the feedback voltage arrives at the base at the correct time, it builds up the oscillation. But if the feedback voltage arrives at the base at the wrong time, it cancels out the oscillation.

We find that "phase shift" means "timing change." In Fig. 17-6, there are three *RC* combinations that produce a phase shift, or that produce a change in timing of the feedback AC voltage. These *RC* combinations are $R_1 C_1$, $R_2 C_2$, and $R_B C_3$. The total phase shift is found by adding up these three phase-shifting actions. Next, there is a very important fact for us to learn. The total phase shift, or timing change that is produced, depends on the operating frequency of the circuit. In other words, *there is only one frequency* at which the phase shift of the feedback voltage is exactly correct to build up the oscillation. Therefore, the *RC* phase-shift oscillator has to operate at this particular frequency.

Notice in Fig. 17-6 that R_B not only acts as a phase-shift resistor, but also provides a path for DC bias voltage from the battery to the base of the transistor. Resistor R_F is also part of this DC bias-voltage path. Since a coil is not used in the collector circuit, resistor R_C is provided to operate as a collector load resistor, and also to provide a path for DC collector current from the battery. Capacitor C_O is called an output blocking capacitor; it acts as an open circuit for DC voltage and current; and it blocks DC voltage out of the oscillator circuit. At the same time, C_O couples the AC voltage that is produced to any load that may be connected across the output terminals.

Instructor Demonstration or Student Experiment

Connect the circuit shown in Fig. 17-6. A transistor with a beta of greater than 100 should be used to assure that the circuit will oscillate. Connect an oscilloscope to the output and observe the frequency, the amplitude, and the shape of the output wave form. If the output wave form is distorted, there is too much gain in the circuit and this problem can be corrected by placing a 0 to 500-ohm potentiometer in the emitter of the transistor between points A and B. The circuit is turned on, and the potentiometer set to the minimum value that will give a goodsine-waveoutput. Replace the three 0.01μF capacitors with three 0.1μF capacitors, and compare the new output to the old signal. Finally, replace the 10K-ohm resistors with three 15K-ohm resistors and observe the output signal.

Questions:

1. What is the relationship of the capacitors to the output frequency?
2. What is the relationship of the output frequency to the values of the resistors?

Next, let us take a look at another important type of *RC* oscillator which is called a *square-wave* oscillator. This type of oscillator does not produce sine waves—instead, it produces square waves, as shown in Fig. 17-7. Mechanical

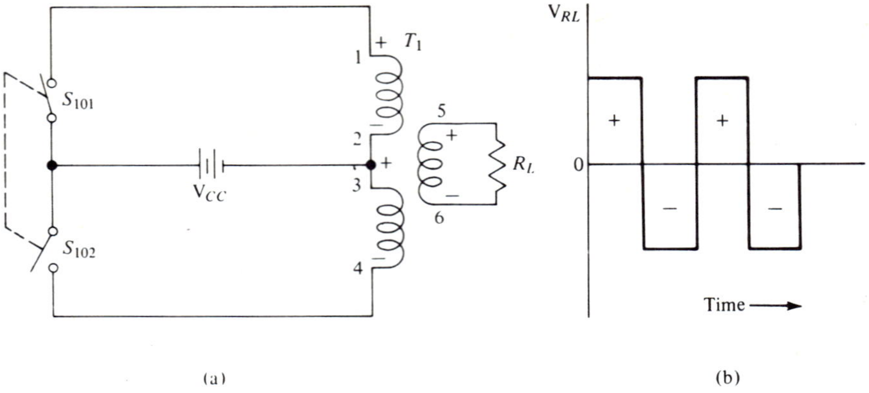

Fig. 17-7
Basic square-wave arrangement. (a) Mechanical switching circuit. (b) Output square-wave voltage.

switching circuits are used to produce square-wave output voltages at low frequencies only. However, the basic mechanical arrangement shows clearly how a square wave is produced. Notice that when switch S_{101} is closed, S_{102} is open. Then, when S_{101} is opened, S_{102} is closed. In turn, the battery current flows first in one direction, and then in the other direction through the primary of transformer T_1. The result is an induced square-wave voltage in the secondary of T_1.

Most square-wave electronics equipment (such as in television receivers) requires comparatively high-frequency square wave forms. Since it is impractical to operate mechanical switches at high speeds, electronic devices, such as transistors, are used as high-speed switches. When transistors are used as switches in a feedback oscillator circuit, the arrangement is called a *multivibrator.* Figure 17-8 shows a simple multivibrator circuit. This circuit produces a square-wave voltage

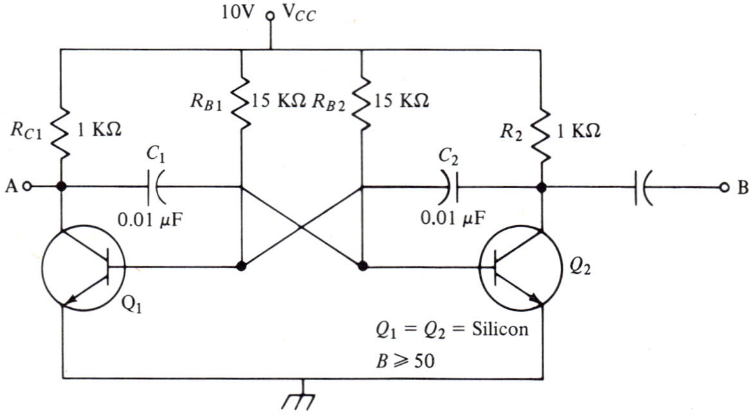

Fig. 17-8
Simple multivibrator.

output because the transistors automatically switch themselves on and off. Let us consider how the multivibrator operates.

Notice in Fig. 17-8 that the collector of transistor Q_2 is coupled to the base of Q_1 through capacitor C_1. Also, the collector of Q_1 is coupled to the base of

Q_2 through C_2. Therefore, each transistor has a feedback path to the other transistor. We find that this circuit has to operate like a seesaw. In other words, let us suppose that the collector voltage on Q_1 starts to go up. Because of the feedback path to Q_2, this rising voltage is coupled to the base of Q_2, which causes the collector voltage of Q_2 to go down. Since the collector voltage of Q_2 is going down, the feedback path through C_2 to Q_1 makes the collector voltage of Q_1 go up still more.

However, this rise of collector voltage on Q_1 can continue only until the supply voltage of V_{CC} is reached. Also, the fall of collector voltage on Q_2 can decrease only to zero. Then the circuit must rest until the charge on C_2 drains away through R_2. At this instant, the feedback conditions are reversed—the collector voltage on Q_1 suddenly falls to zero and the collector voltage on Q_2 suddenly rises to V; this is due to the feedback action of the circuit. This switching action continues back and forth, thereby producing a square-wave output. This square-wave voltage is coupled through blocking capacitor C_O to any load that might be connected across the output of the multivibrator. The repetition rate of the multivibrator is determined by C_1, C_2, R_{B1}, and R_{B2}.

Instructor Demonstration or Student Experiment:

Connect the circuit shown in Fig. 17-8 using silicon transistors that have at least a 6 V reverse breakdown voltage base-to-emitter. Connect an oscilloscope to either the *A* or *B* outputs and observe the frequency and the amplitude of output. Now, replace C_1 with a 0.1 μF capacitor and observe the change in the outputs. Finally, replace C_1 with a 0.1 μF capacitor and again observe the output on the oscilloscope.

Questions:

1. What effect does increasing the value of each capacitor have on the output frequency?
2. What effect does changing the value of the capacitors have on the output wave form?

17.5 Sawtooth Oscillators

Another class of oscillators is the *sawtooth* oscillator group. This type of oscillator works on the principle that a capacitor in series with a resistor takes some amount of time to charge to a value of voltage (Fig. 17-9). The rate of charge

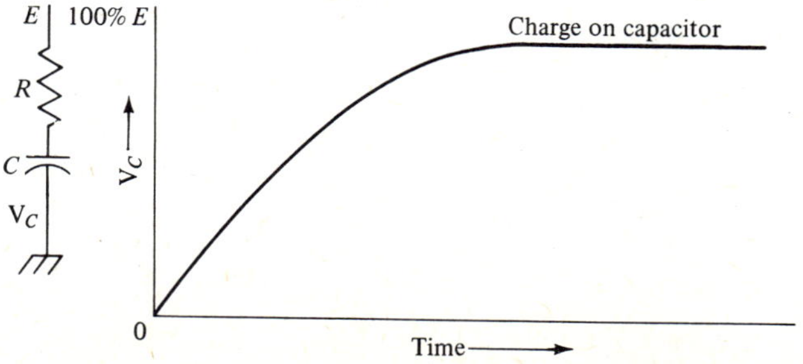

Fig. 17-9. The charge on a capacitor increases with time.

across the capacitor is determined by the value of the series resistor and the value of the capacitor. We find that the capacitor charges to 63 percent of the supply voltage in one time unit where time $= RC$. For example, a 1 μF capacitor in series with a 1 MΩ resistor would charge to 63 percent of the supply voltage in one second. On the other hand, the same capacitor in series with a 1 Ω resistor would charge to the same voltage in one-millionth of a second.

The circuit in Fig. 17-10(c) uses a neon bulb to discharge the capacitor each time it charges to approximately sixty volts. The bulb appears as an open as the capacitor charges toward the supply voltage. When the capacitor charges to the firing voltage of the neon bulb (approximately sixty volts), the bulb becomes a very low value of resistance that begins to quickly discharge the capacitor. At a value of approximately fifty volts, the bulb again appears as an open, and the capacitor begins to recharge toward the supply voltage. The output wave form is a sawtooth.

The sawtooth wave form has many applications in the field of electronics. As an example, the sweep on a cathode-ray tube in an oscilloscope is caused by an electron beam being pulled across the inside of the screen. In the chapters concerning television, we will find that a sawtooth wave form is used to develop the vertical and horizontal scanning of the picture on the picture tube.

Instructor Demonstration or Student Experiment

Construct the circuit in Fig. 17-11, using a four-layer diode or unijunction transistor. As a last resort, a neon (Ne 2) bulb may be utilized as in Fig.

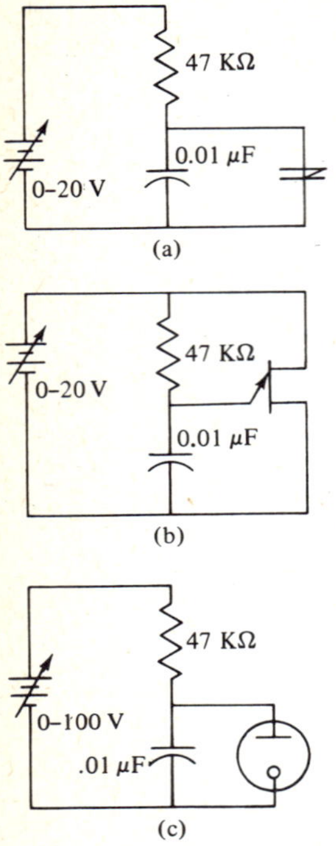

Fig. 17-10
Sawtooth oscillator. (a) Using a 4-layer diode. (b) Using a unijunction transistor. (c) Using a neon bulb.

Notes:
Rate-adjustable from 40 (low largo) to 220 (high presto) beats per minute
R_1-adjusts high rate limit
R_2-adjusts low rate limit
R_3-IRC type Q13-328
potentiometer (log taper)

SPKR-Utah model SP358 3.4 Ω V.C.
Size 3 1/2″ magnet 1 oz.
(However, any speaker
can be used).

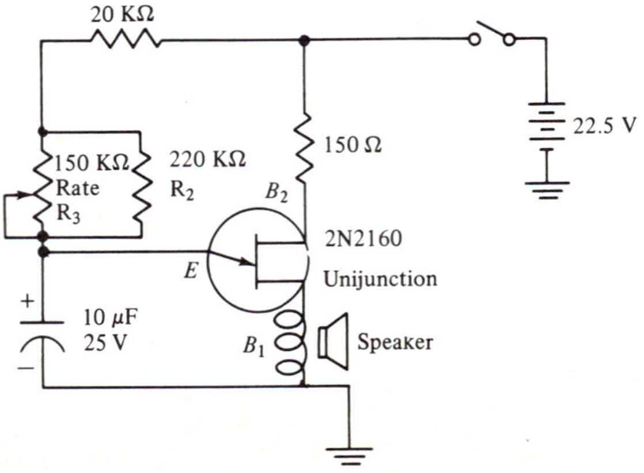

Fig. 17-11
Sawtooth oscillator demonstration circuit.

17-10(c), but the supply voltage must be increased to approximately 100 volts. With an oscilloscope connected across the capacitor, increase the supply voltage until the sawtooth appears. Note the frequency, the amplitude, and the shape of the wave form. Then vary the setting of the rate control while continuing to observe the wave form on the oscilloscope. Now replace the 0.01 μF capacitor with a 0.1 μF capacitor, and repeat the steps in the above procedure.

Questions:

1. What is the principle of operation of a sawtooth oscillator?
2. Explain the relationship between the frequency of oscillation and each of the variables; voltage, resistance, and capacitance.

Questions: Chapter 17

1. What is an oscillator?
2. Discuss the reason for the resonant circuit in an oscillator.
3. What is the basic difference between a shunt-fed and a series-fed oscillator?
4. Explain the term "tickler" as applied to an oscillator.
5. What are the frequency requirements of an oscillator that is to be used in a code-practice oscillator?
6. Which of the components in the phase-shift oscillator in Fig. 17-6 determine the frequency of oscillation?
7. Why does a phase-shift oscillator circuit oscillate at only one frequency?
8. What is the function of a multivibrator?
9. Which components determine the frequency of a multivibrator?

True-False:

1. An oscillator is basically a vibrator.
2. An oscillator cannot produce a sine-wave voltage.
3. Oscillators that depend upon inductors and capacitors to develop a resonant frequency usually produce sine waves.
4. The purpose of the transistor in an oscillator is to supply the amplification of the signal to overcome circuit losses.
5. Series-fed oscillators are used in high-power transmitters for reasons of safety.
6. A Colpitts oscillator uses two capacitors and a coil in the resonant circuit.
7. A code oscillator must operate in the audio-frequency range.
8. A phase-shift oscillator uses resistors and inductors in the frequency selective network.
9. A multivibrator is a sine-wave oscillator.
10. The frequency of a multivibrator is determined by the coupling capacitors and the base resistors.

Multiple choice: choose the best answer.

1. The primary requirement of an oscillator is that the output
   a. frequency be constant
   b. voltage be constant
   c. voltage be variable
   d. frequency be variable
2. Sinewave oscillators usually make use of ____________ to produce oscillations.
   a. *RC* circuits
   b. inductors
   c. capacitors
   d. resonant circuits
3. The shunt-fed oscillator circuit is used in high-power transmitters
   a. to save power
   b. for safety
   c. for extra stability
   d. for simplicity
4. An oscillator that uses two capacitors and an inductor for the resonant circuit is called a
   a. Hartley oscillator
   b. multivibrator
   c. Colpitts oscillator
   d. tickler oscillator
5. An oscillator that produces a square-wave output is called a
   a. multivibrator
   b. Hartley oscillator
   c. Colpitts oscillator
   d. phase-shift oscillator
6. An oscillator that produces a sine-wave output and uses resistors and capacitors as the selective components is called a
   a. multivibrator
   b. phase-shift oscillator
   c. Colpitts oscillator
   d. Hartley oscillator
7. Oscillators may be designed to produce
   a. sine-waves
   b. square-waves
   c. sawtooth waves
   d. all the above
8. The output frequency of a sawtooth oscillator is determined by the
   a. resistance in the charge path
   b. capacitance in the charge path
   c. supply voltage
   d. all the above

18

RADIO WAVES AND BASIC RADIO THEORY

18.1 Radio-Wave Principles

Radio operation is based upon the travel of waves through space. We know that when a stone is thrown into a pond, waves travel across the water from the place where the stone strikes. These waves travel in circles that become larger and larger, as shown in Fig. 18-1. The number of waves that strike the shore in one second is called the *frequency* of the wave motion. A general principle of waves is that they are strongest at their starting point, and become weaker as they travel out. The distance from the top to the bottom of the wave is called its *amplitude*, as shown in Fig. 18-2.

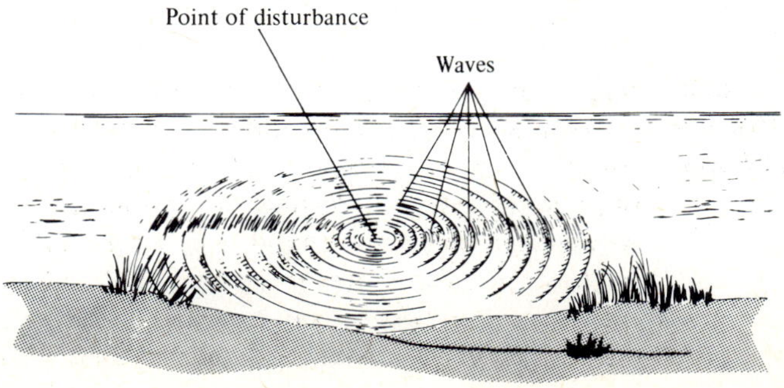

Fig. 18-1
Waves produced by throwing a stone into a pond.

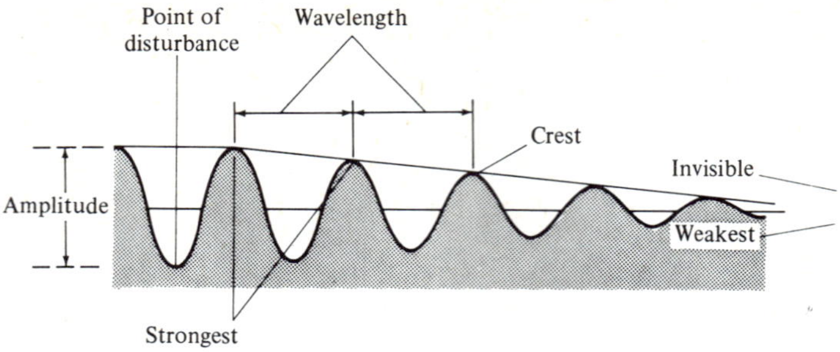

Fig. 18-2
The wavelength is the distance from one peak to the next.

We find that, if vibrations or waves are produced more rapidly, their frequency is higher. For example, if a wave is produced twice as rapidly, its frequency will be twice as great. Figure 18-3(a) shows that *a higher frequency has a shorter wavelength.* And of course, *a lower frequency has a longer wavelength.* We also see

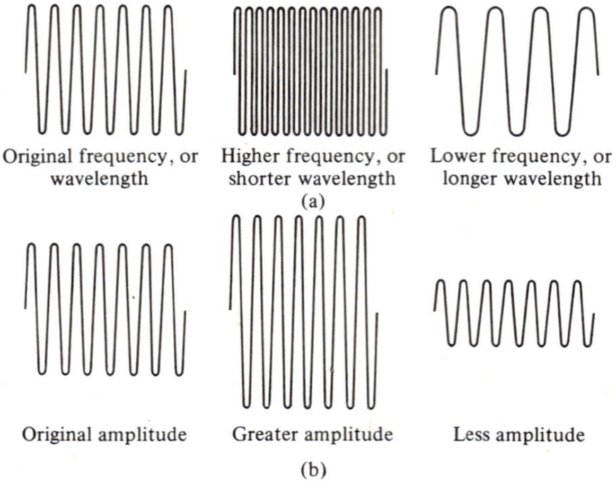

Fig. 18-3
Wavelength, frequency, and amplitude examples. (a) Sine waveforms with same amplitude. (b) Sine waveforms with same frequency or wavelength.

from Fig. 18-3(b) that amplitude does not depend upon frequency or wavelength. In other words, if we change the frequency of a wave, we change its wavelength also, but we do not change its amplitude. This means also that, if we change the amplitude of a wave, we do not change its frequency or its wavelength.

18.2 Electromagnetic (Radio) Waves

Radio waves are also called *electromagnetic waves.* They are similar to water waves in many ways, but there are some important differences. Radio waves consist of *electric* and *magnetic fields* traveling through space. The magnetic lines of force cross the electric lines of force in a radio wave, as shown in Fig. 18-4. As we mentioned earlier, another important difference is in the speed of travel of an electromagnetic wave—this is the speed of light, or 186,500 miles per second. It is a fact that although we cannot see or feel a radio wave, the radio wave is nevertheless closely related to light and heat waves. This fact is shown in Fig. 18-5(a).

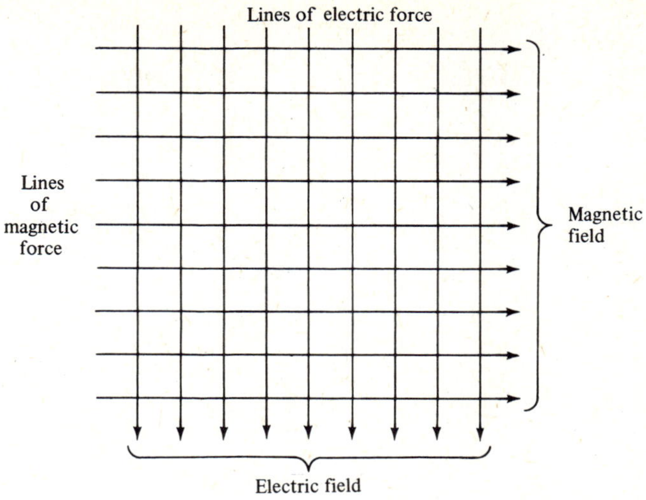

Fig. 18-4
Electric and magnetic lines of force in a radio wave.

It is interesting to note that radio waves travel through the air at 186,500 miles per second.

Scientists often speak of wavelengths in terms of microns, millimicrons, or Angstrom units. However, electronics technicians measure wavelengths in terms of meters—a meter is equal to 39.37 inches. That is, a meter is somewhat longer than a yard. However, we usually speak of the *frequency* of radio waves, rather than the *wavelength.* As seen in Fig. 18-5(a), radio waves range in frequency from 10 KHz to 30,000 MHz. This range of frequencies is lower than the frequencies of light and heat. Except for having a lower frequency, a radio wave is the same as a light wave or a heat wave.

We note in Fig. 18-5(a) that radio frequencies are grouped into VLF, LF, MF, HF, VHF, UHF, SHF, and EHF bands. The meanings of these abbreviations are listed in Fig. 18-5(b). Ordinary radio broadcasting uses the medium-frequency (MF) band. Frequency-modulation, or fm broadcasting, uses the very-high-frequency (VHF) band. Amateur radio stations are licensed to operate on various bands in the range from 3.5 MHz to 40,000 MHz.

18.3 Transmission and Reception of Radio Waves

To start an electromagnetic wave traveling through space, energy must be *radiated* from a suitable conductor arrangement which is called an *antenna.* In general, the higher up an antenna is placed, the better its radiation becomes. A longer antenna will also radiate electromagnetic waves more efficiently than a short antenna. At low and medium frequencies, various types of standard antennas are used, such as shown in Fig. 18-6. The height of an antenna is more important than its length—that is, a short antenna erected at a considerable height will be a better radiator of electromagnetic waves than a long antenna installed near the ground level. An antenna can be compared with a capacitor in that it becomes charged

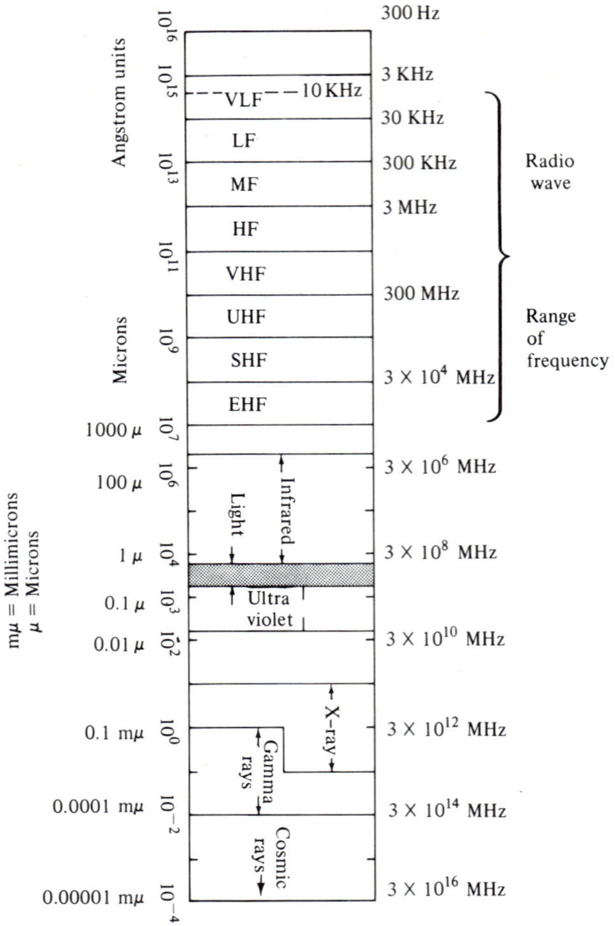

(a)

Abbreviation	Frequency band	Frequency range
VLF	Very low frequency	below 30 KHz
LF	Low frequency	30–300 KHz
MF	Medium frequency	300–3000 KHz
HF	High frequency	3000–30,000 KHz
VHF	Very high frequency	30–300 MHz
UHF	Ultra high frequency	300–3000 MHz
SHF	Super high frequency	3000–30,000 MHz
EHF	Extremely high frequency	30,000–300,000 MHz

(b)

Fig. 18-5 (a) Frequency range of electromagnetic waves. (b) Bands of frequencies.

when it is connected to a source of radio-frequency (rf) voltage. If there is considerable distance between the antenna and the ground below, the charge and discharge action results in efficient radiation of electromagnetic waves.

Radio receivers which operate at low and medium frequencies may use a sin-

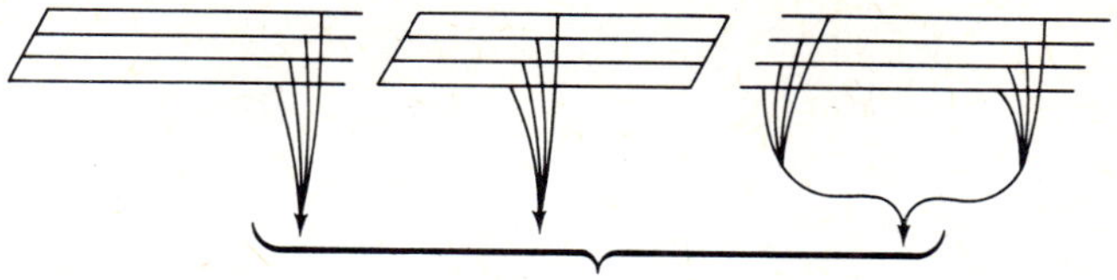

Fig. 18-6
Typical low-frequency radio transmitting antennas.

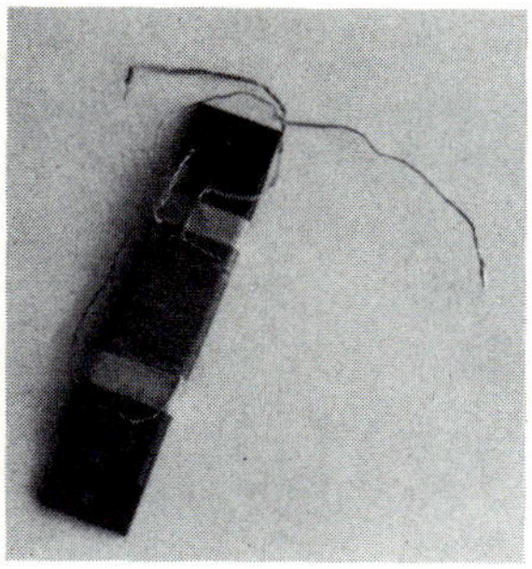

Fig. 18-7
Loop and ferrite-core antenna.

gle short wire for an antenna. Although more energy can be received by using a large and high antenna, this expense and complication is not necessary because a receiver operates at low power, and more transistors can be used to obtain high *sensitivity.* Therefore, medium-frequency receivers, such as ordinary broadcast receivers, generally use loop antennas, or ferrite-core coil antennas, such as shown in Fig. 18-7. When additional sensitivity is needed from a coil antenna, the spiral of wire seen in the figure can be unwound and hung behind the radio receiver.

At higher radio frequencies, roof-top antennas are generally used. A VHF antenna has the basic *dipole* construction shown in Fig. 18-8. "Dipole" means

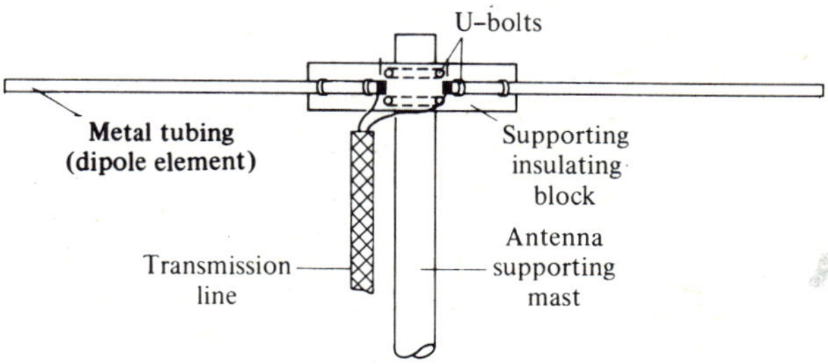

Fig. 18-8
Basic dipole construction.

"two arms." That is, the antenna conductor is separated in the center, and is connected to a pair of wires that run down to the receiver. These connecting wires are called a *transmission line.* Technicians often refer to this type of line as *twin lead.* Simple dipole and twin-lead installations are in wide use for frequency-modulation (fm) reception. The advantage of a dipole antenna is its ability to reduce interference such as ignition noise from automobiles, and random interference from neon signs, electric power lines, and other sources.

A dipole antenna is also directional. It receives best from the front and rear directions—but it has little or no response in its end-to-end direction. Radio amateurs use a more complicated type of antenna called a *beam* antenna, such as shown in Fig. 18-9. This arrangement consists of a simple dipole antenna with another rod in front, and still another rod in back. The front rod is called a *director,* and the rear rod is called a *reflector.* The director is a little shorter than the dipole, and the reflector is a little longer than the dipole. The spacing of the director and reflector rods from the dipole is adjusted carefully in order to obtain the best *front-to-back ratio.* This means that the beam antenna will receive best in the forward direction, and will have very little response from the rear. Not only

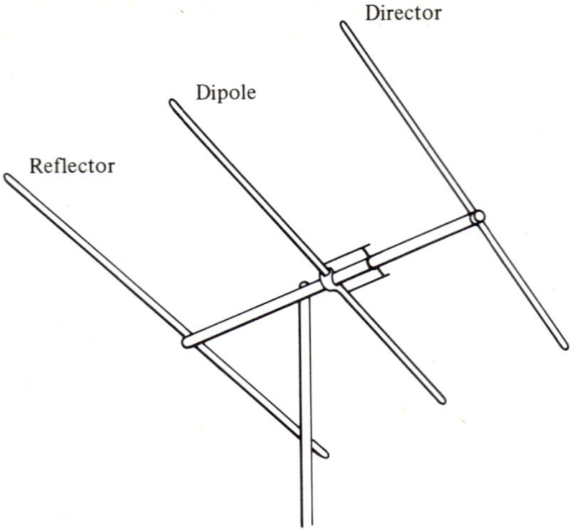

Fig. 18-9
A typical beam antenna.

is a beam antenna more directional than a simple dipole, but it is also more sensitive.

High-frequency radio waves are *propagated* differently from low-frequency radio waves. That is, HF waves do not necessarily travel from transmitter to receiver over the same path that is taken by LF waves; VLF and LF waves travel along the earth's surface as *ground waves.* (See Fig. 18-10.) These waves tend to follow the curvature of the earth, and can be used to transmit halfway around the world provided the transmitter uses very high power. One of the advantages of a ground wave is the fact that its strength is about the same in the day as in the night, and remains the same in summer and winter. The main disadvantage of a ground wave is that an LF long-distance transmitter is quite costly.

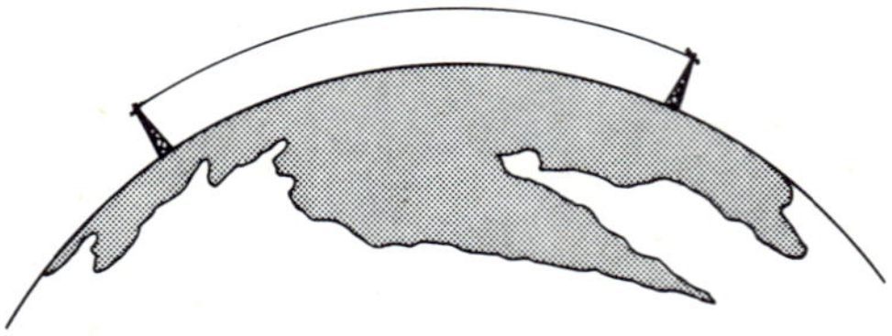

Fig. 18-10
Ground waves tend to follow the surface of the earth.

Medium-frequency waves also travel as ground waves in the daytime and nighttime. However, nighttime transmissions can be received over much greater distances, because a *sky wave* is reflected back to earth, as shown in Fig. 18-11. This reflection is caused by electrified particles high up in the atmosphere. This reflecting layer is called the *ionosphere.* In the daytime, the ionosphere is in a condition which produces little or no reflection of MF waves. Therefore, MF waves escape into outer space during the day. However, at night, the ionosphere reflects MF waves back to the earth's surface; therefore, good reception can be obtained far beyond the ground-wave range at night.

As shown in Fig. 18-11, MF waves must strike the ionosphere at a suitable

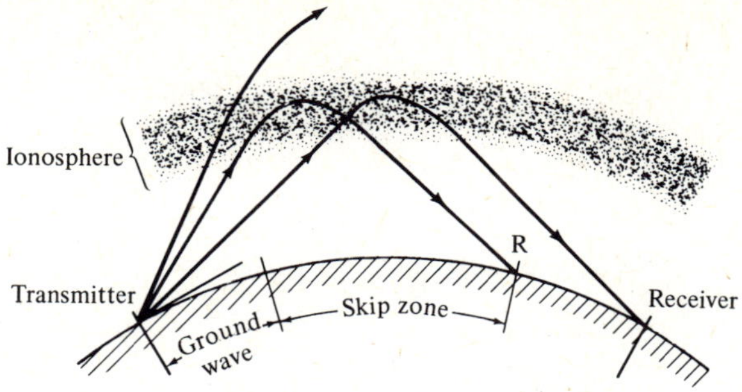

Fig. 18-11
Bending of sky waves.

angle, if they are to be reflected back to earth. This means that there is a *skip zone* between the end of the ground wave and the region where the sky wave comes back to earth. Therefore, during both day and night, there is no reception of MF waves in the skip zone. High frequencies travel in the same general way as medium frequencies, except that the skip zone is much longer. Also, HF waves are reflected from the upper ionosphere both during the day and during the night. In turn, HF waves can be received halfway around the world at any time of the day or night. The main disadvantage of HF waves is that their strength tends to vary from time to time—this is called *fading.* Medium-frequency waves also tend to fade during the night when sky waves are being received.

Very high frequency waves are seldom reflected from the ionosphere. Therefore, only the ground wave can be used. Very-high-frequency reception is sometimes called *line-of-sight* operation, as shown in Fig. 18-12. However, VHF ground

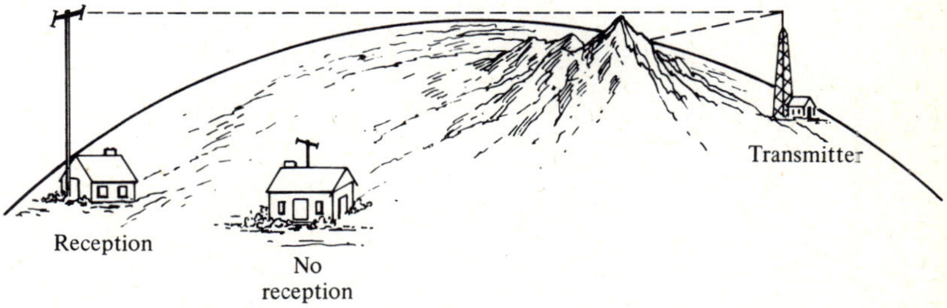

Fig. 18-12
Line-of-sight transmission.

waves tend to follow the surface of the earth to some extent, so that reception is possible even though the receiving antenna cannot "see" the transmitting antenna. The main difference between LF ground waves and VHF ground waves is that the VHF ground waves weaken much more rapidly as they travel out from the transmitting antenna. Since VHF waves are easily reflected by buildings, or by objects such as airplanes, the receiving antenna may pick up both a direct signal and a reflected signal at times. This condition is shown in Fig. 18-13. When this *multipath* reception occurs, the signal is likely to become distorted. For example, an fm broadcast signal can become badly distorted in case of multipath reception.

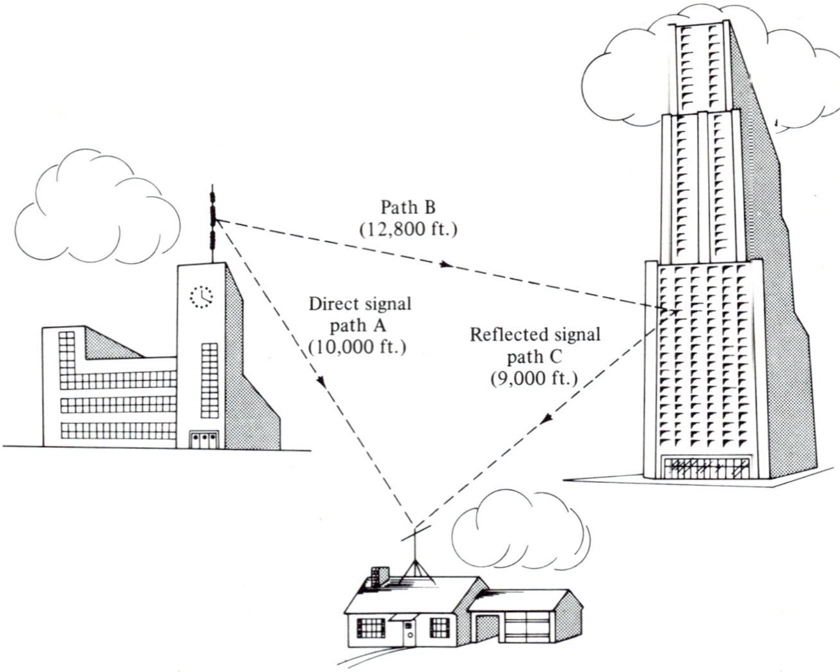

Fig. 18-13
Example of multipath reception.

18.4 Basic Radio Receivers

A basic radio receiver consists of a detector (radio-frequency rectifier) and an earphone, connected to an antenna and ground installation, as shown in Fig. 18-14. Let us see how this arrangement works. In an electrical or electronic cir-

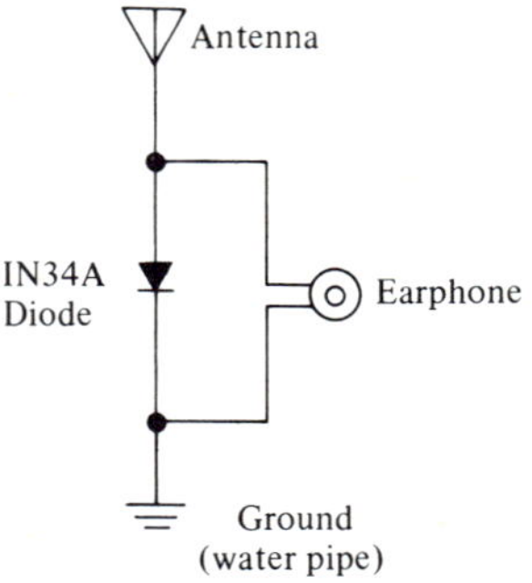

Fig. 18-14
Simplest radio receiver arrangement.

cuit, any sound or tone has a certain *wave form*, such as shown in Fig. 18-15. This is the wave form of the sound "or" that we would find in a telephone circuit, for example. However, in the case of a radio receiver, the antenna voltage is somewhat more complicated. Broadcast radio waves consist of tone (or audio) waves *modulated* on a radio-frequency *carrier* wave, as shown in Fig. 18-16. Unless modulation is used, sound waves cannot be transmitted through space. The radio-

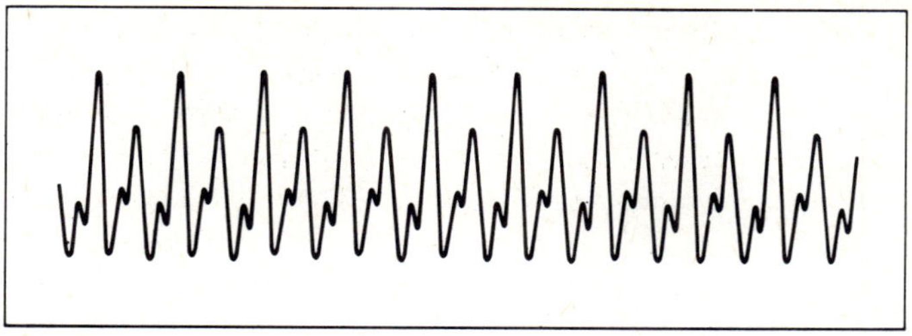

Fig. 18–15
Waveform of the sound "or."

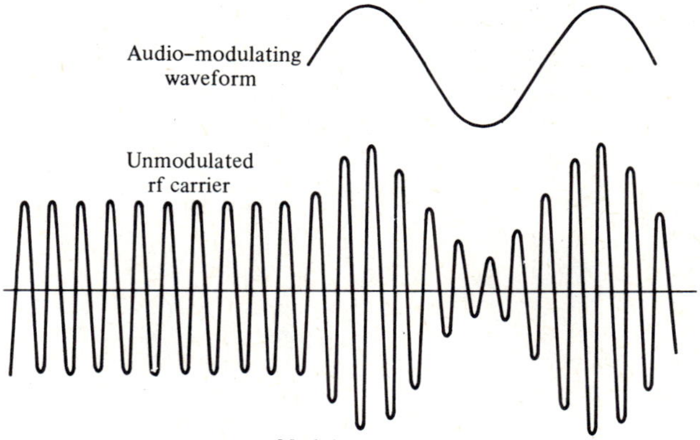

Fig. 18–16
A radio–frequency carrier modulated by an audio wave.

frequency wave is called a carrier, because it actually carries the sound or tone wave through space in the form of electromagnetic energy.

Figure 18–17 shows the construction of an earphone, or telephone receiver.

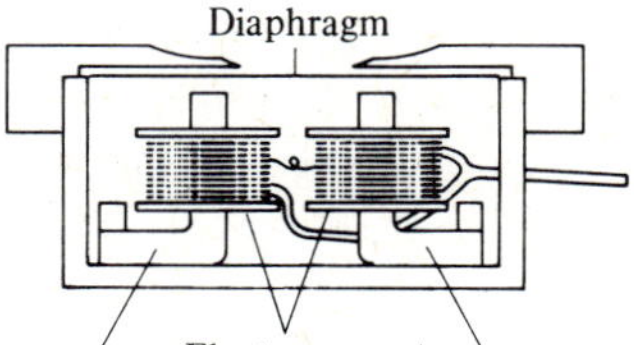

Fig. 18–17
Construction of an earphone.

It consists of a thin iron diaphragm placed near the poles of a pair of electromagnets. The electromagnets are mounted on a permanent magnet. When an audio-voltage wave form, such as that in Fig. 18–15, is applied to the windings of an earphone, the rising and falling audio-frequency current causes the field of the permanent magnet to become stronger and weaker. In turn, the diaphragm of the earphone is pulled more or less, and it vibrates in accordance with the sound—in this example, in accordance with the sound "or."

If we look carefully at the modulated wave form in Fig. 18–16, we see that this wave form cannot operate an earphone. That is, the rf carrier has such a high frequency that the diaphragm cannot vibrate so rapidly, even if we could hear such a high tone (which we cannot). Therefore, some device must be used in a radio receiver to change the modulated rf wave form back into the original audio wave form. This is the job of the 1N34A *detector diode* in Fig. 18-14. A detector is simply an rf rectifier. Figure 18-18 shows the detection process. The detector

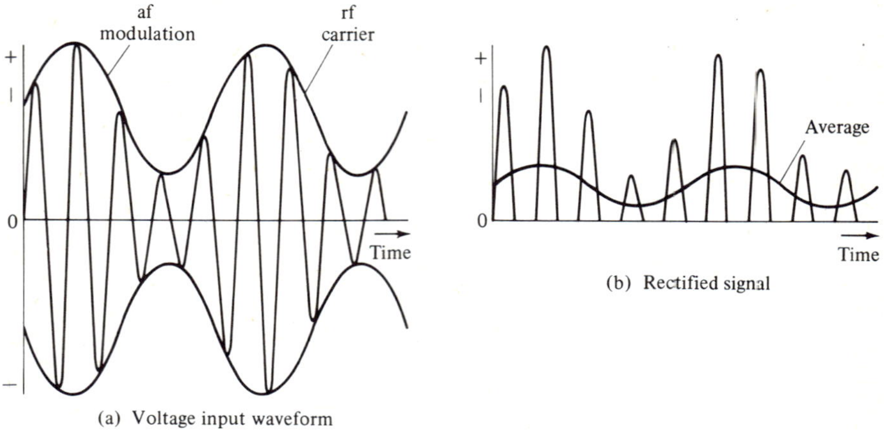

Fig. 18-18
How a detector works.

rectifies the modulated wave form, leaving half cycles of the same polarity (positive, in this example). These half cycles have an average value, which follows the original sound wave form. In turn, the earphone in Fig. 18-14 vibrates in accordance with the sound wave form, and we hear the broadcast program.

An arrangement, such as the one in Fig. 18-14, has the disadvantage that it cannot separate, or "tune out," undesired broadcast stations. That is, this simple receiver usually produces considerable interference in addition to the desired station program. Therefore, you will probably want to construct a receiver that has a *tuner.* A tuner is a resonant-circuit arrangement which responds only to the frequency to which it is set or *tuned.*

We have often had our students build the two-stage radio shown in Fig. 18-19, to give them practice in reading schematic diagrams, constructing circuits, and testing circuits. To construct this radio you will need the parts that can be ob-

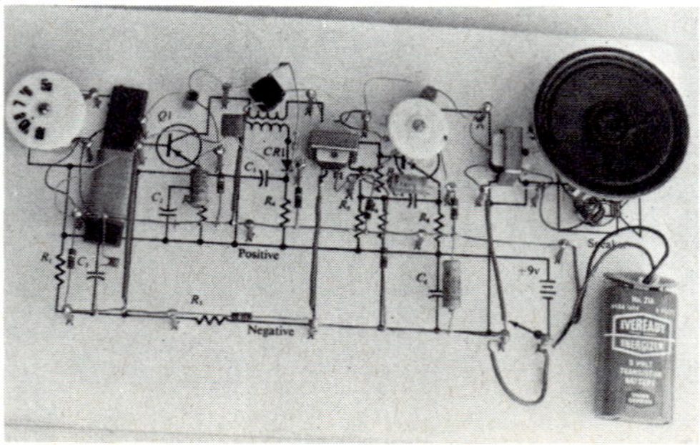

Fig. 18-19
Simple portable radio.

tained from the manufacturer in kit form, a soldering iron (37½ watts), a scrap of wood (approximately 5 inches by 10 inches), and patience. We might note here that the radio may be purchased as a kit, in which case all the parts are included.

Before beginning construction of the radio you should have your instructor test your soldering skills. Proper soldering techniques are discussed in the Appendix on page 256.

When you are satified with your ability to make proper soldering connections, begin construction of the breadboard as follows:

1. Trace the schematic in Fig. 18-20 and glue the tracing to the 5 by 10 inch board (Fig. 18-19).

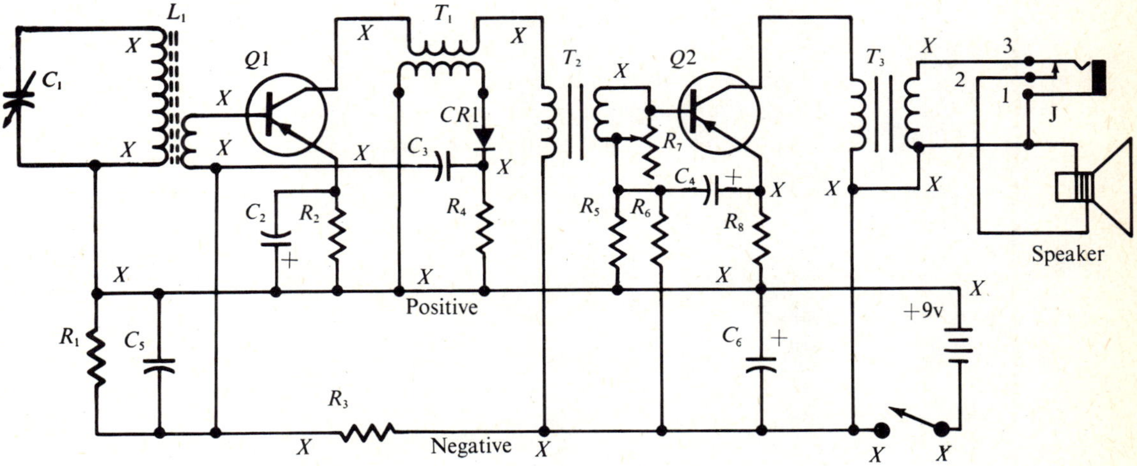

Fig. 18-20
Schematic of a two-transistor radio
Courtesy: Graymark, Inc.

2. Drive copper nails into each of the large X's in the schematic. Leave these nails protruding about ½ of an inch from the board.
3. Tin (coat with solder) the tops of each copper nail with rosin core solder.
4. Solder a bare copper wire along the COMMON POSITIVE and the COMMON NEGATIVE lines as shown in Fig. 18-19.
5. Mount the transformers upside down, as shown in Fig. 18-19. Be sure to place the colored marks on the transformers in the direction indicated on the schematic. Note, that extreme care should be taken when bending the leads of the transformer T_1. A heatsink, such as shown in the Appendix (Fig. A-2 [f]), should be used when soldering to this transformer, the diode, and the transistors. The heat sink is placed between the component and the soldered connection.
6. Identify each of the parts and mount them on the breadboard and double-check each soldering connection. The speaker is connected to transformer T_3 as shown in Fig. 18-21.

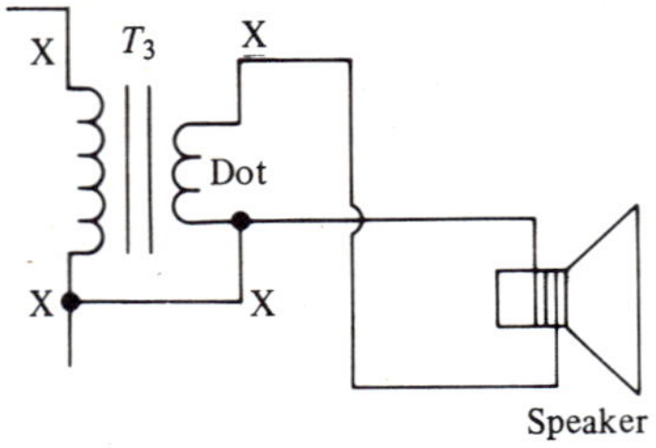

Fig. 18-21
Speaker connection.

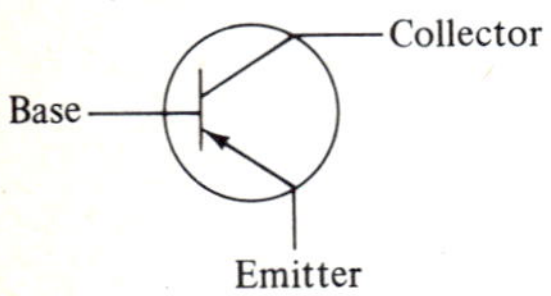

Fig. 18-22
PNP transistor.

7. The transistors used in this circuit are the PNP type made of germanium. The circuit symbol of a PNP transistor is shown in Fig. 18-22. The transistors may be constructed in any of several types of packages or cases as shown in Fig. 18-23. The operation of the transistor is the same, regardless of the type of case. In fact, you may substitute any PNP germanium transistor that has a frequency cutoff of over 2 megahertz and a breakdown voltage of over 10 volts. Ask your instructor about these terms.

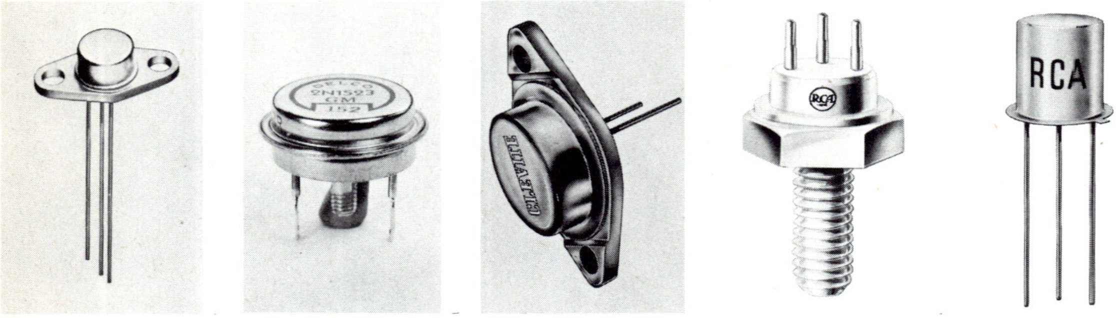

(a) (b) (c) (d) (e)

Fig. 18-23
Typical transistor cases.

The radio shown in Fig. 18-19 is a reflex type; that is, the first transistor $Q1$ is used to amplify the radio frequency signal that is selected by the setting of the tuning capacitor C_1 and the antenna coil L_1. The signal is coupled through the transformer T_1, the radio frequency transformer. The radio frequency portion of the signal is eliminated and the audio frequency saved by the detection action of diode $CR1$. The audio signal is then coupled to transistor $Q2$ through transformer T_2. The 5,000 ohms potentiometer serves as a volume control to control the amount of audio coupled to the base of transistor $Q2$. The audio signal is amplified by $Q2$ and coupled through the output transformer T_3. The purpose of transformer T_3 is to match the output of transistor $Q2$ to the low impedance value of the speaker.

When you have completed your radio, have someone check all the connections before connecting the 9-volt battery. If the connections seem to be in order, you are now ready to test your radio. First, rotate the volume control R_7 fully clockwise, connect the battery, and rotate the tuning capacitor through its entire range from one extreme to the other. You should be able to receive several stations. If you have no reception, rotate the volume control fully counter-clockwise and repeat the above steps. If your radio still does not receive a station you have a wiring error or a defective component. At this point you should seek the aid of your instructor who may have you perform the following test to locate the problem.

1. Recheck all connections and parts placement. Common problems are improper connection of transformer T_1, or antenna coil L_1.
2. Test the voltage across capacitor C_6 with the battery connected and the switch closed. The voltage should measure only slightly less than 9 volts. If the voltage is low, the battery may be defective or there may be a short circuit in the radio.
3. The front or left part of the radio in Fig. 18-19 acts as the station selector or tuner. The operation of this section can be tested with a crystal earphone or audio amplifier as shown in Fig. 18-24. If

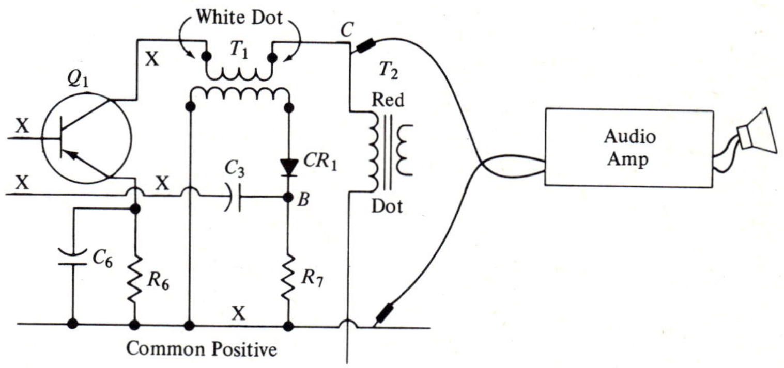

Fig. 18-24
Testing the finished radio.

there is reception at this point the problem is further along in the radio.
4. Placing the audio amplifier or crystal earphone between the base transistor $Q2$ and the COMMON POSITIVE (Fig. 18-25) checks the operation of transformer T_2.

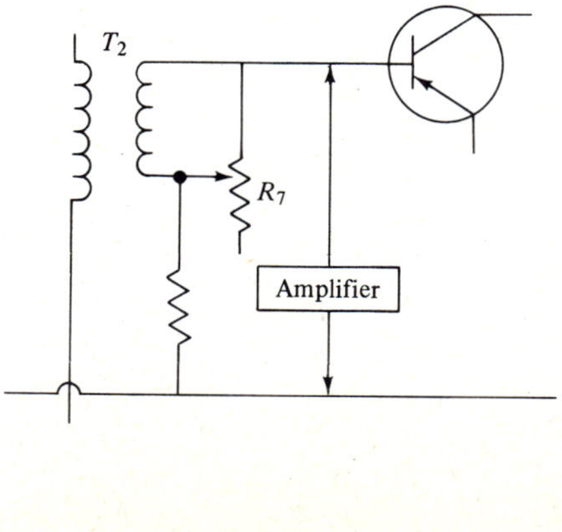

Fig. 18-25
Testing transformer T_2.

5. The operation of transistor $Q2$ can be tested by placing the audio amplifier or crystal earphone between the collector lead of $Q2$ and the COMMON POSITIVE.
6. Finally, transformer T_3 can be tested by placing an audio amplifier or crystal earphone across the output leads of transformer T_3.

We might also note that the radio can be tested from the output to the input by placing a finger tip at the base of each transistor and listening for noise in the speaker. Once the trouble has been localized to a specific part of the radio, that section should be examined for wiring errors or defective parts. Defective parts can be determined by substitution of a new part or testing of the suspected part.

Questions: Chapter 18

1. Define the terms wavelength, amplitude, and frequency as they apply to a radio system.
2. What is the measurement in which wavelength is usually given?
3. What is an antenna?
4. Which is more important, a long antenna or a high antenna?
5. Define each of the following words: *transmission line, dipole, twin-lead, beam, sensitivity, director, reflector, propagated, ground wave, sky wave,* and *ionosphere.*
6. What causes fading of a distant radio signal?
7. What is the distance limitation of VHF radio transmission?
8. What are the components of a basic radio receiver, and what is the function of each component?
9. What is a detector?
10. What is the purpose of the tuner in a radio?
11. What do we mean by breadboard construction?

True–False:

1. The operation of a radio is based on the travel of sound waves through space.
2. We can change the amplitude of a radio wave without changing its frequency.
3. Increasing the frequency of a radio wave results in an increase in wavelength.
4. Energy is radiated from the transmitting antenna.
5. Increasing the height of a receiving antenna increases its range of possible reception.
6. The radio broadcast band uses the VHF band of frequencies.
7. The sensitivity of a radio has to do with its signal to interference level.
8. A dipole antenna is not directional.

9. An antenna may be made to be directional.
10. The ionosphere reflects some radio waves.
11. At night the ionosphere produces little or no reflection of radio waves.
12. Reception of a radio signal can be obtained much farther from the transmitting station at night than during the day.
13. Fading is caused by variation in the signal strength of a radio signal.
14. A detector is a rectifier.
15. The audio signal is separated from the carrier by a detector circuit.

Multiple choice: choose the best answer.

1. The wavelength of a radio signal is determined by its
   a. amplitude
   b. frequency
   c. phase
   d. location
2. The wavelength of a radio wave is measured in
   a. radians
   b. inches
   c. feet
   d. meters
3. Energy is transmitted through space from
   a. an antenna
   b. a receiver
   c. a transmission line
   d. a wavelength
4. The higher up an antenna is placed the
   a. more power required for transmission
   b. farther the signal can be received
   c. longer the wavelength of the signal
   d. shorter the wavelength of the signal
5. The waves transmitted from an antenna are called ________ waves.
   a. electric
   b. magnetic
   c. audio
   d. electromagnetic
6. An antenna that is rather directional is called a ________ antenna.
   a. dipole
   b. beam
   c. long
   d. short
7. Night time transmissions can be received over a much greater distance because

a. the air is cooler
b. the sun has set
c. of the sky wave
d. of the ground wave

8. A detector is a
   a. radio frequency rectifier
   b. type of amplifier
   c. type of antenna
   d. type of speaker
9. The information transmitted in a radio wave is in the
   a. carrier
   b. amplitude of the wave
   c. sensitivity
   d. modulation
10. The device in a radio that selects the desired signal from the interference is called a
    a. modulator
    b. tuner
    c. selector
    d. detector

19

PRINCIPLES OF TELEVISION

19.1 Basic Arrangement

Television transmission and reception make use of electromagnetic waves to send and receive pictures over considerable distances. Either still pictures or motion pictures may be used at times, but most television programs provide images of live scenes. Figure 19-1 shows the basic plan of a television system. A television

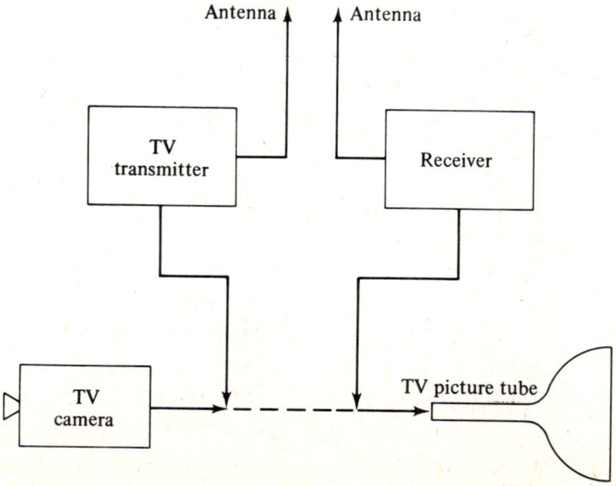

Fig. 19-1
Basic plan of a television system.

camera picks up light images. These images are sent to the transmitter and are changed into electromagnetic waves, which are transmitted through the air by the antenna. At the television receiver, these waves are picked up by the receiving antenna, processed by the receiver, and changed into signal voltages which are fed into a *cathode-ray tube.* The *CRT,* also called a picture tube, changes the signal voltages back into the original light images. Of course, the system used in this process is rather complicated.

Since sound is also transmitted at the same time as the picture, a television transmitter has both picture and sound channels. The receiver also, has both picture and sound channels, as shown in Fig. 19-2. To transmit a picture image by

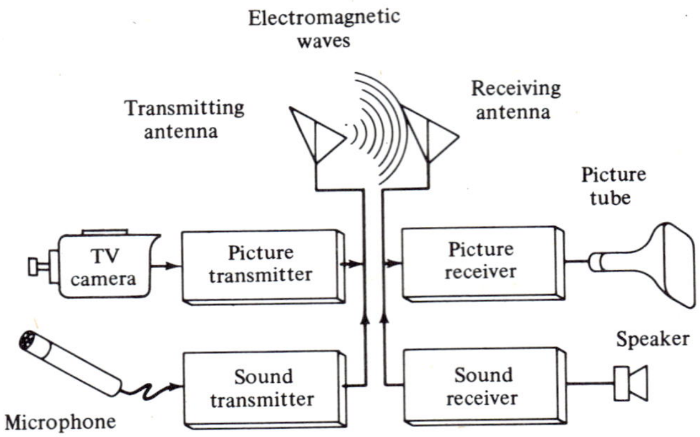

Fig. 19-2
Plan of a television system.

television, the image must be broken down into small "elements" which are changed into signal voltages as shown in Fig. 19-3. This is called the *scanning*

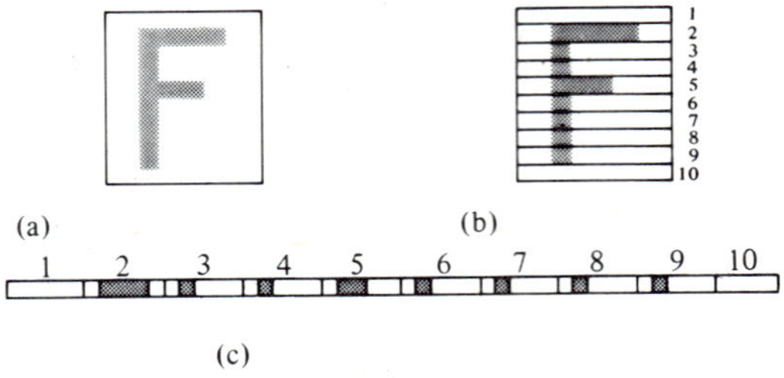

Fig. 19-3
Development of signal voltages by the scanning process. (a) Pattern to be scanned. (b) Scanning sequence. (c) Sequential signal voltage.

process. It can be compared with reading a printed page. In the example of Fig. 19-3, we have broken the pattern down into 10 strips, called *scanning lines.* The scanning process starts at the upper left-hand corner, and moves left to right. When the end of the first scanning line is reached, the scanner moves back quickly to the left-hand end of the second scanning line. This process is repeated until all 10 scanning lines have been completed. In the meantime, the black-and-white picture "elements" are changed into signal voltages. A picture element is simply the spot which is being scanned at a given instant. In a television system, the picture is scanned by 525 lines. The vertical scanning rate is thirty times per second, and the horizontal scanning rate is 15,750 times per second.

Thus, the television camera scans a light image as it changes its light and dark "elements" into signal voltages. Figure 19-4 shows how an image can be built up from black and white dots. A *flying-spot scanner* used in television servicing pro-

Fig. 19-4
An image can be built up from black and white dots.

cedures is a simple form of television transmitter which scans a slide transparency, and changes its picture "elements" into signal voltages. Figure 19-5 shows how a phototube is used to produce a signal voltage as the light beam scans the transparent slide. A phototube has a light-sensitive plate which emits electrons when

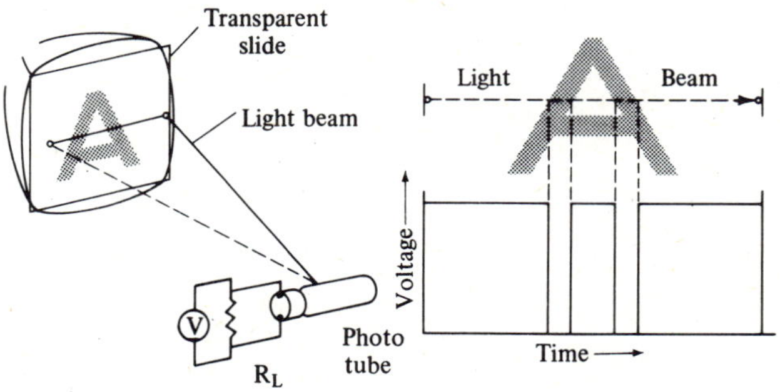

Fig. 19-5
Phototube converts light to a signal voltage.

the plate is struck by light. A bright light causes more electrons to be emitted than a dim light. When the phototube is dark, no electrons are emitted. A flying-spot scanner is a simple type of television camera.

At the television receiver, the picture tube is operated as shown in Fig. 19-6. Three main circuit sections are used. The first section is simply a pair of power supplies that provide DC voltage for operation of the picture tube and its operating circuits. The second section includes the receiver signal circuits and the *synchronization* circuits. Let us see how this system operates.

The picture tube in Fig. 19-6 has a fluorescent screen which glows at any point where it may be struck by the electron scanning beam. This electron beam is produced by a heated cathode. In turn, the electron beam scans the picture-

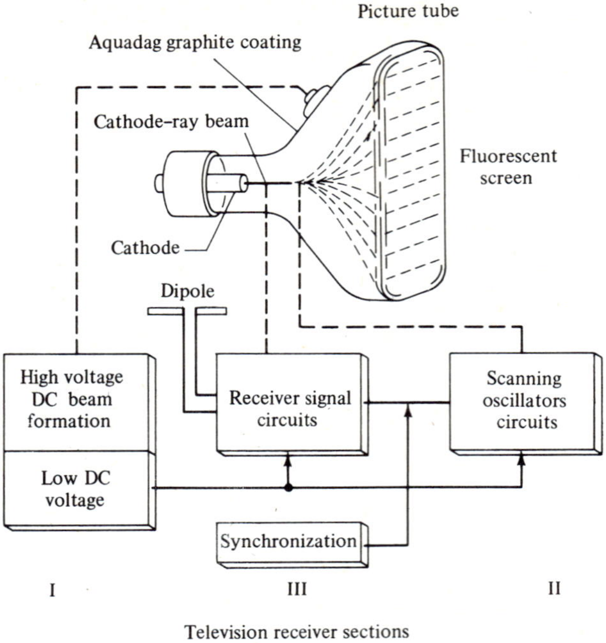

Fig. 19-6
Simplified plan of a picture tube and its operating circuits.

tube screen as it is attracted or repelled by the scanning voltages produced by the scanning oscillators. These scanning voltages have a *sawtooth* wave shape, as shown in Fig. 19-7(a). The *horizontal oscillator* produces a sawtooth voltage which moves

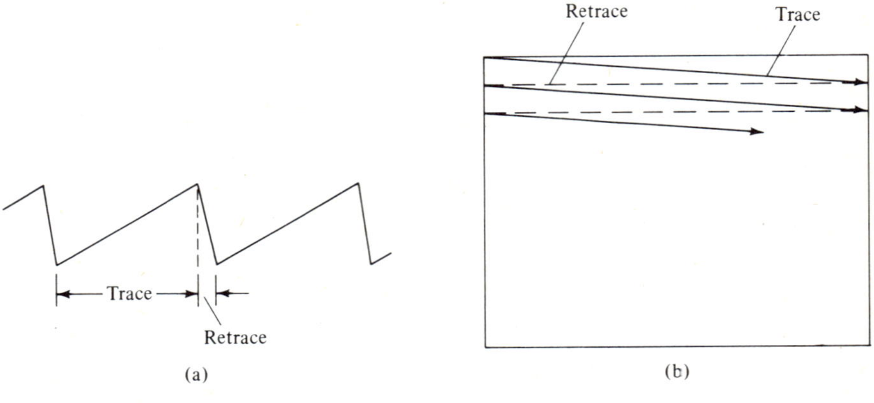

Fig. 19-7
(a) Saw-tooth scanning waveform. (b) Screen scanning pattern.

the electron beam back and forth between the left and right sides of the screen. Also, the *vertical oscillator* produces a sawtooth voltage which moves the electron beam up and down between the top and bottom of the screen. In this way, the screen is scanned line by line in much the same way as we read a printed page.

The receiver signal circuits in Fig. 19-6 step up the small signal voltages from the television antenna in order to control the strength of the electron beam in the picture tube. In other words, when the scanning beam comes to a spot on the screen that is to be white, the receiver signal circuits "turn on" the electron beam to its full strength. On the other hand, when the scanning beam comes to a spot on the screen that is to be black, the receiver signal circuits "turn off" the electron beam so that the screen does not glow at this point. Notice in Fig. 19-6 that a synchronization section operates between the receiver signal circuits and the scanning oscillator circuits. This *sync* section obtains timing signals from the receiver signal circuits. In turn, these timing (sync) signals are used to keep the scanning oscillators in step with the picture signal so that the image on the screen does not become bent, torn, or scrambled.

19.2 Television Receiver Circuit Arrangements

A simplified diagram of the television receiver system is shown in Fig. 19-8. No-

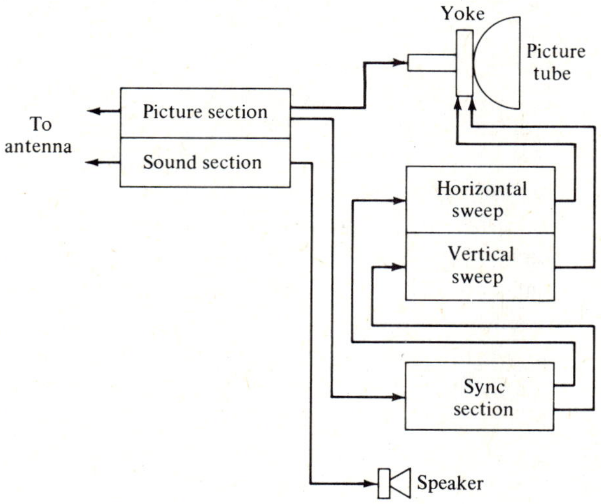

Fig. 19-8 Simplified diagram of a television receiver arrangement.

tice that the outputs from the *horizontal sweep* and the *vertical sweep* sections are fed to a *yoke*. A sweep section consists of a sawtooth oscillator and an amplifier to step up the sawtooth voltage. A *deflection yoke,* or simply yoke, is a pair of coils on the neck of the picture tube. The output from the horizontal sweep section is fed to the *horizontal deflection coil.* Also, the output from the vertical sweep section is fed to the *vertical deflection coil.* When the sawtooth deflection currents flow through these coils, the electron beam inside the picture tube is scanned over the screen. This scanning action is produced by the magnetic fields of the deflection coils. Since the electron beam is moving through the center of the yoke, this electron motion is the same as an electric current. Therefore, the electron beam is surrounded by a magnetic field so that the beam is deflected by the magnetic fields from the yoke.

Figure 19-9 shows the details of a television picture tube. The *control grid*

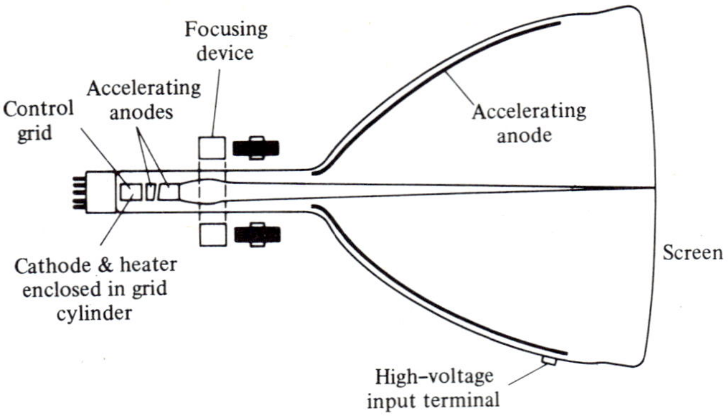

Fig. 19-9
A magnetically-controlled cathode-ray tube.

is a metal cylinder which encloses the cathode and heater. This entire assembly is called an *electron gun.* An electron gun not only emits electrons from the cathode part but also forms these electrons into a thin beam. Moreover, the strength of this electron beam can be changed by changing the signal voltage on the control grid. When the signal voltage is negative, the electrons are repelled back toward the cathode. On the other hand, when the signal voltage is positive, the electrons flow out the end of the control grid in large numbers.

Next, the electron beam from the control grid must be speeded up. This is the job of the *accelerating anodes.* These are metal cylinders which are connected to a fairly high positive DC voltage. This positive voltage attracts the electrons and speeds them toward the screen. In order to focus the electron beam to a small spot, a *focusing device* is placed on the neck of the picture tube. This may be a permanent magnet. Next, the deflection coils are mounted on the neck of the picture tube, to scan the electron beam across and up and down the screen. Notice that high DC voltage is applied to the accelerating anode. This is a graphite coating inside the picture tube, and gives the final speed-up to the electron beam before it strikes the fluorescent screen.

Let us consider the television signal shown in Fig. 19-10—this is also called a

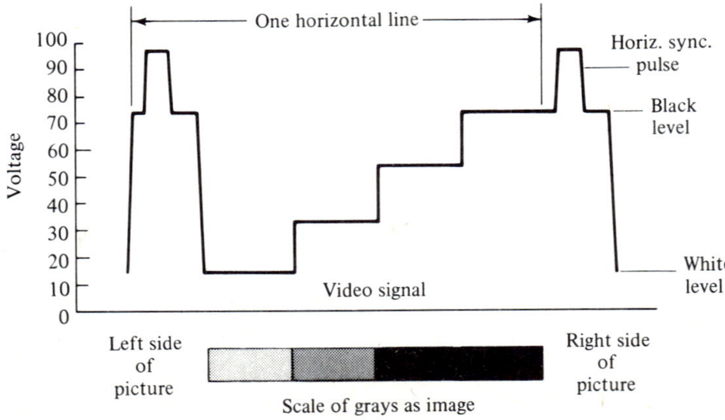

Fig. 19-10
A television video signal voltage.

video signal. In this example, four gray blocks are being scanned. The grays increase in darkness from left to right. In turn, the video signal has the appearance of stair steps. That is, each darker shade of gray produces a higher video signal voltage. Notice that a *horizontal sync pulse* is included at the beginning and at the end of the signal. The sync pulses are put into the signal automatically at the television transmitter. At the receiver, the sync pulses are picked out and fed to the horizontal sweep section as shown in Fig. 19-8, thereby keeping the horizontal scanning lines in step with scanning process at the television transmitter. When the bottom of the picture is reached, a comparatively long *vertical sync pulse* is automatically put into the video signal at the television transmitter. The beginning and end of each vertical sweep is then kept in step with the scanning process at the television transmitter.

Instructor Demonstration or Student Experiment:

Turn on a black-and-white television receiver and adjust the set for a good picture. Take the north pole of a strong permanent magnet and move it across the front of the screen while observing the distortion of the picture. Now use the south pole of the magnet and repeat the above action. Move the magnet at different speeds and observe the screen.

Questions:

1. Why does the magnet cause distortion of the picture?
2. Why does the speed at which the magnet is moved have an effect on the picture?
3. How do changing poles of the magnet affect the results?

19.3 Television Receiver Adjustments

A television receiver has several controls which must be properly set to obtain normal operation. One group of controls is called *operating adjustments,* and another group of controls is called *service adjustments.* The operating adjustments may be used each time a television station is tuned in. On the other hand, the service adjustments seldom require attention. Therefore, the operating adjustments are placed on the front, side, or top of the television receiver where they are easily reached. However, the service adjustments are usually on the rear of the receiver; some service adjustments are on the neck of the picture tube.

Figure 19-11 shows a typical group of operating controls or adjustments.

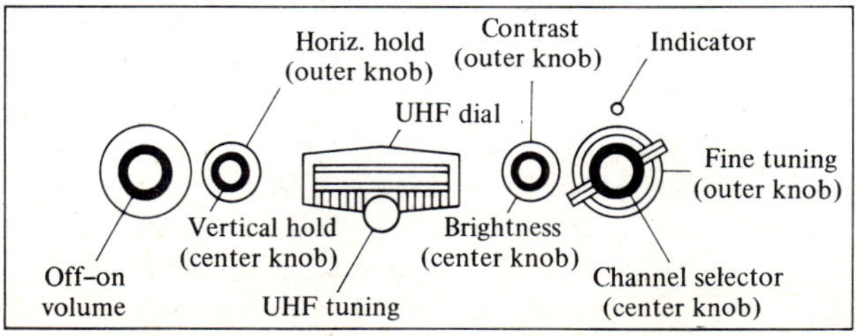

Fig. 19-11
Typical operator controls.

However, we find that some receivers differ in that the *horizontal-hold* control may be included with the service adjustments, for example, or the *brightness control* may be included with the service adjustments. In summary, most receivers tend to provide the following operating controls:

1. Off-on switch, and volume control
2. Channel selector
3. Fine-tuning control
4. Contrast control
5. Brightness control
6. Horizontal-hold control
7. Vertical-hold control

In some cases, it might be necessary only to turn the channel-selector control when changing stations. In other cases, however, it may be desirable to also adjust the fine-tuning control. Sometimes the contrast control or the brightness control or both may need to be reset. The volume control is less likely to need resetting. Usually the horizontal-hold and the vertical-hold controls will stay in adjustment for days or weeks at a time.

The off-on switch is used to apply and remove power from the receiver circuits. It is usually combined with the volume control, which makes the sound output louder or weaker. Next, the channel-selector is a switch which is turned to tune in the television station that is desired. This control has twelve positions, numbered from two to thirteen. These are the channel numbers of the twelve VHF bands. All modern television receivers also have a UHF tuning control, as shown in Fig. 19-11. The UHF channels are numbered from fourteen through eighty-three; "VHF" means very high frequency, and "UHF" means ultra high frequency, as noted earlier.

The fine-tuning control trims up the tuning to exact frequency so that the best picture and sound operation are obtained. If the picture does not look quite normal or if the sound seems distorted after a channel has been selected, the fine-tuning control is turned for best operation. Notice that the fine-tuning control is used only on the VHF channels. When the tuner is set for UHF reception, the fine-tuning control has no effect—only the UHF tuning knob is effective in tuning UHF stations.

Next, the contrast control has the effect of making the dark grays darker or even black and of making the light grays lighter or even white. In turn, this control is set to make the shades of the image or picture as natural in appearance as is possible. Although the brightness control might seem to have some of the effect of a contrast control, its action is basically different. In effect, the brightness control changes all the gray shades of the picture by the same amount and in the same direction. For example, if the brightness control is turned up, all the gray shades will become lighter (see Fig. 19-12). Or if the brightness control is turned down, all the gray shades will become darker. A good rule to follow is to set the brightness control for the most natural background, such as sky or hills.

Notice that the horizontal-hold control changes the frequency of the horizontal-scanning oscillator. If this control is adjusted incorrectly, the picture will

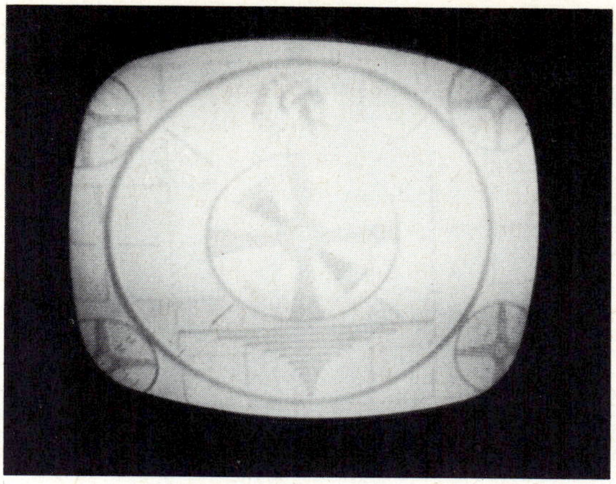

Fig. 19-12
Test pattern showing the effect when the brightness control is set too high.

break up into slanting black-and-white bars, as shown in Fig. 19-13. The hori-

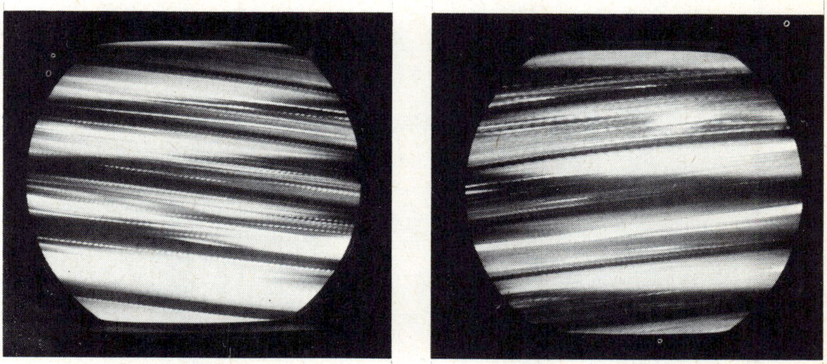

(a) (b)

Fig. 19-13
Loss of horizontal sync. (a) Oscillator running too fast. (b) Oscillator running too slow.

zontal-hold control should be set to a position such that the picture reappears, and stands still, with no tendency to pull either to the left or to the right side of the screen. In much the same way, the vertical-hold control changes the frequency of the vertical-scanning oscillator. If the picture rolls up or down, the vertical-hold control should be set to a position such that the picture stands still with no tendency to pull either up or down.

Figure 19-14 shows a typical group of service controls or adjustments. As noted above, we also find some of the service adjustments on the neck of the pic-

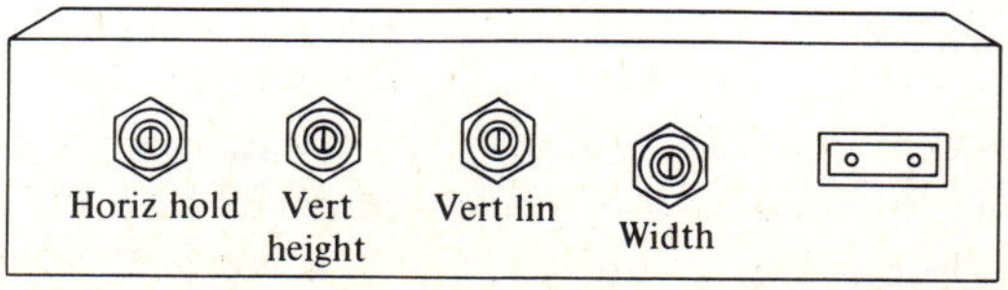

Fig. 19-14
Typical service adjustments.

ture tube. The service adjustments usually consist of the following:

1. Focus control or connection board
2. Vertical-size (or height) control
3. Vertical-linearity control
4. Horizontal-width (or size) control
5. Horizontal-linearity control
6. Automatic-gain control
7. Buzz control

Some service adjustments, such as the horizontal-linearity control or the buzz control, may be omitted from certain types of receivers. Let us see how the service adjustments work. As its name indicates, the focus control is used to produce a sharp and clear picture. The focus control may be a flexible shaft extending from a permanent magnet on the neck of the picture tube. Or it may be a knob on the rear of the chassis. Again, a terminal strip may be provided with a choice of three or four different voltages. The focus lead is then connected to the terminal that gives the sharpest picture.

The vertical-height control is adjusted to make the picture fill the screen exactly from top to bottom. However, the picture might appear cramped at the top or bottom. Figure 19-15 shows top-cramping, and the vertical nonlinear-

Fig. 19-15
Test pattern showing the effect of misadjustment of the vertical-linearity control.

ity control must be adjusted. Notice that adjustment of the vertical-linearity control usually requires a little readjustment of the vertical-height control also. In case the picture is too short or too long from left to right, the horizontal-width control should be adjusted. In some receivers, this is a control knob on the rear of the chassis. However, in other receivers, the width control is a thin brass cylinder that slides on the neck of the picture tube. It is slid more or less under the yoke until the picture exactly fills the screen from left to right. Still other receivers provide a trimmer capacitor which operates as a horizontal-width control.

Figure 19-16 shows the appearance of horizontal nonlinearity. If the television receiver has a horizontal-linearity control, it is adjusted for perfect circles in the picture. We find that a horizontal-linearity control may need readjustment after the width control is changed. (The horizontal-width control may be called a horizontal-drive control in some receivers). Sometimes the horizontal-linearity

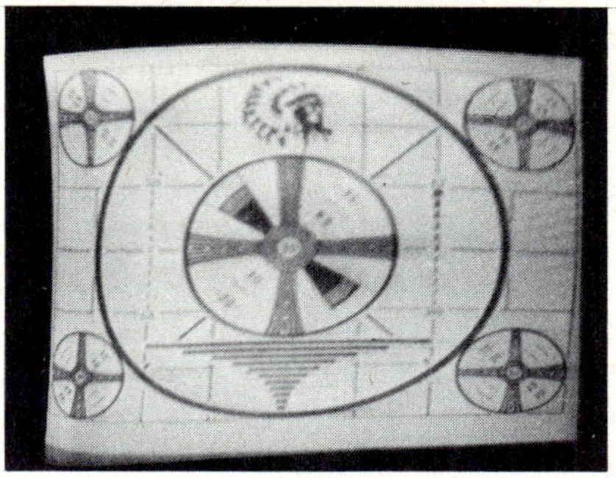

Fig. 19-16
Test pattern showing the effect of misadjustment of the horizontal-linearity control.

control is a knob on the rear of the chassis. Or it may be a slug (adjustable core) in a coil. In still other receivers, small magnets are provided on the neck of the picture tube for adjustment of horizontal linearity.

The automatic gain *(AGC)* control is set to a point such that weak stations show up best on the screen. However, the AGC control must not be set so high that strong stations produce a "muddy" or "filled up" appearance on the screen. In finding the best setting for the AGC control, the contrast control should also be turned up or down a little, as may be needed for best operation.

Finally, the buzz control that we find on some television receivers has the effect of quieting sound interference. That is, if we hear a continuous buzzing in the sound, or if we only hear a buzz at certain times when particular scenes are being programmed, the buzz control should be adjusted. The setting of this control is correct when no buzz is heard in the sound. However, the control must not be turned so far that the sound volume is seriously weakened.

Many television receivers also provide centering controls. In most cases, we will find the horizontal and vertical centering controls in the form of a pair of levers on the back of the deflection yoke, as shown in Fig. 19-17. These levers

Fig. 19-17
Neck of a picture tube in a TV receiver which employs an unsecured deflection yoke and two magnetized rings for a centering device.

will rotate a pair of magnetized rings. When each lever is rotated to the correct position, the picture will be framed properly on the screen. Occasionally, we may find an *ion trap* on the neck of the picture tube. Most picture tubes do not have ion traps. However, when we are working with a tube that has this device, it is adjusted by turning and sliding the magnet on the tube neck until the picture is brightest, and any corner shadows are eliminated.

Questions: Chapter 19

1. Draw a block diagram of a basic television system.
2. What is the purpose of the CRT in a television receiver?
3. Why must a television receiver have both sound and picture channels?
4. Why is it very important that the scanning sequence that produces the picture at the receiver be in synchronization with the scanning sequence at the transmitter?
5. What component on the CRT produces the scanning action of the electron beam?
6. Explain the purpose of the electron gun.
7. What is the purpose of the yoke on the neck of the picture tube?
8. Describe the functions of the electron gun in the CRT.
9. What is the purpose of the accelerating anode in the picture tube?
10. Outline the composition of the television video signal.
11. What is the basic difference between the service adjustment controls and the operating adjustments?
12. List the operating controls that are provided on most television receivers and state the purpose of each.
13. Name the service controls that are provided on most television receivers and state the purpose of each.
14. What is the purpose of the ion trap in a CRT?

True–False:

1. The cathode-ray tube changes electrical signals to light images.
2. A television set is made up of two separate receivers, one for sound and and one for picture.
3. The process of breaking a scene down into small elements for transmission is called scanning.
4. Scanning of the picture on the CRT screen is controlled by both the vertical and the horizontal oscillators.
5. A yoke on a CRT neck is also called an electron gun.
6. The purpose of the yoke is to cause the electron beam to scan the screen of the picture tube.
7. A focusing device on the neck of the picture tube may be a permanent magnet.
8. Loss of timing (sync) signals in the receiver can cause the picture to be torn, bent, or scrambled.
9. One function of the electron gun is to assist in the focus of the electron beam.
10. The accelerating anode serves the function of a scanning device.
11. A video signal received by a television receiver carries the picture in-

formation as well as the sync pulses.

12. The off-on switch on a television is usually combined with the volume control.
13. UHF means unusually high frequency.
14. The fine-tuning control on the VHF channel is used for tuning UHF stations.
15. A horizontal-hold control changes the frequency of the horizontal scanning oscillator.
16. The vertical-hold control is adjusted to prevent picture roll up and down.
17. Not all receivers are furnished with a horizontal-width control.
18. The AGC control adjusts the height of the video signal.
19. An ion trap is furnished on all picture tubes.

Multiple choice: choose the best answer.

1. The television camera breaks a scene down into small segments for transmission, and this process is called
   a. dotting
   b. focusing
   c. scanning
   d. screening
2. The three main sections of a television receiver are
   a. antenna, CRT, and screen
   b. power supplies, receiver circuits, and synchronization circuits
   c. synchronization circuits, power supplies, and antenna
   d. receiver circuits, focusing circuits, and CRT
3. The purposes of the scanning oscillators are
   a. to generate the deflection voltages
   b. to prevent shadows on the screen
   c. to develop the vertical and horizontal synchronization signals
   d. to develop the black-to-white ratio on the screen
4. The electron gun serves to
   a. emit electrons and form them into a thin beam
   b. emit electrons
   c. focus the electron into a beam
   d. shoot ions at the screen to produce the black area
5. The signal containing the sync pulses and picture information is called a ____________ signal.
   a. picture
   b. complete
   c. composite
   d. video
6. The control that is used to set the ratio of the light to dark areas in a scene is called the ____________ control.
   a. contrast
   b. brightness
   c. tuning

d. darkness

7. The fine-tuning control is used to
   a. select UHF stations
   b. select VHF stations
   c. set the horizontal-scanning rate
   d. set the vertical-hold rate
8. To correct a slow rolling of the picture, we would adjust the
   a. contrast control
   b. vertical-hold control
   c. horizontal-hold control
   d. fine-tuning control
9. To increase the light level of the picture on a CRT, we would adjust the ____________ control.
   a. contrast
   b. brightness
   c. darkness
   d. fine-tuning
10. A picture that is pulled to the right or left indicates a misadjusted
    a. antenna
    b. horizontal-hold control
    c. vertical-hold control
    d. contrast control
11. The vertical-linearity control is interactive (must be adjusted) with the
    a. horizontal-width control
    b. vertical-height control
    c. volume control
    d. contrast control
12. To quiet sound interference, we would adjust the ____________ control.
    a. ion
    b. sound
    c. fuzz
    d. buzz
13. To center the picture on most television receivers, we would
    a. set the centering control
    b. adjust the ion trap
    c. adjust the yoke
    d. set the horizontal sync
14. A circular shadow on one edge of the picture would be caused by a misadjusted
    a. shadow guard
    b. brightness control
    c. yoke
    d. ion trap

20

BASIC COLOR TELEVISION

20.1 General Principles

Color-television transmitters and receivers are more complicated than black-and-white transmitters and receivers, because additional circuits and parts are necessary to send and receive the color signal. The basic plan of a color-television system is shown in Fig. 20-1. A color-television camera produces the color signal voltages, and sends them to the transmitter where they are amplified and sent through the air from the antenna. At the receiver, the signals are picked up by the antenna, amplified, and a color picture tube is used to reproduce the scene that is being scanned at the transmitter. There are a large number of elec-

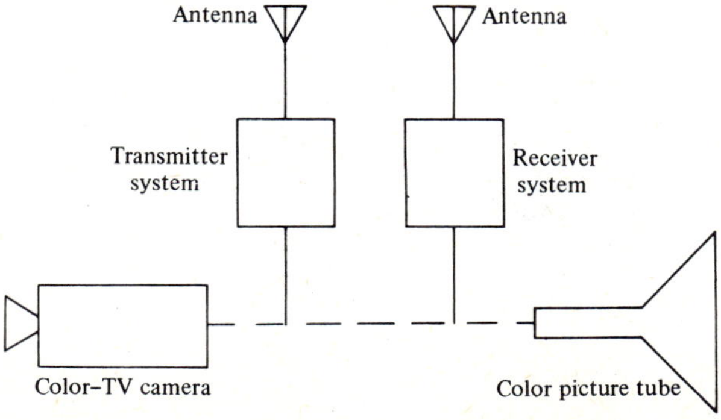

Fig. 20-1
Basic plan of the color-TV system.

tronic circuits and devices in between the color-television camera and the color picture tube.

Although there are hundreds or even thousands of different colors in a scene such as a garden, only three colors are actually transmitted in the color-television system. These are the primary colors: red, green, and blue. It is practical to use only three colors in the television system, because any desired color can be obtained by means of a suitable mixture of the three primary colors. To transmit these three colors, a color-television camera (Fig. 20-2) consists of three separate cameras, each of which responds to one of the primary colors.

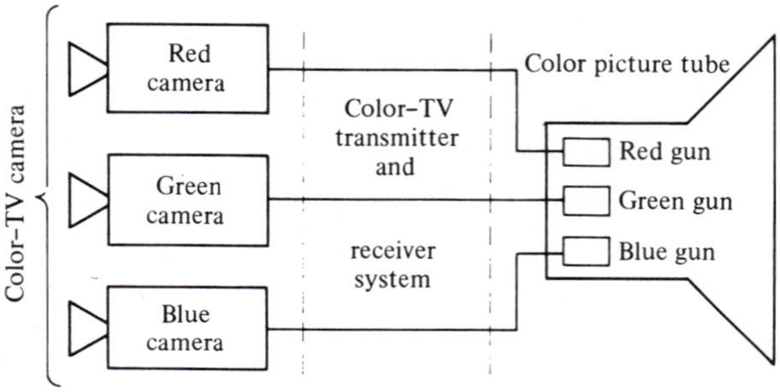

Fig. 20-2
Operating principle of the color-TV system.

That is, one camera responds to red light, another camera responds to green light and still another camera responds to blue light.

At the color-television receiver, the color picture tube contains three separate electron guns. These are called the red gun, the green gun, and the blue gun. A signal voltage fed to the red gun will produce a red color on the picture-tube screen, and likewise with the other two electron guns. Figure 20-3 shows the locations of the three main parts in a color picture tube. These are the electron gun structure, the shadow mask, and the phosphor-dot screen. A glowing image is produced by the phosphor-dot screen when it is struck by the electron beams from the guns.

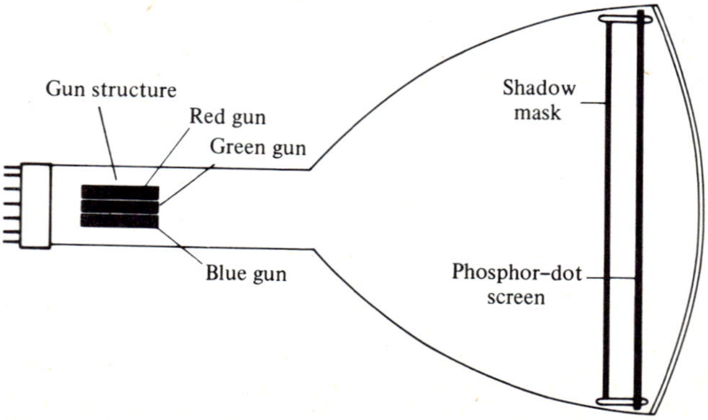

Fig. 20-3
Locations of the three main parts in the color picture tube.

An enlarged view of how the fluorescent screen is made is shown in Fig. 20-4; the screen consists of many tiny phosphor dots. A red phosphor dot will

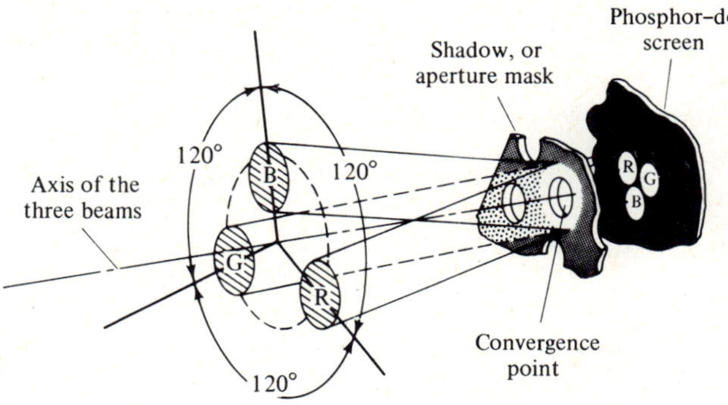

Fig. 20-4
The relationship between beams, aperture mask, and phosphor screen in the color picture tube.

glow with a red color when struck by an electron beam. In the same way, the screen will glow with a blue color when the blue phosphor dots are struck by a beam. Or the screen will glow with a green color when the green phosphor dots are struck by a beam. If all the red, green, and blue phosphor dots are struck at the same time by the electron beams, the screen will glow with a white light. That is, white is produced by a mixture of the three primary colors. If yellow is to be produced, the red and green dots are struck. Yellow is a mixture of red and green.

Since there are three electron guns in a color picture tube, there are three electron beams scanning the screen. These three electron beams must pass through the shadow (aperture) mask before they can strike the phosphor dots. Careful study of Fig. 20-4 shows that the electron beam from each gun arrives at the shadow mask from a different angle. This mask has many tiny holes, or apertures. Since the three electron beams arrive at different angles, each beam can strike only one color dot. For example, the "red" beam can strike only the red dots, and so on. If the "red" beam is turned on, and the green and blue beams are turned off, the scanning of the screen will produce red light only.

Color-television receivers are *compatible,* which means that if a color-television receiver is tuned to a black-and-white broadcast program, the black-and-white picture will be displayed on the screen of the color picture tube. From what has been said previously about mixtures of the primary colors, we can understand that all three electron guns are operating continuously while a black-and-white picture is being displayed. On the other hand, during reception of a color-television program, one or two of the electron guns may be turned off

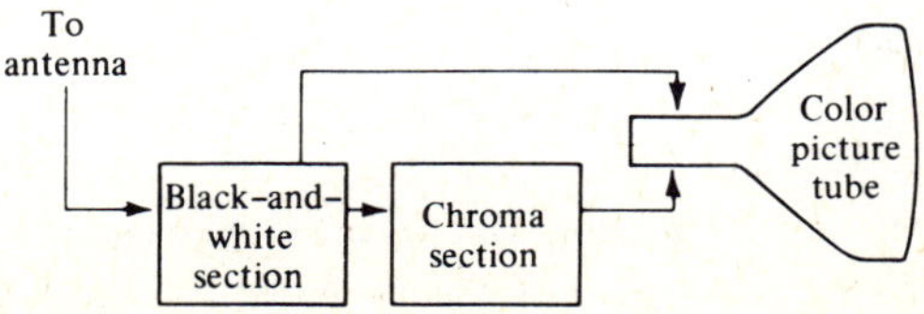

Fig. 20-5
Two main sections of a color receiver.

more or less from one part of the picture to another. This unbalance in gun outputs results in display of various colors, tints, and shades.

20.2 Color-Television Receiver Arrangement

A color-television receiver must have all the circuits and parts used in a black-and-white receiver, plus additional circuits and parts for the color signal. Figure 20-5 shows that a color-television receiver has a black-and-white section, followed by a *chroma* section. Both the black-and-white and the chroma sections feed signal voltages to the color picture tube. It is important to know that, in the operation of a color receiver, only the black-and-white section operates when the receiver is tuned to a black-and-white program. However, both the black-and-white and the chroma sections operate when the receiver is tuned to a color program. That is, the incoming color signal automatically turns the chroma section on. If the color signal stops, the chroma section automatically turns off.

Now let us see how a color signal differs from a black-and-white signal. The first fact that we notice is that the horizontal sync-pulse of a color signal has a *color burst signal,* as shown in Fig. 20-6. This burst signal is absent in a black-

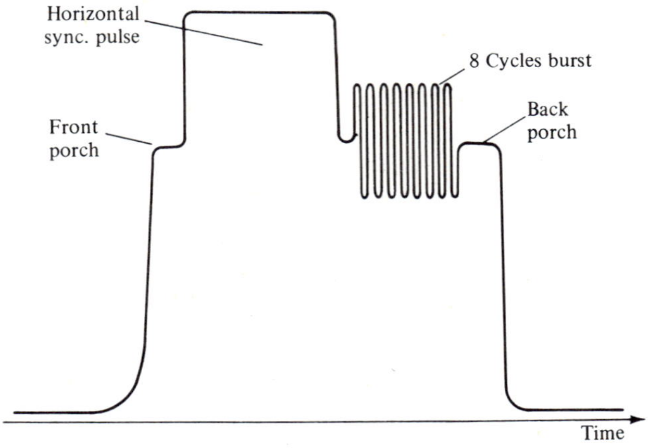

Fig. 20-6
Horizontal sync pulse with the color burst.

and-white signal because it is this signal that turns on the chroma section in a color-television receiver. The frequency of the burst signal is 3.58 MHz. When a black-and-white receiver is tuned to a color signal, the burst has no effect on its operation. This is because a black-and-white receiver does not have any chroma circuits tuned to 3.58 MHz. On the other hand, the chroma section in the color receiver in Fig. 20-5 has circuits which are tuned to 3.58 MHz so that the color burst can be separated from the horizontal sync pulse and used to turn on the chroma circuits.

The complete color signal is made up as shown in Fig. 20-7. Notice that the complete color signal has a black-and-white part, and a chroma part. The

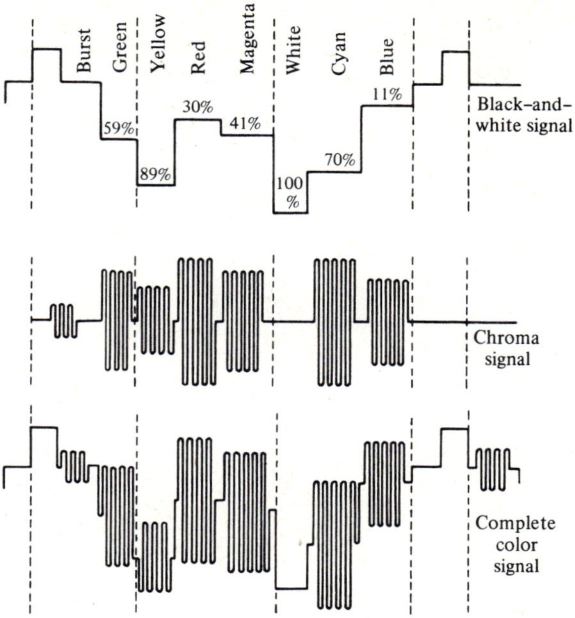

Fig. 20-7
Elements of TV color bars.

chroma portion contains the color burst and, in this example, the 3.58 MHz portions of the primary and complementary colors. This is the complete color signal that is produced by one type of *color-bar generator* that is used to test and adjust color-television receivers. Remember that the primary colors are red, green, and blue, while the complementary colors are magenta, cyan, and yellow. White is not a color in the strict sense of the word and black is simply the absence of light. Figure 20-8 shows a color-bar pattern as it appears on the screen of a color-television receiver.

Fig. 20-8
Color-bar pattern.

In color-television receivers, tuned circuits are used to separate the black-and-white signal and the chroma signal, as shown in Fig. 20-9. Notice that the

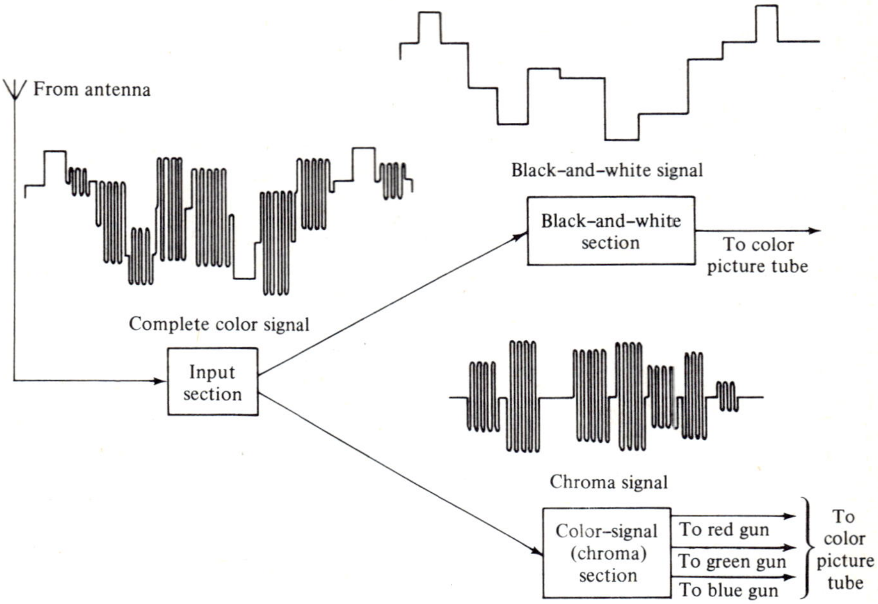

Fig. 20-9
Separation of the black-and-white and chroma signals in the color-TV receiver.

television antenna is connected to an input section, which feeds the complete color signal to the black-and-white and to the chroma sections. The chroma section is tuned to 3.58 MHz. This tuning action allows the chroma signal to go into the chroma section, and keeps out the black-and-white signal. In the black-and-white section, the circuits are tuned to pass the black-and-white signal, and a trap circuit is included to keep the chroma signal out. Therefore, only the black-and-white signal can pass through the black-and-white section.

The black-and-white part of the signal is fed to all three electron guns in the color picture tube. On the other hand, the chroma section breaks down the chroma signal into its primary red, green, and blue portions, and feeds the signals by three separate paths to each of the electron guns. The red signal voltage is fed to the red gun, the green signal voltage is fed to the green gun, and the blue signal voltage is fed to the blue gun. The result is that any one of the electron guns can be turned off or on by the chroma section. However, the black-and-white section turns all three guns on or off at the same time. The end result is that the chroma signal can produce a color image, but the black-and-white signal can produce only a black-and-white image.

Comparatively simple and less costly color-television transmitters and receivers were used in the early days of color-television broadcasting, but they had a serious disadvantage. A simplified system operated with a simplified color signal which could not be displayed by an ordinary black-and-white television receiver. In addition, a simplified color-television receiver could not display an ordinary black-and-white program transmission. In other words, the simplified color-television systems were *incompatible.* This means that if a viewer wished to receive both color and black-and-white television programs, he would have had to purchase two separate television receivers. The total cost would have been greater in that case. Public opinion eventually voted against the simplified systems, and the *compatible* system was legalized.

20.3 Color-Television Receiver Adjustments

A color-television receiver has all the controls and adjustments that have been explained in the previous chapter. In addition, a color receiver has certain controls that affect only the color portion of the picture. These can be grouped into *operating adjustments* and into *service adjustments* (Figure 20-10). The two operating adjustments are called the *color control* and the *tint control.*

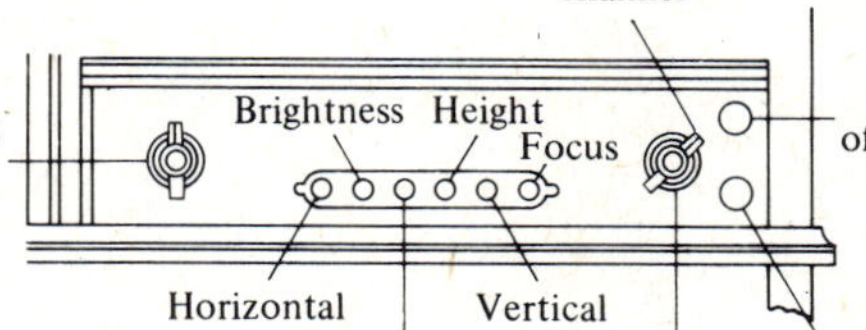

Fig. 20-10
Typical control arrangement for a color-TV receiver.

Notice that the color-control setting puts more or less color into the image when a color program is being received. That is, if we turn the color control all the way down, there is no color in the picture—only the black-and-white portion of the picture is displayed on the screen. But if the color control is turned too far up, there is too much color in the picture—the image looks like a color cartoon. Therefore, we turn the color control up just far enough to display a normal amount of color in the picture.

Next, the tint control makes the colors look natural when the control is correctly adjusted. If the tint control is incorrectly adjusted, "purple people" may be displayed on the screen. It is usually best to set the tint control for the most natural face color because the actors or performers are watched closely. We find that both the color control and the tint control may need to be readjusted when another station is tuned in. Sometimes the tint control may need to be reset slightly during the course of a program from the same station. For example, if the program switches from an indoor scene to an outdoor scene, another camera will be used; and this may require resetting of the tint control of the receiver.

We find that the setting of the fine-tuning control must be made more exactly in the case of color reception than in black-and-white reception. For example, if you tune in a color program, and no color appears in the picture even when the color control is turned up all the way, it is most likely that the fine-tuning control is not correctly adjusted. The best way to set the fine-tuning control is to turn it until sound interference appears in the picture. Then, back off on the control setting until the sound interference just disappears. This setting of the fine-tuning control will provide the best reception of the color signal.

There are more service adjustments in a color-television receiver than in a black-and-white receiver. Most of these additional service adjustments are used to obtain proper operation of the color picture tube. Since there are three electron beams in the tube, all three beams must be focused to the same spot on the screen. This is called *convergence* of the color picture tube. If one or two of the beams do not strike the screen at the same spot, we observe color fringing in the picture. This is called *misconvergence.* Convergence adjustments are usuallly made with a *crosshatch pattern* on the screen of the color picture tube. This pattern is produced by a test instrument called a crosshatch generator.

An example of a misconverged crosshatch pattern is shown in Fig. 20-11.

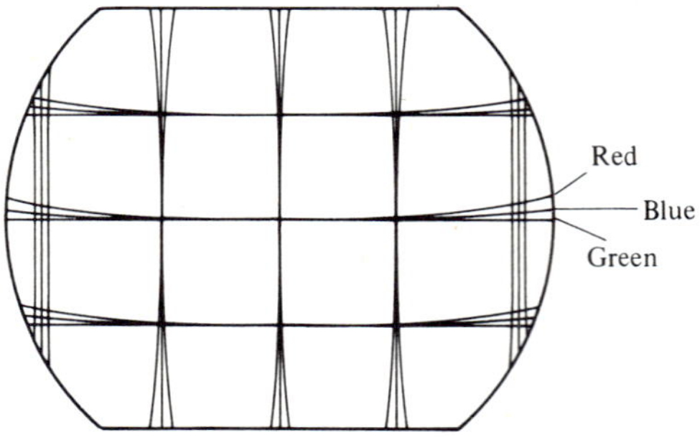

Fig. 20-11
Crosshatch pattern on screen of misconverged color picture tube.

The red, green, and blue lines do not fall together at all points in the pattern. In this example, the greatest misconvergence appears at the screen edges. There are about a dozen convergence controls provided in a typical color receiver. These controls must be adjusted according to a standard procedure, and should never be adjusted without the proper instruments. When the convergence adjustments are correctly made, the red, blue, and green lines fall together at all points in the pattern. Then there is no color fringing in the pattern, and the crosshatch pattern consists of white lines only.

Questions: Chapter 20

1. Why are color-television transmitters and receivers more complicated than black-and-white transmitters and receivers?
2. What basic colors are transmitted by a color-television system?

3. How are the thousands of colors in an outside scene presented on a television screen?
4. Explain the differences between a color picture tube and a black-and-white picture tube.
5. How does a color picture tube present a black-and-white television scene?
6. What is the function of the shadow mask in a color picture tube?
7. How many more color dots are there on a color-picture-tube screen than holes in the shadow mask?
8. Discuss the word "compatible" as it is applied to a color television receiver.
9. How is the chroma section of a color television turned on to receive a color signal?
10. What is the name given to the test instrument that produces the vertical bars of different colors on a color television?
11. How is the color signal prevented from entering the black-and-white section of a color television?
12. Why is the compatible color-television system more complex than the simplified color-television system?
13. Why was the compatible color-television system adopted as the standard system?
14. How should the color and the tint controls be adjusted on a color television?
15. Why does the tint control need to be adjusted so often?
16. What is the reason for color fringing in a picture?
17. State the name of the color-television generator that produces vertical and horizontal lines on a screen.

True–False:

1. Color-television transmitters are less complicated that black-and-white transmitters.
2. A color camera is really three separate cameras.
3. Only three basic colors are transmitted in a color television system.
4. A color picture tube contains three separate electron guns.
5. The screen of a color picture tube is made up of many colors of phosphor.
6. If the electrons from the red gun strike the blue or green phosphor dots, a red color will be produced.
7. The position of the shadow mask is very important to the production of the correct colors on the screen of a color television.
8. There are one-third as many holes in the shadow mask as there are phosphor dots on the screen of a color-television picture tube.
9. Only one electron gun operates to produce a black-and-white picture on a color-television screen.

10. Both the black-and-white and the chroma sections operate when the the receiver is tuned to a color program.
11. The horizontal sync pulse of a color-television signal has a color burst signal of 3.58 MHz.
12. Traps allow the 3.58 MHz color burst to enter the black-and-white section of a television receiver.
13. The chroma section of a color television is turned off during black-and-white signal reception.
14. A color-bar generator produces a cross-hatch pattern on the screen.
15. The composite color-television signal can be received by either a color or a black-and-white television.
16. The cross-hatch generator is used to check convergence of the three electron gun beams.
17. The tint control should be adjusted for the correct face color of the actors.
18. The adjustment of the fine-tuning control is less critical on a color television than on a black-and-white set.

Multiple choice: choose the best answer.

1. The color-television camera contains
   a. only one camera tube
   b. three camera tubes
   c. many camera tubes
   d. one camera tube and three filters
2. The primary colors transmitted by a color transmitter are
   a. red, green, and black
   b. red, green, and white
   c. red, green, and yellow
   d. red, green, and blue
3. The color-television picture tube contains
   a. an electron gun and shadow mask
   b. a phosphor screen and an electron gun
   c. an electron gun, a phosphor-dot screen, and a shadow mask
   d. three electron guns, a shadow mask, and a phosphor-dot screen
4. When a color television receives a black-and-white program
   a. all the electron guns are on
   b. only the black-and-white gun operates
   c. only the green gun operates
   d. only the green and blue guns operate
5. The colors of the dots that make up the phosphor dot screen are
   a. white, blue, green, and red
   b. white, black, green, blue, and red
   c. green, blue, and red
   d. green, red, and yellow

6. When a color television is tuned to a black-and-white program
   a. the chroma section is turned off
   b. all circuits are in operation
   c. only the chroma section is in operation
   d. the set does not receive the program
7. A color-bar generator is used to
   a. calibrate the transmitter
   b. check the color camera
   c. align the color circuits of the transmitter and camera
   d. adjust the color circuits of the television receiver
8. The color-television system that is used in the United States is a/an ______________ system.
   a. incompatible
   b. compatible
   c. linear
   d. universal
9. To make the colors look natural on a television screen, we would adjust the
   a. natural control
   b. color control
   c. vivid control
   d. tint control
10. To adjust a television receiver for convergence, we would use a
   a. color-bar generator
   b. local station
   c. crosshatch generator
   d. distant station

APPENDIX: SOLDERING

Solder is an alloy, generally composed of tin and lead, which is used to join metals together without actually melting them. Solder makes this possible because its melting point is lower than that of the metals to be joined.

When heat is applied to a solder connection and solder is applied, the solder flows around the joint and mixes with the molecules on the surface of the metals to be joined. Upon cooling, the connection becomes a continuous metal surface.

For example, if two copper wires are to be soldered together, they are, before soldering, two separate wires. They can be tied together mechanically (wrapped or twisted together) but are still two wires. When the joint is heated to the melting point of solder and solder is applied, both a chemical and metallurgical reaction takes place. The solder causes a small amount of the copper to be dissolved at temperatures well below the melting point of copper. This causes the formation of a new alloy which is neither solder nor copper. Upon cooling, a continuous metal connection is formed. The two copper wires are fused together and are no longer two separate wires. This fusion of the base metals must take place for us to have a good electrical connection.

A simple mechanical connection between two wires, or between component leads and wires, is not acceptable in electronics equipment for two reasons: First, when copper is exposed to air, oxides form on the surface of the copper (like rust on iron). This oxidation is greatly increased by the application of heat and increases with age. Since a coat of oxides exist on the surfaces of the copper wires when the wires are connected together they really aren't touching. The wires are separated by the coating of oxides. This causes the conductivity at the joint to be unstable. That is, the resistance will be higher than if the wire were continuous, and it will change as the coating of oxides changes. Second, the connection is not solid, and the wires will move easily at the joint and may even come apart causing an equipment failure.

The following rules should be followed when soldering:

1. Prepare the soldering iron by applying flux and solder to the *clean* tip the moment it is hot enough to liquefy solder. This coats the tip with solder and protects it from oxidation (tinning the iron). Leave a spot of solder on the tip between uses to help keep the tip clean and prolong tip life.
2. Inspect the terminals and the wires for cleanliness. Clean with solvent or lead cleaner.
3. Attach the wire(s) to the terminal(s) in a manner suitable to the type of terminal used.
4. Apply a rosin or resin flux *only*. When cored solder is used, the flux is applied automatically just before the solder liquefies, because the flux melts at a lower temperature than the solder. It is often desirable to use extra flux for best results.
5. Clean the tip of the heated soldering iron of old solder by wiping it lightly on a kimwipe or rag. The hot tip may be cleaned by using a sponge which has been moistened with a water-soluble flux.
6. Place the soldering iron tip on the terminal and immediately melt a *small* amount of solder at the point of contact. This forms a solder bridge through which the heat can flow more rapidly to the terminal.
7. When the terminal has reached the melting temperature of the solder, apply solder to the terminal (NOT the soldering iron). This must be done quickly, melting as much as is needed without delay.
8. As soon as enough solder is applied, remove the solder and then remove the iron.
9. Allow the connection to cool *without disturbance* until the solder alloy solidifies. After all soldering is completed, the flux residue can be removed by cleaning with a solvent. Never scrape or wire-brush to remove the flux residue, since this would weaken the connection.

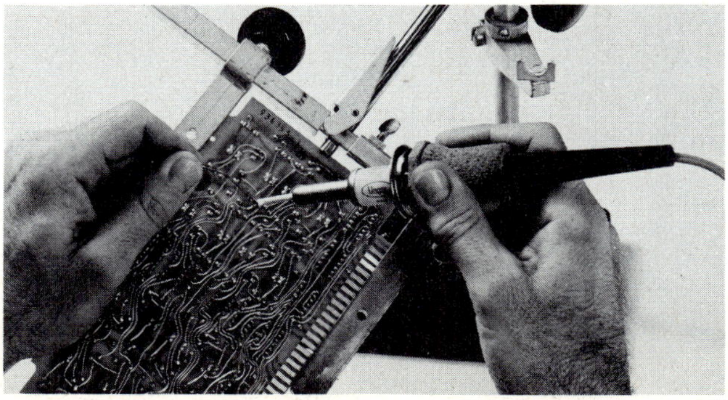

Fig. A-1
Method of soldering to printed-circuit board.

Printed-circuit (PC) assemblies are more likely to be damaged by excessive heat than ordinary chassis assemblies. Therefore, we must learn to be careful in the handling of PC boards, and in the soldering and unsoldering procedures. Two of the most important precautions are to use the right amount of heat and not too much solder. Figure A-1 shows a student making a solder connection on a PC board. As shown in Fig. A-2, it is very helpful to use some special tools such as the following:

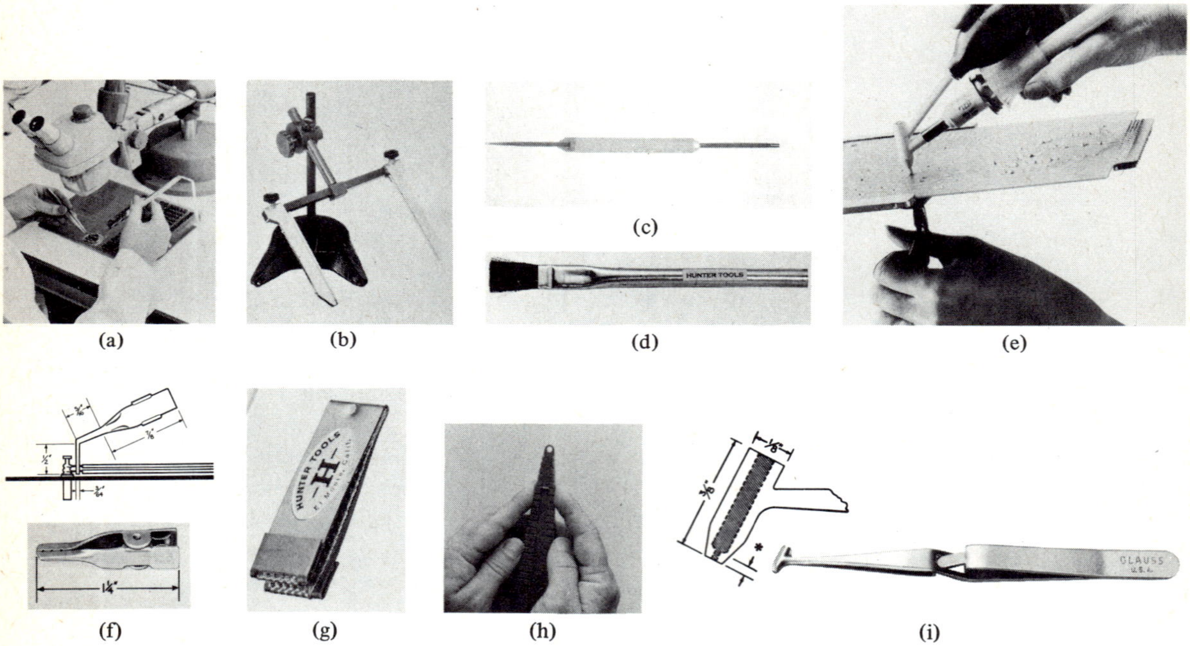

Fig. A-2
Special tools for PC board work.
(a) Small soldering iron. Courtesy: Ungar, Inc.
(b) PC board vise. Courtesy: Hunter Tools
(c) Soldering pick.
(d) Bristle brush.
(e) Desoldering tool. Courtesy: Ungar, Inc.
(f) Antiwicking heat sink. Courtesy: MacDonald & Co.
(g) Component lead cleaner.
(h) Component lead bender. Courtesy: Hunter Tools
(i) Antiwicking tool. Courtesy: Clauss Cutlery

Pencil-type soldering iron, 25 to 37.5 watts
PC board vise
Soldering pick
Bristle brush
Desoldering tool
Heat sink
Component lead cleaner
Component lead bender
Antiwicking tool

Isopropyl alcohol is useful for removing excess flux after soldering, and for cleaning the joint before soldering.

We are using a small soldering iron to avoid excessive heat, and also to keep the heat pinpointed where it is needed.

To install a component on a PC board, first clean the leads with a lead cleaner or abrasive eraser to remove oxides (Fig. A-2[g]). The leads are then carefully bent so that they will enter the terminal holes in the PC board easily. A lead bender (Fig. A-2 [h]) is useful. The new component is then mounted, and the leads can be clinched or swaged close to the pad. This forms a mechanical basis for a good solder joint. Next, a small amount of rosin flux is applied to the lead and pad. A type 60/40 solder is recommended. A small amount of solder is used, and the heat is removed as soon as the solder flows into the joint. Remember to keep the lead motionless until the solder has cooled. If any excess flux remains, it should be removed with a cloth moistened with isopropyl alcohol or other solvent.

After the solder cools, the joint should be inspected to make sure that it is solid and that the solder did not flow over or drop where it can cause a short circuit.

Now let us see how a component is removed from a PC board. We first dip the tip of the metal braid into the soldering flux, and then place the braid over the pigtail lead on the part and the tip of the soldering iron to the braid—we never apply the heat directly to the trace. As soon as the solder melts, the braid sucks up the excessive solder. These steps can be repeated if the area is not free of solder. Special suction devices may also be used to suck up the solder. If we must release two or more terminals to remove the defective component, it may be necessary to heat all the terminals at the same time. This can be done by means of soldering irons that have special tips. Never use force or pressure in an attempt to remove the defective component.

Sometimes we find it difficult to remove a defective component. If the ends of the leads have been squeezed out of shape, it is necessary that they be cut with a pair of diagonal cutters to allow withdrawal from the board. In case the leads are bent, it is usually possible to straighten them before removal. If some solder still remains on the leads, the soldering iron can be momentarily applied, and the leads then withdrawn before the solder cools. In any case, it is important to pull the component straight up from the PC board without forcing or twisting it.

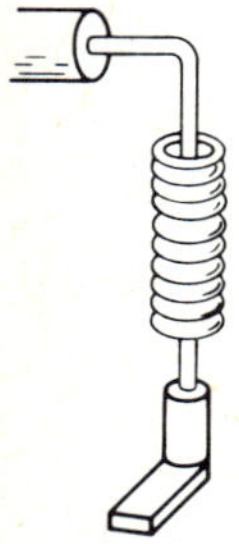

Fig. A-3
Helix-type soldering aid.

Some technicians prefer to leave the cut ends of the component leads in place, and to install the new component with a helix-type soldering aid, as shown in Fig. A-3. This is a small coil of copperweld wire with an outer coating of solder and flux. It is slipped over the end of a cut lead, and the pigtail lead of the new component is inserted into the other end. The coil is then heated with a soldering iron, and a solid connection is made without danger of damage to the PC board.

When component leads are easily removable from a PC board, it may still happen that molten solder will flow into the lead holes while the component is being removed. The excess solder must be removed before the new component is installed. Otherwise, the end of the lead may tear the pad loose from the board. If the solder cannot be pulled out with braid or a solder sucker, use a twist drill of the proper size and carefully drill out the solder from the pad side of the board. Never drill out solder from the component side of the board, or the pad is likely to be torn loose.

When a trace is damaged, it is not always necessary to discard the PC board. For example, it may be practical to remove the loose or charred portions of the trace and replace them with pieces of tinned copper wire bent at both ends. The wire must be long enough to bridge the damaged section of trace; then drill two holes of correct size to pass the wire through the PC board. These holes should not be drilled through the trace, but near the trace. Insert the wire and bend the ends over the trace. Then solder the wire to the trace and the conduction path will operate as it did before the section of trace was damaged.

Remember that transistors and diodes are very sensitive to heat, and will be destroyed if they are overheated. Therefore, the soldering technique that we use to replace semiconductor devices is very important. A heat sink of some sort should be used between the semiconductor case and the point of contact on the lead by the soldering iron. A standard heat sink is shown in Fig. A-2(f). This small clamp is placed between the component and the solder joint. A pair of long-nose pliers can also be used as a heat sink. A heat sink absorbs excess heat before it travels up the lead and into the transistor or diode. The use of a heat sink to prevent damage to insulation is very important and is necessary whenever a soldering iron touches a semiconductor lead.

Large transistors, such as the power transistor shown in Fig. A-4, produce considerable heat in normal operation. This heat must be absorbed to avoid dam-

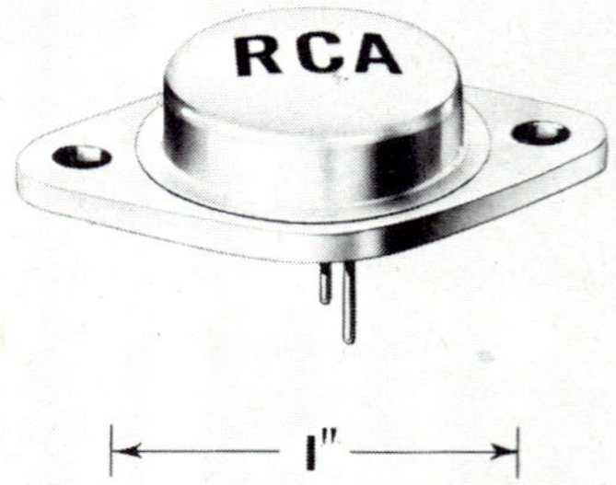

Fig. A-4
Typical power transistor
Courtesy: Radio Corp. of America

age to the transistor. A power transistor may be bolted to a metal plate with a number of radiator fins. This arrangement is called a cooler, or a heat sink. Notice that transistor sockets are usually unmarked and unkeyed, thereby permitting a replacement power transistor to be inserted into the socket backwards. To avoid this risk, the emitter terminal of the original transistor should be marked on the cooler or socket before the transistor is removed.

Remember that many soldering irons, when plugged into an AC line, have some value of voltage on the tip with respect to ground. This means that leakage currents from the soldering iron to a transistor can cause damage from excessive current flow. Therefore, it is good practice to use an isolation transformer or to connect a jumper lead from the metal body of the soldering iron to the common ground bus or to the chassis of the equipment that is being worked on. *Never hold solder in your mouth when soldering.*

GLOSSARY

A

AC, a. c., a–c. Abbreviation for alternating current.

AC spectrum. Conventional range of alternating–current frequencies; power frequencies are from 25 to 60 Hz; audio frequencies are from 15 to 20,000 Hz; radio frequencies extend from 10 KHz to 100,000 MHz.

AF, a. f., a–f. Abbreviation for audio frequency.

AM, a. m., a–m. Abbreviation for amplitude modulation.

ammeter. An instrument for measurement of electric current values.

ampere. The basic unit for measurement of electric current values.

amplification. Increase in magnitude of an electronic signal.

amplifier. A device for providing amplification.

amplitude modulation. A method of communication that involves change in amplitude of an alternating current called the carrier.

anode. An electrode that has positive polarization.

antenna. A device consisting of spatially–arranged electrical conductors designed to radiate or to intercept electromagnetic wave energy.

attenuation. Diminution of the effectiveness of a radio (or other) signal. It is commonly caused by increase in transmission distance, passage through the exhaust flame of a rocket engine, or other opposing situation.

audio frequency. An AC frequency in the range from 15 Hz to 20,000 Hz.

B

ballistic missile. A vehicle that is guided or controlled during the initial flight phase only, its flight during later phases being governed only by the characteristics of a freely falling body.

battery. Two or more electric cells connected to provide more power than a single cell. A single cell is sometimes referred to, but incorrectly, as a battery.

binding post. An electrical terminal device for securing a connecting wire.

blip. A spot of light or similar indication on a radarscope indicating the position of a reflecting object such as a missile in flight.

blockhouse. A building, generally with heavy reinforcement, that houses electronic equipment and controls for firing a missile.

C

cable. One or more wires used to conduct electricity.

capacitance. A characteristic of a device or arrangement which permits it to store electric charges in the form of an electrostatic field.

capacitor. A device that provides capacitance.

cathode. A negatively polarized electrode, or one that supplies electrons.

cathode ray. A stream of electrons moving in an evacuated enclosure.

cathode-ray oscilloscope. An electronic test instrument for visible display of electrical wave forms on the screen of a cathode-ray tube.

cathode-ray tube. An electron tube with a glass face coated with a phosphor that glows when struck by a beam of electrons. The electron beam is produced by an internal structure called an electron gun.

charge. Electrostatic charge refers to the quantity of electricity of a specified polarity on the surface of an insulator.

chassis. A metal structure on which the component parts of an electronic unit are mounted.

circuit. A conductive path for the flow of electric charge (usually electrons).

code. Formalized electrical signals employed in electronic communication.

coil. Turns of an electrical conductor, often wound as a helix, for concentration of the magnetic field produced by flow of electric current.

command. A transmission by radio of data to a missile in flight in order to correct its position and velocity.

command-destruct signal. A radio signal that is operated to actuate a destruction device contained in a missile.

conductor. A substance that permits comparatively free flow of electric charge.

cooled detector. A device that indicates the presence of electromagnetic-wave energy, cooled to reduce internal noise voltages, and thus responsive to weak signals.

continuous wave. An uninterrupted flow of radio energy. In tracking, the continuous wave from a radar is reflected by the target, but because of changing distance between the tracker and target, the frequency of the reflected wave also changes (Doppler effect). Comparison of the transmitted and received frequencies provides range measurements.

coulomb. The conventional unit of electric charge.

countermeasures. Electronic countermeasures have the purpose of impairing the operational effectiveness of enemy activity, and operate in the area of electromagnetic-wave energy.

current. Rate of charge flow past a point in ampere units.

cycle. One complete excursion of an alternating current. A cycle starts at zero, rises through positive values to a peak, then decreases to zero, descends through negative values to a minimum, and finally increases to zero again.

D

DC, d. c., d–c. Abbreviation for direct current.

detection. A process involving radio systems, radar, infrared, optical, or other means for location of objects in space.

detector. A device that recovers the signal information in a modulated wave form. A semiconductor diode is commonly used as a detector device.

device. A physical construct designed to carry out a specified function. An active device is exemplified by a transistor.

dielectric. Insulating substance such as mica, used as a separator between the plates of a capacitor.

diode. A one–way electronic device (rectifier); a rectifier can be compared with a valve.

direct current. An electric current that consists of a charge flow in one direction only.

Doppler effect. (See continuous wave.)

drain rate. Rate of consumption of electrical power in a system.

dry cell. A primary cell that employs zinc and carbon electrodes with chemical action provided by ammonium salts. It changes chemical energy into electrical energy.

E

E. Symbol for electromotive force, or voltage. An alternating voltage is also symbolized by *e.*

Edison effect. An electronic process discovered by Thomas Edison, wherein a metal plate inside of an incandescent light bulb becomes negatively charged when the filament glows.

electricity. A fundamental entity of the physical universe, called charge; it is not mass or energy, although it may be associated with either or both; electricity occurs as subatomic particles, and has either positive or negative polarity. Electricity is not magnetism, although it is often associated with magnetism.

electromagnet. A coil of wire, often wound around an iron core, that develops a magnetic field when an electric current flows through the coil. One end of an electromagnet has north polarity, and the other end has south polarity.

electromagnetic waves. An electromagnetic wave comprises electric– and magnetic–field energy traveling through space with the velocity of light, or 186,500 miles per second.

electromotive force. A potential difference that is provided by a source of electricity when no current is demanded from the source. Abbreviated EMF, it is measured in volt units.

electron. A fundamental subatomic charge of negative electricity. All atoms contain electrons and other subatomic particles.

electron gun. A metallic construction, as in a cathode–ray tube, for the thermal emission of electrons and their concentration into a narrow beam.

electronics. The science of generation and control of electric current consisting of electrons in a vacuum, gas, or solid–state device. Electric current in liquids is also covered.

electron tube. An evacuated or gaseous control device designed to supply electrons and control their flow for utilization by an external circuit.

electrostatic field. The space in the vicinity of an electric charge in which electric field energy is distributed.

emission. A process of ejecting electrons, as from a filament in a vacuum tube by means of thermal energy.

F

farad. The standard unit of capacitance.

field. An altered condition of space in the vicinity of an electrically charged object, or of a magnet.

filament. In a vacuum tube, a wire which is heated by electric current to a sufficiently high temperature that electrons are emitted.

FM, f. m., f–m. Abbreviations for the frequency modulation.

free electron. An electron which is not bound by any force to the vicinity of an atom.

frequency. The number of complete vibrations or alternations of an electric current or wave per unit time.

frequency modulation. A method of communication involving the change in frequency of an alternating current called the carrier.

frequency spectrum. A range of frequencies from a high to a low limit expressed in terms of the number of vibrations or cycles in a unit of time.

G

gamma rays. Electromagnetic radiation with a frequency higher than that of X–rays.

generator. An energy converter; an electronic generator commonly converts DC energy into high–frequency AC energy.

germanium. A gray–white brittle metallic element called a semiconductor. It is the basic substance used in fabrication of solid–state devices.

grid. The control electrode (control element) in a triode vacuum tube.

ground. Originally, a ground path forming part of an electrical circuit; now, the term usually refers to the metal chassis of an electronic unit.

H

headphone. Syn., earphone; a small telephone receiver worn by various electronic personnel. Usually mounted on a headband.

headset. A pair of earphones with a headband.

henry. The standard unit of inductance.

Hertz. Abb. *Hz;* denotes one complete vibration or alternation per second.

HF, h. f., h–f. Abbreviation for high frequency.

high frequency. A frequency within the range of from 3 to 30 MHz (megahertz).

I

I. Symbol for electric current. Alternating current is also symbolized by *i.*

image intensifier. An amplifying device used with an image orthicon to increase the energy of photoelectrons and thereby increase the contrast between an object being tracked and the sky background.

image orthicon. A design of television camera tube often used in tracking, wherein an electron image is produced and then scanned with an electron beam.

inductance. The property of reaction exhibited by a current–carrying coil, due to storage of magnetic–field energy.

information. A coded structure impressed upon an electric current symbolizing letters or numerals, or corresponding to light levels, tones, and so on.
infrared. The range of electromagnetic radiation frequencies next above the visible spectrum.
interferometer. An instrument that separates a radio wave (or other radiation) into two or more parts, and causes them to travel by different paths to a given point. Thus, radio waves from a missile may be received by separated ground antennas.
instrument. An electrical or electronic device that measures a value of voltage, current, resistance, power, etc. Or, a generator of a specified electrical output, such as a signal generator, pulse generator, or square-wave generator.
insulation. A comparatively non-conducting substance used to prevent the escape of electricity from a conductor.
interference. Electrical or electromagnetic disturbances that cause undesirable responses in electronic equipment.
ion. An atom or molecule which is not electrically neutral, having acquired or lost one or more electrons.
ionization. The process of ion formation.
ionization current. A charge flow (electric current) between two oppositely polarized electrodes in an ionized gas.

J

jack. A plug-in type of electrical spring-terminal device.
jamming. Interference with the reception of radio or radar signals, as by countermeasures to distort or prevent the guidance of a missile.
joule. The standard unit of electrical energy.
junction transistor. A transistor design that utilizes a layer arrangement of semiconductor substances.

K

K. Abb. from 1,000; thus, 20,000 volts - 20 KV.
KHz. Abb. for 1,000 Hz. (Kilohertz)
kilo. Metric prefix for 1,000; thus, 10 kilowatts – 10,000 watts.
KV. Abb. for kilovolt; 1,000 volts.
KW Abb. for kilowatt; 1,000 watts.
KWh. Abb. for kilowatt-hours; 1,000 watt-hours.

L

L. Symbol for inductance; the unit of inductance is the henry.
leakage. Undesired escape of electricity, e. g., from a conductor through poor insulation to ground.
lines of force. Imaginary field-structure lines emerging from the north pole of a magnet and entering the south pole. Imaginary field-structure lines emerging from a negative electric charge and entering a positive electric charge. Lines of force are used to map a field.
loran. An acronym denoting long-range navigation.

M

M. Abb. for 1,000,000; thus, 1 MHz = 1,000,000 alternations per second.

magnet. A device or object that provides a magnetic field. The earth is a huge, although weak, magnet.

magnetic field. The space in the vicinity of a magnet in which magnetic-field energy is distributed.

magnetic lines of force. Imaginary field-structure lines that are employed to map the distribution of magnetic energy in the vicinity of a magnet.

magnetic poles. Regions of a magnet at which the field strength is concentrated.

magnetism. A fundamental physical property associated with electricity in motion.

magnetohydrodynamics. The characteristics of ionized gases when acted upon by electric and magnetic fields.

maser. A type of microwave amplifier capable of amplifying very weak signals without interference from internally generated noise.

MHz. Abb. for megahertz.

mercury-vapor tube. A gas tube that operates with vaporized mercury.

meter. An electrical or electronic measuring instrument.

microfarad. One millionth of a farad; abb. μF.

Microlock. The tracking and communication system of the Explorer satellite. It is based on the interferometer principle.

micromicrofarad. One millionth of a microfarad; also termed a picofarad; abb. pF.

microphone. A transducer for converting acoustic energy into electrical energy.

microwaves. Electromagnetic waves with a frequency greater than 300 MHz.

Minitrack. The tracking and communication system of the Vanguard satellite. It is based on the interferometer principle.

multimeter. An electrical instrument that measures various electrical quantities, usually DC and AC voltage, DC current, and resistance values. Also called a volt-ohm-milliammeter, abb. VOM.

multiplex operation. Transmission of two or more individual signals in the same communication channel.

N

negative charge. The electron is the elementary charge of negative electricity. A body with an excess of electrons with respect to ground or to some other reference point is said to be negatively charged.

neutron. A fundamental subatomic particle; it is electrically neutral, and is a constituent of all atoms with the exception of the hydrogen atom.

north pole. The magnetic polarity corresponding to that of the end of a compass needle which points toward the polestar.

O

ohm. The standard unit of electrical resistance; defined as the resistance that permits the flow of 1 ampere of current when 1 volt is applied across the resistance.

Ohm's law. A basic electrical law that expresses the relation among voltage, current, and resistance values. The law states that current flow is directly proportional to the applied voltage, and inversely proportional to the resistance value.

orbit. From the electronic aspect, an orbit is a closed path in which an electron revolves around the nucleus of an atom.

oscillation. A mode of operation of an electrical or electronic circuit wherein free electrons surge back and forth in the conductors.

oscilloscope. An electronic instrument that displays AC wave-forms on the screen of a cathode-ray tube.

P

P. Abb. for power; *W* also symbolizes power.

permanent magnet. Abb. PM; a device or substance that exhibits a magnetic field.

phosphor. A substance such as Willemite that glows when struck by an electron beam.

phosphorescence. Afterglow resulting from activation of a phosphor.

photoelectric. Process of converting light into electricity by a transducer.

photoelectric cell. An electronic device that operates as a photoelectric transducer.

picture tube. A cathode-ray tube employed in a television receiver.

plasma rocket. A rocket that employs a partially ionized gaseous propellant, by discharge of a powerful arc through it, producing very high temperatures and forces.

polarity. A physical distinction between the two kinds of electric charge and the two kinds of magnetic poles. Denoted by plus or minus, and by north or south.

pole. A region of a magnet at which the magnetic field force is concentrated.

positive. Symbolized by a + sign; denotes a point that has a deficiency of electrons with respect to a reference point.

potential. Electric potential is a measure of the work that must be done to move an electron through a given distance.

potential difference. Equivalent term for electromotive force, measured in volts.

primary cell. A chemical generator of electricity; it develops an excess of electrons at one terminal, and a deficiency of electrons at the other terminal.

proton. A subatomic particle with a polarity opposite to that of an electron.

pulse. Radio energy transmitted for a short time, usually in millionths of seconds.

Q

Q. Symbol for quantity of electricity, as the number of electrons that flow in a conductor for a given period of time.

R

R, r. Symbol for electrical resistance.

radar. An acronym denoting radio detection and ranging.

radiation. Electromagnetic energy, such as radio waves, which travel through space at the speed of light (186,500 miles per second).

radio. (English eq., wireless); a communication system that employs electromagnetic waves.

radioactive decay. Spontaneous transmutation of a complex element such as uranium into simpler elements, radiation, and subatomic particles such as electrons.

radio astronomy. A branch of astronomy that employs radio telescopes to study electromagnetic-wave radiations from outer space.

radio frequency. Frequencies in the range from 10 kHz to 10,000,000 MHz.

radio-inertial guidance system. A command type of basic guidance system. It includes a radar tracking unit consisting of radar equipment on the ground, one or more transponders (an electronic device that receives a challenging signal and automatically transmits a response) in the missile, and necessary communication links to the guidance station; a computer that accepts missile position and velocity data from the tracking unit and furnishes the command link an appropriate signal to steer the missile; the command link consists of a transmitter on the ground and an antenna and receiver on the missile (actually, the command link is contained in the tracking unit); an inertial system is also provided for partial guidance in the event of radio failure.

radio-ranging. Process of radio beam transmission in the form of directional signals.

radio receiver. Apparatus for interception and processing of electromagnetic-wave energy, and its conversion into acoustic energy.

radiosonde. Electronic system for transmission of meteorological instrument data from a weather balloon or other vehicle to recording instruments located on the ground. Also, radio meteorograph.

radio telegraph. An electronic system that employs electromagnetic waves for communication by means of a code, such as the Morse code.

radio telephone, radiophone. An electronic system that employs electromagnetic waves for communication by speech, or transmission of musical programs.

radio wave. An electromagnetic wave, utilized for communication.

rate transmitter and rate receiver. A special-purpose radio transmitter and receiver installation used for determination of the rate of change in missile range during flight.

rectifier. A device that conducts electrons in one direction only.

resistance. A physical property that opposes the flow of electrons.

S

scanning. Progressive sweeping of a surface, line-by-line, as by an electron beam.

secondary electron. An electron ejected or emitted from a substance as a result of bombardment by an electron stream.

secondary emission. The process of secondary electron emission.

semiconductor. An electrical classification of substances which are neither good conductors nor good insulators.

short-circuit. A very low-resistance path for current flow, usually accidental in occurrence.

short-wave transmitter. A radio transmitter with an operating frequency above 1,500 KHz.

signal. An alternating current or electromagnetic wave which has been systematically varied in some characteristic for purposes of information transfer in communication.

sky wave. An electromagnetic wave that proceeds to the ionosphere and is then reflected back to earth.

solid state. Descriptive term for devices (usually semiconductors) in which charge flow occurs in crystal lattices.

south pole. The magnetic polarity corresponding to that end of a compass needle which points in a direction opposite to that of the polestar.

spark. A lightning-like electric discharge between separated terminals in air, resulting from ionization of the intervening air molecules by intense electrostatic forces.

speech-frequency range. The narrowest band of audio frequencies within which intelligible speech transmission and reception is practical; it extends approximately from 100 to 2,000 Hz.

standard. An established value for a physical unit, such as the electron charge, the volt unit, and so on.

static. Random interference in communication systems, caused by undesired electromagnetic waves produced by natural sources, such as lightning storms.

subatomic. Pertaining to characteristics of the fundamental particles, such as electrons and protons, which are smaller than atoms.

super-high frequency. Frequencies in the range from 3,000 to 30,000 MHz.

symbol. A formalized sign, letter, or design that represents a concept or physical property, as the polarity symbol "–" which denotes an excess of electrons.

T

telecommunication. Communication (information transfer) by means of electron flow or electromagnetic waves.

telegraphy. Communication over wire circuits by means of a code such as the Morse code.

telemetering. A system that takes measurements within a missile during flight and transmits them by radio to a ground station, where the data are used to evaluate the performance of the missile and its components.

telemetry. Transmission and reception by means of electromagnetic waves of instrument data from a remote location to a receiving station on the ground.

telephone receiver. An electrical transducer for the conversion of electrical energy into acoustic energy.

telephone transmitter. Also termed a *microphone*; an electrical transducer for the conversion of acoustic energy into electrical energy.

television. The transmission and reception of scenes and sounds by means of electronic devices and electromagnetic waves.

thermocouple. A transducer composed of two different metals that generate an electric current when heated.

tracking. The process of observing the progress of an object moving through space, usually by means of a radar system.

transducer. A device for the conversion of one form of energy into another form.

transistor. A solid-state electronic device that performs the basic functions of an electron tube.

transmission. Transfer of electrical or electromagnetic energy or signals from one location to another.

triode. A three-element electron tube, usually employing a filament, a grid, and a plate.

tube. Abbreviated expression for vacuum tube or for electron tube (there is no technical distinction between the terms).

U

UHF, uhf. Abb. for ultra–high frequency.

ultra–high frequency. Frequencies in the range from 300 to 3,000 MHz.

ultra–violet radiation. A range of electromagnetic radiation that extends beyond the visible violet limit of the solar spectrum.

V

V, v. Symbol for voltage.

vacuum. A space from which gas molecules have been pumped. Outer space approximates a perfect vacuum.

vacuum tube. Syn., electron tube; a device that provides rectification, amplification, or other functions by means of electron flow in a vacuum.

Van de Graaff machine. An electrostatic generator used to accelerate electrically charged particles for initiating nuclear reactions.

voltaic cell. An electrochemical device that changes chemical energy into electricity.

voltmeter. An electrical instrument that measures voltage values.

W

W, w. Symbol for electrical energy (sometimes used also to symbolize power).

watt. Standard unit of electrical power.

watt–hour. A unit of electrical energy.

watt–hour meter. An electrical instrument that measures watt–hour values.

wave. In electronics terminology, a structure of electrostatic and electromagnetic energy traveling through space at the speed of light.

wave length. The distance from one peak to the next of an AC wave.

X

X–rays. Electromagnetic radiation with a very high frequency which penetrates many substances such as wood, human tissue, and so on.

X–ray tube. An electron tube that generates X radiation.

Z

Z. Symbol for electrical impedance (opposition to flow of AC current).

zero level. Horizontal axis in a wave form (as displayed on the screen of a cathode–ray tube) which corresponds to the zero–voltage level.

INDEX

JAMES HILSINGER